# Petroleum Engineering

## Sustainable Geoenergy Engineering and Technology

The objective of this series is to establish a link between traditional hydrocarbon science and engineering and the development of sustainable engineering to achieve net-zero and sustainability goals in energy production. The series aims to promote and expedite the dissemination of new research results, innovations, and technology in the field of exploration, production, and management of oil and gas, aiding the oil and gas industry in navigating through the energy transition and driving it towards the net-zero target. Additionally, the series covers developing areas related to the geological subsurface, such as geothermal energy and underground storage. It includes monographs, lecture notes, and edited volumes, with a subject focus on upstream processes. Coverage extends to all theoretical and applied aspects of the field, encompassing both traditional and modern topics, including but not limited to:

Exploration Formation evaluation (well logging) Drilling Economics Reservoir simulation Reservoir engineering Well engineering Artificial lift systems Facilities engineering Sustainable energy exploration and production Carbon capture, utilization, and storage (CCUS) Geothermal energy development Underground Hydrogen Storage (UHS) Information technology (IT)

Contributions to the series can be made by submitting a proposal to the responsible Springer contact, Anthony Doyle anthony.doyle@springer.com.

Pablo Druetta · Remo Frabotta ·
Gustavo Luis Cavallo

# Transport by Pipelines and Its Optimization

## A Sustainable Approach to the Design

Springer

Pablo Druetta
Chemical Engineering
University of Groningen
Groningen, The Netherlands

Remo Frabotta
Mechanical Engineering
UTN-FRBA
Buenos Aires, Argentina

Gustavo Luis Cavallo
Mechanical Engineering
UBA-IGPUBA
Buenos Aires, Argentina

ISSN 2366-2646 ISSN 2366-2654 (electronic)
Petroleum Engineering
ISBN 978-3-031-66565-3 ISBN 978-3-031-66566-0 (eBook)
https://doi.org/10.1007/978-3-031-66566-0



This Springer imprint is published by the registered company Springer Nature Switzerland AG
The registered company address is: Gewerbestrasse 11, 6330 Cham, Switzerland

*To Magda, Mili and Ola, who taught me how to write a technical book while playing horse-ride.*

*Pablo Druetta*

*To Dr. Ing. Ezio Carlos Lorenzelli (1910–2001), who introduced us to the fascinating world of the study of fluids, from his undergraduate classes taught at the Universidad Tecnológica Nacional—Regional Buenos Aires, during the 1970s.*

*Remo Frabotta & Gustavo Luis Cavallo*

# Prologue

The fundamental objective of this text is to introduce a calculation methodology to obtain the *optimal project* of a pipeline for long-distance transport of a certain fluid, either liquid or gaseous; which is strictly related to the primary goal of its transport, i.e., a continuous supply of material to consumers or industries, at the lowest cost.

During the 1st Argentine Congress of Engineering Policies, held in Buenos Aires at the Argentine Center of Engineers in 1977, the following was defined as the Mission of the Engineer:

> The Engineer's first and last mission is to master, transform and take advantage of the forces and things of nature, preserving the environmental ecosystem, producing goods and services with a technical-economic concept, maximum performance, minimum costs, increasing and optimizing the productivity of the system required by scarce resources and growing demand, thereby satisfying social interests.

Theodore von Karman once said, "scientists study the world as it is, engineers create the world that never was", and we think this summarizes nicely the job engineers do. The job they do is a thin balance between the comprehension of the natural laws that determine how things behave, and the constant need of improving the quality of life and the use of the scarce resources we count with. This book was written with this concept as one of its foundation stones.

Considering these premises and our own convictions, the Pipeline Project Optimization methodology, presented in this book, was elaborated. It was developed and improved by the authors from 2001 to date, and has been applied to the design of numerous fluid transportation lines (oil pipelines, multi-product lines, gas pipelines, etc.) in Latin America, as part of their professional engineering tasks aimed at to the oil and gas industry. The courses "Transport of Fluids through Pipelines" and "Fluid Mechanics" developed at the National Technological University (UTN) in Buenos Aires, Argentina; "Transport of Natural Gas" corresponding to the Specialization Course in Natural Gas, postgraduate from the Institute of Gas and Oil (IGPUBA) dependent on the Faculty of Engineering of the University of Buenos Aires (UBA) were created by the authors to teach this method and the design of pipeline systems, who have more than 30 years of university teaching activity.

The target audience is students who find themselves at the end of their B.Sc. or M.Sc. programs in engineering and want to understand some more details on the design of pipelines from a techno-economic perspective. It contains examples of different systems, which are aimed also to make students used to the orders of magnitude usually found in the pipeline industry. The required background is then a good knowledge of basic sciences (e.g. multivariable calculus, algebra, numerical methods) as well as some basic (mechanical) engineering science courses, such as thermodynamics, fluid mechanics and mechanics of materials.

With the theoretical and experimental tools of fluid mechanics, resistance of materials and heat transmission from the thermodynamic perspective, cost functions corresponding to all the elements of the installation (line pipe, compression or pumping plants, accessories, necessary energy, initial product load, etc.) are defined and then, with appropriate mathematical methods, the variables are optimized, in such a way that the installation results in a *minimum cost*. The fact of looking for a minimum cost, forces a detailed evaluation of the costs of all the components involved in the system. Once the optimal variables of the pipe have been determined, it is advisable to carry out a more precise calculation, considering all the variables and variations in play. Given the complexity that an integral calculation would require, the calculation scheme is facilitated by making an approximate calculation by a finite difference approach, thus obtaining for any section along of the pipe, all the parameters that define the state of the fluid (pressure, density, temperature, etc.). Furthermore, to complement these (quasi-)static studies, dynamical analyses of the pipeline should be evaluated in order to determine, e.g., if the pressure at the reception point is within the tolerances established by the client.

To focus on the problem, it is first necessary to carry out a limited market investigation in order to obtain all the costs of the elements that intervene in the installation, i.e., materials, labor and equipment, and then elaborate adequate mathematical regressions to obtain the functions of costs required by the method. The proposed method aims to meet the requirements of effective use of resources with rationality and effective calculation possibilities, since most of the data used is usually available in the companies of the sector or industry in question. Given the crucial importance of liquid transportation systems through pipelines within the hydrocarbons industry, the efforts made to achieve the greatest economic rationality when projecting a new transportation system will never be enough.

This development is not the first treatment given to the subject. The optimization methods, applied to the transport of fluids and in the hydrocarbons industry in particular, are already long-standing. However, this new approach is that of an updated review in its mathematical language and fundamentally in the number of variables optimized simultaneously.

Chapters 1 to 3 present the introduction of the basic knowledge needed to understand the optimization process developed in this book. Chapters 4 to 6 contain the study and application of the optimization technique to different types of pipelines, i.e., from an incompressible oil flow, to the analysis of pipelines of refined products, and then the case of compressible gas pipelines. Chapters 7 and 8 present the study of flow with two (or more) phases: firstly the case of liquid/gases; and finally the mechanics behind the two-phase flow of solids with liquids or gases. Finally, Chap. 9 makes a brief introduction to the study of transient in this pipelines, presenting the most common numerical techniques used to solve the differential equations governing the flow of (in)compressible flows.

Groningen, The Netherlands Pablo Druetta
Buenos Aires, Argentina Remo Frabotta
Buenos Aires, Argentina Gustavo Luis Cavallo
2024

# Acknowledgements

Our deepest gratitude goes to the National Technological University in Argentina and the University of Groningen in the Netherlands. Without the patience of our colleagues, this book would not have been possible. Pablo Druetta especially thanks to his friends, Profs. Patrizio Raffa and Daniele Parisi, for their support during this time and their valuable advices while having our important meetings in ACLO. We also want to thank the countless students, colleagues, professors, engineers and many others who helped us through the years, giving their valuable opinion to make this technique what it has currently become.

We would also like to thank to our families, without their patience and support this project had been impossible to materialize.

Finally, to the people from Springer Nature, specially to Anthony Doyle, Amudha Vijayarangan and Sharmila Anbu for the opportunity they gave us to write this long-awaited book.

Buenos Aires, Argentina
Groningen, The Netherlands
2024

# Contents

# Chapter 1
# Introduction

> Science can amuse and fascinate us all, but it is engineering that changes the world
>
> *Isaac Asimov*

The problem of the transport of fluids by means of ducts comes from old data and there are plenty of historical examples showing the concern of these first engineers of the Roman Empire and ancient Egypt regarding its design (See Fig. 1.1).

It is undeniable the role that raw materials and energy have, how they have radically changed the way our society lives and the economy works. The changes generated in both over the last 200 years, since the beginning of the Industrial Revolution, are more than obvious (Druetta 2018; Hopkins 2007). Nevertheless, focusing on the raw materials used to generate useful energy, they are most of the times far away of their final consumption points, leading to a major conundrum: how do we transport them in an effective and safe way, from their production points to their final destinations?

Generally speaking, human kind counts nowadays with five different transportation media, namely: land transport (road and railway), maritime, pipelines and air transport. Even though many factors should be considered in the analysis of how to transport these fluids, which will be developed during this book, one of the major goals in any engineering project is to reduce the cost (and the use of material) to a minimum, leading to a sustainable approach in the design. If one analyzes the transport costs, one immediately realizes that air transport is not an attractive way, leaving the other four as possible alternatives. Figure 1.2 shows the approximate costs to transport one barrel of oil between two points by using these four alternatives (Hafner and Luciani 2022; Hansen and Dursteler 2017; Cookenboo Jr 1954). As it can be seen, maritime and pipelines render the minimum values, mainly due to the fact that large volumes can be transported per cycle. Going into a more detailed analysis, the transport costs will depend on the total distance to be transported and mainly the



P. Druetta et al., *Transport by Pipelines and Its Optimization*, Petroleum Engineering,
https://doi.org/10.1007/978-3-031-66566-0_1

**Fig. 1.1** Aqua Claudia, regarded as one of the four great aqueducts of Rome, by Tabora (Tabora 2006)

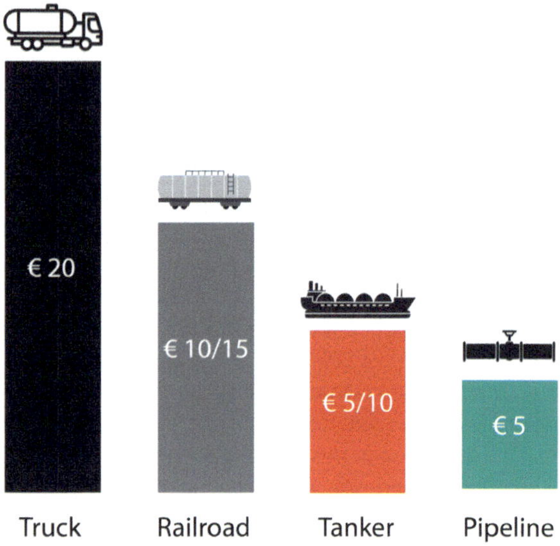


**Fig. 1.2** Comparative costs per barrel of oil for the different transportation systems

location of these two points, being affected by many factors, e.g., climate, political conditions, price of commodities, etc. This does not apply only to the classical fluids seen in the market, i.e., oil (and refined) and gas, but also to other fluids, such as $CO_2$, as Fig. 1.3 shows Nilsson et al. (2011), Weihs et al. (2014). In this case, it is evident that for shorter distances, pipelines take the lead as preferred mean of transportation, whilst above 1000 km, the maritime transport option renders the minimum costs. A similar trend can also be observed in Figs. 1.4 and 1.5 for the transportation of natural gas: at longer distances, involving crossing oceans, the option of building a pipeline is no longer considered, and the use of LNG tankers become the favorite option. It is noteworthy to point out that these numbers are relative and they depend on many factors, namely: price of commodities, origin and destination points, geopolitical

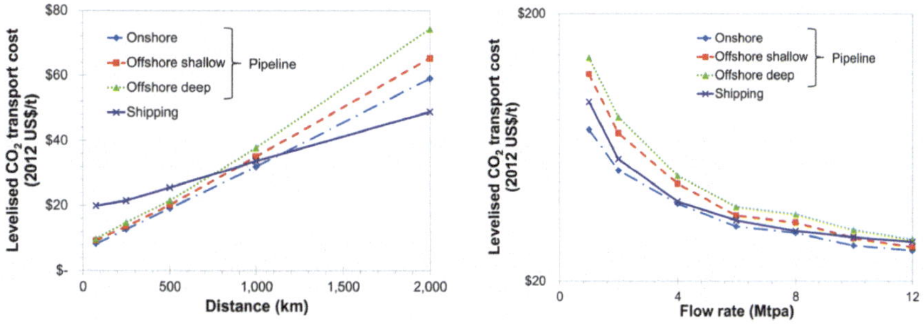


**Fig. 1.3** Comparative costs for transport of $CO_2$ for 1000 km (left) barrel of oil for a mass flow rate of 6 Mtpa (left) and 1000 km (right) (Weihs et al. 2014)

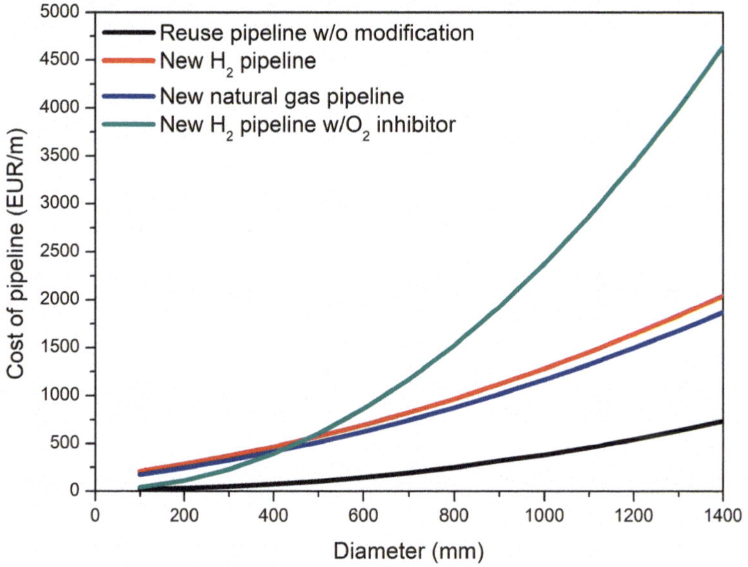


**Fig. 1.4** Comparative costs for natural gas and hydrogen pipelines, reusing existing ones or using oxygen as inhibitor for the hydrogen transport, as a function of the pipeline diameter

reasons, etc. In general lines, it can be said that the investments costs for pipelines can vary from 30 to 200 €/m/inch, with this wide range depending on several factors, such as terrain and climate, geographical location, local legislation, geopolitical issues and/or safety/environment factors (Hopkins 2007).

Based on the previous analysis, it is considered that the use of pipelines for the transport of fluids represents an interesting choice, both from an economic and safety point of view. In its major usage, pipelines transport the oil and gas necessary to fulfil the steadily increasing demand for energy, being used in power generation, refining, and synthesis of derived products (e.g. plastics). Nonetheless, different products are also being transported nowadays, with possible different purposes than those

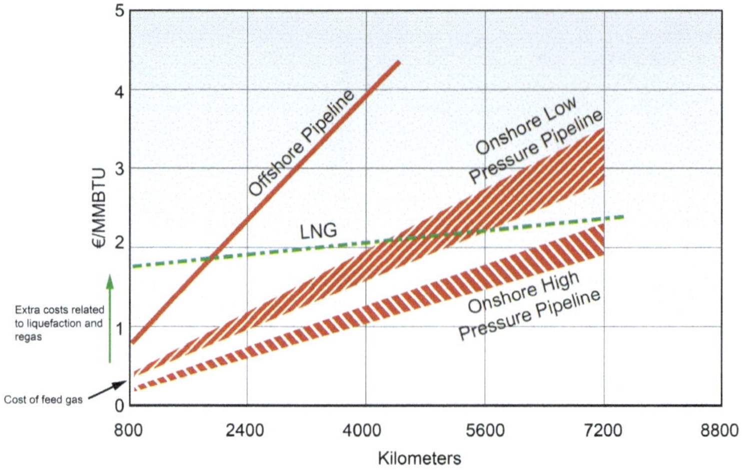


**Fig. 1.5** Comparative costs to transport 1 tcf/year of gas by different pipelines and by sea transport (LNG), including the liquefaction and regasification costs (Mokhatab and Purewal 2006)

mentioned previously, namely: water for human consumption; hydrogen for energy generation and storage in caverns (Erdener et al. 2023; Kazi et al. 2024); $CO_2$ for carbon capture and enhanced oil recovery; district heating; and finally, the transport of minerals in a carrier (i.e. water or air) for energy generation (e.g. coal) of further processing (e.g. iron ore) (Mokhatab 2014). One can categorize a pipeline according to several factors, depicted in Fig. 1.6.

Focusing mainly on the oil/gas trunk pipelines, there are as of 2023 in the world around of 2,193,250 km of trunk pipelines (i.e. main lines), of which crude oil constitutes the 18%, refined products the 13%, natural gas liquids (NGL) the 6% of the total, and finally, natural gas pipelines the 63% (Figs. 1.7 and 1.8) (Global Data Energy 2022). Within Europe, it is evident the inflow to the continent coming from Africa, via the Mediterranean Sea crossings such as Medgaz; those coming from the North Sea, such as the Europipe; and those coming from Russia, such as the Yamal Europe crossing Belarus and Poland.

## 1.1 Pipeline Design and Project

It is evident that any project for large-distance transport of fluids as diverse as petroleum, natural gas, water, refined products, or multiphase systems, by means of pipelines must comply with certain fundamental premises imposed in the design bases. These are basically the required supply flow rate, the pressure available in the header of the line, the minimum arrival pressure at the point of consumption

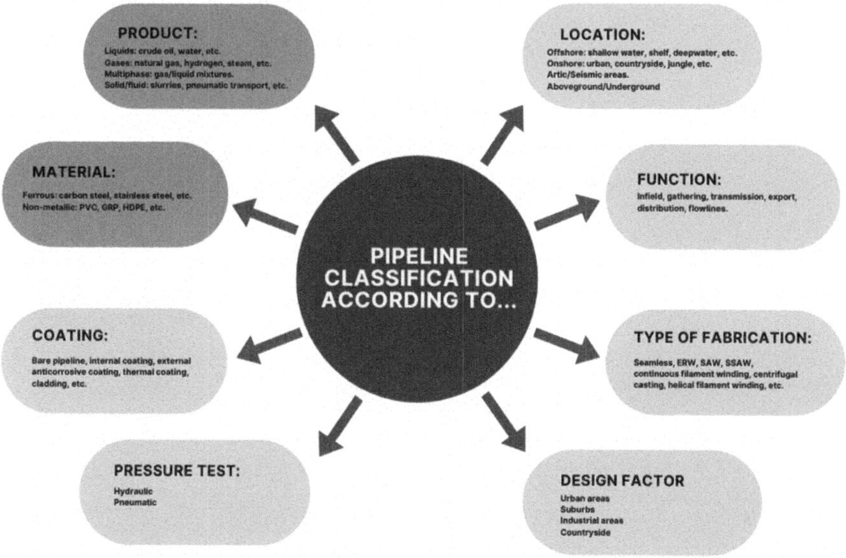


**Fig. 1.6** Pipeline classification according to different perspectives

(if any), the geography where the trace of the duct must cross, and other conditions pre-established by the project owner (Robinius et al. 2017; Jenkins and Murray 2005).

Considering these conditions, and with the respective mathematical formulations and calculation procedures, engineers ensure that the basic design is suitable for transporting the flow of the fluid to the pressure provided in the design bases. However, for the purposes of a comprehensive project this is not enough, since this technical calculation method does not provide any information whatsoever about the cost of the project, namely: costs of fixed assets (property, plant and equipment, *PP&E*, also known as CAPital EXpenditure (CAPEX) and the OPerating EXpenses (OPEX), which are the costs of the day-to-day operations. Both together form the final cost of transportation (Fig. 1.9) (Mohitpour et al. 2007).

In general, and as mentioned previously, the final objective of this type of systems is to transport the fluid in question to a "minimum cost". The methodology developed in this text will just search and find the minimum cost project, which it will be called the Optimal Project. Nevertheless, it is worth mentioning that the project should consider other facets besides the techno-economical, applying a concept that has gained more and more relevance during the last decades: the *sustainability*. This involved more than a minimization of costs, but comprises societal, environmental and economical impacts: the project is no longer good only if minimizes costs. It should also minimize its environmental footprint and maximize its impact in the society, improving the economy, education and/or standard of living of the place in question.

**Fig. 1.7** World map showing the operating (brown), in construction (red) and proposed (orange) oil pipelines (Global Energy Monitor 2022a)

**Fig. 1.8** World map showing the operating (brown), in construction (red) and proposed (orange) gas pipelines (Global Energy Monitor 2022b)

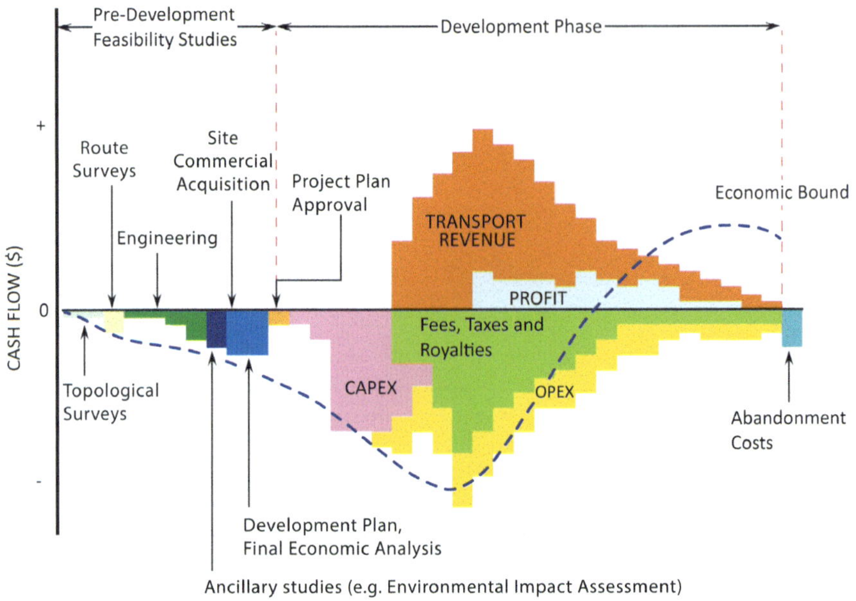


**Fig. 1.9** Cash flow of a pipeline project, with the dashed blue line representing the instantaneous value (Druetta et al. 2016)

When analyzing the technical aspect of the project, it is convenient to do the transport of fluid by a pipe at high speeds, because in this way the pipe size becomes relatively smaller and hence the costs are reduced. However, the transport of a fluid at a relatively high speed results in the need to boost the fluid with a higher pressure; consequently, a greater consumption of energy is required for the impulse, increasing the CAPEX and OPEX because more complex facilities are required, with bigger size compressors/pumps.

Another variable at play is also the amount of plants (compression or pumping) needed to transport the fluid at the required rate and pressure, from the header (e.g. gas/oil field) to the main distribution point (e.g. cities). The greater the number of compressors or pumps, the smaller the pressure jump in each of them will be for the same transport distance. On the one hand, the cost of the project increases by increasing the number of compression or pumping plants (CAPEX), and on the other hand, the operational costs decrease for each plant by decreasing its pressure variation (i.e., OPEX, the power involved).

When trying to solve this problem mathematically, the fact that the amount of equations defining the problem is lower than the amount of unknowns to be solved, rendering the system under-determined, and consequently the system admits multiple solutions. Among this set of possible solutions, and with the help of cost functions obtained on the basis of all the elements that constitute the installation and its energy consumption, the optimal project can be found, using the mathematical optimization method of the Lagrange[1] multipliers.

Applying the Lagrange Multipliers Method to the Transport of Fluids through Pipes is the final objective of the engineering design proposed in this book. It is obtained then a set of relatively simple equations, applicable to many practical cases such as: oil pipelines, aqueducts, multi-product lines, gas pipelines, hydrogen transport, $CO_2$ transport, mineral pipelines, etc. These optimum values of the variables under study correspond to the minimum cost of installation and operation. Once the theoretical optima corresponding to the diameter have been obtained, the pipe thickness calculated, and the number of pumping/compression stations (most certainly a non-integer number), it will be necessary to select commercial diameters and thicknesses (close to the theoretical ones) and a round number of plants.

Notwithstanding, this is not the end of the design process, but it must be developed in more detail. To do this, it is necessary to superimpose the analysis carried out by the altimetry of the route, the particular characteristics of the areas that the pipeline will cross (important when defining the heat transfer) and the effect of the temperature and pressure of the fluid on its own characteristics. In order to carry out this analysis, a finite approximation methodology is developed in this book, that is, a basic diagram of the flow of fluids in pipes by small sections or segments, applicable to the different types of fluids analyzed (liquids, gases, and mixtures between them, including solids). This is the last of the (quasi-)static analyses; after this, a transient study should be carried out in order to understand the dynamics behavior of the system and to verify whether it will also comply with the transport requirements when one (or more) of the variables change as a function of time. A short study of this will also be presented in Chap. 9 of this book.

Finally, it is highlighted that it is not the claim of this book to have found a definitive solution to the transport of fluids through pipes, already developed by numerous authors, but that it is one more effort to solve the "optimization" problem of distant transport of fluid or flowable products, by means of an appropriate mathematical methodology.

## 1.2 General Information About Fluids, Transport Systems and Flowable Mixtures

The basic question that an engineer must answer when developing a project for a transport system for a certain fluid is why the transport the fluids through a pipelines should be done instead of using other means? Indeed, at the beginning of the 20th

[1] de LAGRANGE, Joseph Louis (1736–1813): Franco-Italian astronomer and mathematician. Lagrange was of French descent, although he was born and raised in Italy. As a child, at school, he came across an essay by Halley on mathematical analysis and immediately decided to pursue mathematics. Lagrange's mathematical ability was recognized by Euler from a memorandum he received from Lagrange on the calculus of variations.

century, the transport of fluid substances was basically carried out by ships, barges, trucks, railways, etc. The change towards the selection of transportation systems through pipelines was initially due to a problem of costs, later security, and more recently, environmental care, which is immersed within the concept of sustainability. Undoubtedly, not only fluids (liquid and gaseous substances) are feasible to be transported by pipelines. The technique has been devised to transport solid substances through pipes by means of their "fluidification", mixing them in adequate proportion with liquids (usually with water as a carrier fluid, known as slurries) or with air (Krieg 2012).

Pipes for transporting fluids through distant lengths (usually called "pipeline" or simply "line", or "duct") are made up, in the case of steel, of a succession of individual pipes of variable lengths between 6 m (single random length or SRL), most commonly 12 m (double random length or DRL), and 18 m (triple random length or TRL).[2] Pipes are built in various materials and joined by various methods, although the most usual—in large pipes (important diameter and length)[3]—is on the basis of shaped steel pipes (generally low carbon content) and joined together by arc welding techniques (e.g. shielded metal arc welding—SMAW).

Although welding is a highly developed metallurgical procedure, one must keep in mind the issue of micro-structural changes in steel derived from sudden heating and cooling of the welded area.[4] It is evident that a pipe line cannot be heat treated in a furnace (e.g. normalized to 600°C to eliminate residual stresses from the welding process). It is therefore necessary to select suitable and compatible welding procedures and pipe materials. When it comes to steel pipes with low carbon content, tempering is not produced, but if the carbon content is increased, weldability and toughness problems may appear. As a general rule, *the greater the resistance/yield stress, the greater the complication at the time of welding*. Another important point to pay attention to is the relationship between the specified minimum yield strength (SMYS) and the ultimate tensile strength (UTS): this gives an idea of the margin of safety against a failure by plastic collapse and the capacity of the pipe to have a plastic deformation before its failure. This is quite important, though not in its design, but during the construction of the pipeline, specifically in the laying down operations. For instance, the American Petroleum Institute specification for Line Pipes (API 5L) states that this relationship should be lower than 0.93 for C-Mn steel pipelines with SMYS of 555 MPa (or 80 ksi, X80) and then increases up to 0.99 for steel grades with SMYS of 830 MPa (or 120 ksi, X120). Another issue to consider in the broad technical spectrum of large pipelines is their manufacturing process. It is known that pipes obtained from shaped (rolled) and welded sheet steel are more suitable for obtaining various thicknesses and resistances, generally for pipes greater than DN 200/250 mm (NPS 8/10). For smaller diameters, the seamless rolling process is more suitable.

---

[2] In certain cases also 24 m as the union of two 12-m pipes, known as "double-jointed".

[3] Though this may vary, it can be considered in cases where diameter >300 mm and lengths > 10 km.

[4] Also known as the heat affected zone or HAZ.

## 1.3 Definitions Over the Fluids Being Transported

### *1.3.1 Homogeneous Fluids—Pure Substances and Mixtures*

In natural sciences, a substance or mixture is called homogeneous when its composition and microscopic structure are uniform. A pure substance is one that is not mixed with anything else, while a mixture is a bond or group of substances that can be homogeneous or heterogeneous from a macroscopic point of view.

In the analysis one will call *pure* or *homogeneous* fluids those that have a specific density and viscosity at any point of their mass as long as the pressure and temperature are the same; for example oil, water, natural gas and also to solutions of liquids that are miscible with each other (such as LPG composed of liquid propane and butane) (Poling et al. 2000).

The term *mixtures* is meant to encompass all those conditions in which immiscible liquids flow within the pipe, or it is a flow of pure or homogeneous substances in different phases (gas-liquid/s flows).

### *1.3.2 Low Compressibility Fluids (Liquids)*

A liquid is a substance at a state of matter, whose molecules have less cohesion than those at a solid state, but in turn greater than those of gases. In a liquid, the volume remains relatively constant due to the cohesive forces between the molecules. As a result, a liquid takes the shape of the vessel that contains it and forms a free surface.

Taking petroleum as an example of a liquid, like any other fluid called "incompressible", it is actually slightly compressible, since it is well-known that incompressibility is merely an ideal concept. On the other hand, the physical properties of a petroleum, as it is the case of the density and the viscosity, change with both the pressure and the temperature. Therefore, as the pressure and temperature change along the pipeline, the flowing characteristics of the oil also change. To optimize the project variables, as a first approximation (sufficient for practical purposes), it is advisable to consider the oil physical properties and the temperature throughout the pipeline as constant. This so-called calculation or optimization temperature will depend on the pipeline's route and its location (e.g. off-/onshore), that reflects in the best possible way the average transport conditions (Monge et al. 2023).

As it was mentioned before, a liquid is defined as a practically incompressible fluid, but in reality it is not, since while flowing through a pipe there will be variation in the temperature caused by the heat transmission as well as a variation of the density by pressure, and these variations must be taken into account when the optimum system is verified by using a finite approximation method.

The case of natural gas (and other gases) is different because, since it is a compressible fluid, the density variations must be taken into account in the calculations, but in this case the optimization will be considered with the simplification of assuming the gas flow at constant temperature, i.e., an isothermal flow.

### Density and Viscosity Change with Pressure and Temperature

The variation of the viscosity and density of liquids due to the effect of pressure and temperature will be analyzed below. It is noteworthy to keep in mind that the objective of this book is to propose optimal solutions for the transport of fluids by applying correlations that allow the reader to build relatively simple calculation programs. There are many correlations, of the polynomial type, in many cases more accurate than those given here, but their application to optimization or calculation by sections, generally complicates the resolution (and even in some cases makes it so complex that it does not allow the engineering calculations), without substantially improving the calculations.

The density is defined as the relationship between mass and volume, assuming the fluid as a continuum,[5] it is mathematically expressed as (Druetta et al. 2016),

$$\rho = \lim_{\Delta V \to 0} \frac{\Delta m}{\Delta V} = \frac{dm}{dV} \tag{1.1}$$

This renders a function of *pressure and temperature* (in short, the **equation of state**, though in general it is customary to apply this definition only to gases), thus $\rho = f(p, T)$. This function relating the density of a liquid with the variations of pressure and temperature in general is not known explicitly as in the case of gases, but must be found experimentally. In order to assemble it, it is possible to take as starting point the shape of a certain general function (i.e. there is no single equation).

An applicable equation is the one defining the isothermal bulk modulus $K$ (or its inverse the compressibility modulus, $\beta$), which relates the pressure variations within a liquid enclosed in a container, and therefore with an invariable mass, when it is subjected to small variation in the volume of the container, while keeping the temperature constant.

$$K = -V\left(\frac{dP}{dV}\right)_T \Rightarrow K = \rho\left(\frac{dP}{d\rho}\right)_T \tag{1.2}$$

If the temperature changes, the variation in volume can be expressed as,

$$\Delta V = \alpha_T V \Delta T \tag{1.3}$$

Being $\alpha$ the volumetric expansion coefficient. Written in differential terms,

[5] A continuum is defined as a substance that, at a macroscopic level, does not discern between the inter-molecular spaces and molecules themselves, which allows applying the derivation rules by defining the limits of space, surface or volume to zero.

$$\frac{dV}{V} = \alpha_T dT = -\frac{d\rho}{\rho} \tag{1.4}$$

Writing now the density as a general function of the pressure and temperature and calculating the total differential of that function, it renders,

$$\rho = f(p, T) \Rightarrow d\rho = \left(\frac{\partial \rho}{\partial p}\right)_T dp + \left(\frac{\partial \rho}{\partial T}\right)_p dT \tag{1.5}$$

From Eqs. 1.2 and 1.4, the partial derivatives of the density with respect to the pressure and temperature are obtained, respectively. Introducing these in the previous equation, it renders,

$$d\rho = \frac{\rho}{K} dp - \alpha_T \rho dT \Rightarrow \frac{d\rho}{\rho} = \frac{1}{K} dp - \alpha_T dT \tag{1.6}$$

For the sake of this demonstration, one can call $\alpha_p$ to the compressibility modulus $\beta$. In reality, the values of $K$ and $\alpha_T$ are also functions of the pressure and temperature. If they variability within the range in study is high, then the integration becomes more difficult, but for a small range they can be considered constant. Thus, integration of the previous equation renders,

$$\ln \rho = \alpha_p p - \alpha_T T + C \tag{1.7}$$

In order to solve this, a reference density is needed, at known pressure and temperature.[6] Then,

$$\ln \frac{\rho}{\rho_0} = \alpha_p (p - p_0) - \alpha_T (T - T_0) \tag{1.8}$$

Consequently, the variation of density in liquids due to pressure and temperature variations can be found from a base value $\rho_0$, with the following expression:

$$\rho = \rho_0 \cdot e^{[\alpha_p (p - p_0) - \alpha_T (T - T_0)]} \tag{1.9}$$

A property that plays a major role in the transport of fluids through pipelines and fluid mechanics in general is the viscosity. In fact, the viscosity is related to the pressure loss, and in this case as well its variations due to pressure and temperature will affect specially the solution verification in the finite approximations study. The viscosity in liquids decreases with increasing temperature, in contrary to what occurs in gases. In general, the influence of temperature is important in the case of crude oils or refined products (See Figs. 1.10 and 1.11).

The dynamic or absolute viscosity $\mu$ is defined as the property of a fluid that offers resistance to the relative movement of its molecules, which is denoted as

[6] In general, obtained in laboratory tests at atmospheric pressure and 15 °C of temperature.

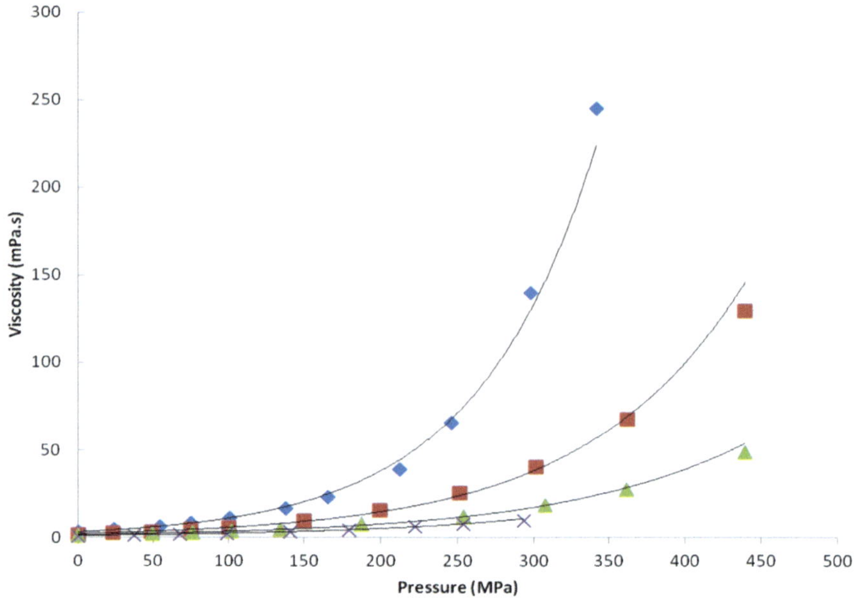


**Fig. 1.10** Variation of viscosity for a Diesel Fuel as a function of temperature and pressure for: blue diamonds at 298 K; red squares at 323 K; green triangles at 348 K; and the crosses at 373 K (Schaschke et al. 2013)

a shear stress. In the International System of Units (SI), the dynamic viscosity is expressed in *Pascal·seconds* (Pa·s).

In order to simplify our study, one use a formulation analogous to that found for the variation of density under pressure and temperature changes. Thus, the viscosity variation in liquids due to pressure and temperature variations can be found, defining a known reference viscosity $\mu_0$ (for the reference pressure and temperatures $p_0$ and $T_0$, respectively), according to the following:

$$\mu = f(p, T) \Rightarrow \mu = \mu_0 \cdot e^{[A_p(p-p_0)-A_T(T-T_0)]} \tag{1.10}$$

## 1.3.3 Compressible Flows

Compressible fluids, commonly called gases, are those that expand until they meet the walls of the container and fill the entire space available to them. This is because the gas molecules are widely spaced and the cohesive forces between them are weak and, unlike liquids, they cannot form a free surface (Mohitpour et al. 2007).

Gas and vapor are often used interchangeably. The vapor phase of a fluid is called a gas when it is above its critical temperature, and by vapor it is usually implied that the gas is not far from its condensation state.

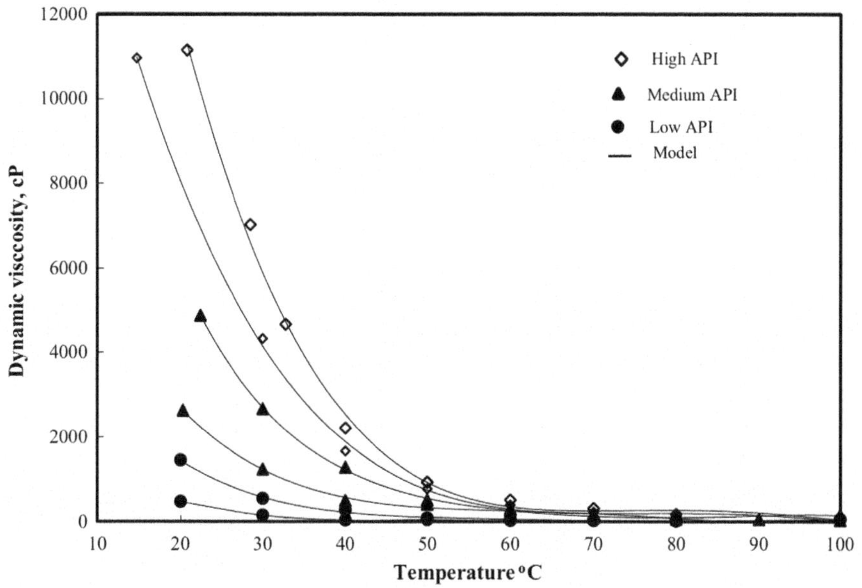


**Fig. 1.11** Variation of the viscosity as a function of the temperature for three different API grade oils (Alomair et al. 2016)

## Real and Ideal Gases—Compressibility Factor

Analogous to liquids, density with pressure and temperature are related via the equation of state.[7] The general form of any equation of state can be written as $f(p, V, T) = 0$, and for the specific case of ideal gases is (Mansour 2020),

$$\frac{p}{\rho} = RT \tag{1.11}$$

in which the absolute pressure of the gas, the density and the absolute temperature are related through the particular constant of the gas $R$. This constant of the ideal gas or mixture of gases (considered as ideal) is obtained by dividing the universal constant $R_u = 8314.5$ J/kmol·K by the molecular weight of the gas (pure or mixed).

In the case of real gases, better, improved correlations must be adapted to described the relationship among the state parameters. There are several functions proposed, for instance, the cubic equations of state (e.g. van der Waals, Redlich-Kwong, Peng-Robinson) (Redlich and Kwong 1949; Peng and Robinson 1976; Soave 1972; Tosun 2021). A different approach can be followed by using the concept of *compressibility factor*, defined as $Z_c$. The equation of state of real gases can be then written as,

[7] This is a thermodynamic equation relating the states variables of a system, which are used to describe the state of matter.

$$\frac{p}{\rho} = Z_c RT \tag{1.12}$$

The compressibility factor shows the deviation between the ideal gas model and its real behavior. It depends on the pressure, the temperature and the composition of the gas. This factor is based on the principle of corresponding states, indicating that any pure gas (or mixture of them) at the same reduced pressure and temperature, should exhibit the same deviation from the ideal gas law, i.e., the same compressibility factor. The reduced parameters are defined by $T_r = \frac{T}{T_c}$ and $p_r = \frac{p}{p_c}$. In these two expressions, the $p_c$ and $T_c$ are the critical pressure and temperature of the (mixture of) gas, respectively. The critical temperature is that above which it is not possible to liquify the gas, while the critical pressure is the pressure required to liquify a gas at the critical temperature. They defined what is known as the critical point of the material, above which there is no clear distinction between liquid and gas phases. This is the area of supercritical fluids, which has become an interesting area of research due to the properties exhibited by the fluids (e.g. supercritical $CO_2$ acting as a green solvent) (Dune and Oriji 2007). This is known as the *two-parameter theory of corresponding states*,

$$Z_c = f(p_r, T_r) \tag{1.13}$$

Although this equation renders nearly exact results for simple gases (e.g. Argon, Xenon), it presents deviations as the molecules become larger and more complex in its structure. Therefore, Pitzer introduced in 1955 a third factor in the equation, the *acentric factor* $\omega$, in what is known now as the *three-parameter theory of corresponding states*,

$$Z_c = f(p_r, T_r, \omega) \tag{1.14}$$

The acentric factor mainly takes into account the non-sphericity of the molecules. In the natural gas, composed primarily of methane, its acentric factor will tend to zero, and thus the Eq. 1.13 is used to described its real behavior. However, when analyzing the flow of different molecules, such as hydrogen or $CO_2$, it can be seen from the table that Eq. 1.14 must be used to calculate their compressibility factors.

For natural gases (hydrocarbon gas mixtures, with a preponderance of methane) it has been found that the compressibility factors can be generalized with sufficient accuracy for most engineering calculations by introducing the concept of pseudo-reduced pressures and temperatures based on pseudo-reduced critical pressures and temperatures (in all cases applying measured values in absolute scales). These can be found using Kay's (1936) mixing rules (Kay et al. 1936). Several publications, such as the GPSA Engineering Data Book (graph 16-3 GPSA) or authors such as Standing and Katz, Hall and Yarborough, and more recently Dranchuk, Purvis and Robinson have established equations and different mathematical methods to avoid the use of graphs, and facilitate the calculation of the $Z_c$ in the calculation programs for flow in gas pipelines (Dranchuk et al. 1973; Standing and Katz 1942) (Fig. 1.12).

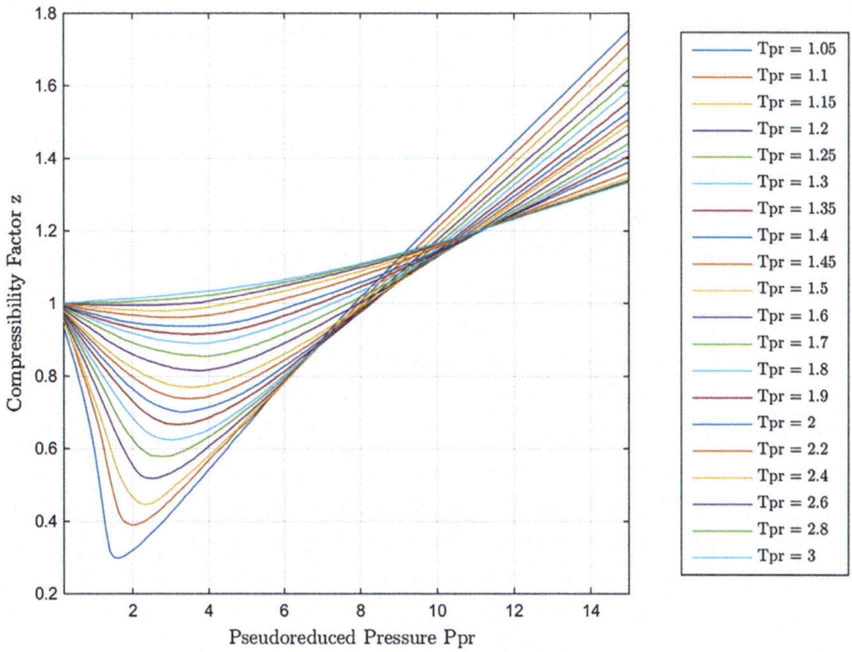


**Fig. 1.12** Compressibility factor for gases with $M_w < 40$ (Kareem et al. 2016)

### Mathematical Estimation of the Compressibility Factor $Z_c$

At a certain pressure and temperature, the models to calculate the thermodynamic properties of real gases are based on the Equation of State of the gas, corrected by the compressibility coefficient $Z_c$. In the case of a gas of any known composition, this factor can be determined with relatively good precision by graphic methods, using the set of curves (GPSA graph) in reduced coordinates, $T_r$ and $P_r$. If it is a mixture of gases, the pseudo-critical reduced temperatures and pressures are obtained by weighing the critical temperatures and pressures of the gas constituents, weighted using their molecular compositions.

A mathematical method for calculating the compressibility factor is generally more convenient than using the graphical method of Standing and Katz (1942), since this allows computer calculations and a better precision in the determination of the compressibility factor. Some are simple correlations derived from the mentioned diagram, while other authors prefer iterative calculations of the compressibility factor. Other methods have proceeded to calculate the factor based on other Equations of State. The precision obtained by these methods in calculating the compressibility factor is around 0.5%. The application of these equations requires the previous calculation of the pseudo-critical coordinates and then based on these, the pseudo-reduced pressure and temperature, knowing the molar composition of the gas mixture.

In the case for natural gas, the first step is then to find the critical pressure and temperature based on the gas composition. Nevertheless, there are approximations based, for instance, on the gas relative density,[8] as defined by Thomas (1970),

$$p_c = 4.892 - 0.405 \cdot \rho_{rel} \qquad p = [MPa] \tag{1.15a}$$

$$T_c = 94.72 + 170.75 \cdot \rho_{rel} \qquad T = [K] \tag{1.15b}$$

Another commonly used relationship to calculate the critical parameter of the mixture of the Sutton correlation (Sutton 1985),

$$p_c = 52.1795 - 9.0321 \cdot \gamma_g - 0.2482 \cdot \gamma_g^2, \qquad p = [bar] \tag{1.16a}$$

$$T_c = 94 + 194.1666 \cdot \gamma_g - 41.1111 \cdot \gamma_g^2, \qquad T = [K] \tag{1.16b}$$

The presence of certain molecules, like sour gases (e.g. $CO_2$ and $H_2S$), affects the critical parameters of the mixture and therefore they must be considered when calculating these values. A good correlation considering the presence of these gases is the following set of equations (Ahmed 1989; Guo 2011),

$$p_c = 46.75 - 3.45 \cdot (\gamma_g - 0.5) - 14.25 \cdot y_{N_2} + 30.34 \cdot y_{CO_2} + 41.83 \cdot y_{H_2S}, \quad p = [bar] \tag{1.17a}$$

$$T_c = 181.11 + 175.39 \cdot (\gamma_g - 0.5) - 133.33 \cdot y_{N_2} - 46.28 \cdot y_{CO_2} + 74.06 \cdot y_{H_2S}, \quad T = [K] \tag{1.17b}$$

Where $y_i$ represents the mole fraction of the species $i$ in the natural gas. With respect to the compressibility factor for natural gas, a simple approximation is,

$$Z_C = 1 + \left(\frac{0.257}{p_C} - \frac{0.533}{p_C}\frac{T_C}{T}\right) p_m \qquad p_C \wedge p_m = \left[\frac{kg}{cm^2}\right] \tag{1.18}$$

$$p_m = \frac{2}{3}\frac{p_1^3 - p_2^3}{p_1^2 - p_2^2} \tag{1.19}$$

If greater accuracy is required, the Dranchuck-Purvis-Robinson equation (DPR) (Dranchuk et al. 1973) can be used to obtain the compressibility factor, with the difficulty that it is an implicit expression for the $Z_c$, requiring successive approximations to be solved.

$$Z_c = \left[1 + \left(A_1 + \frac{A_2}{T_r} + \frac{A_3}{T_r^3}\right)\rho_r + \left(A_4 + \frac{A_5}{T_r}\right)\rho_r^2 + \frac{A_5 A_6}{T_r} \cdot \rho_r^5 + \right.$$
$$\left. + A_7 \frac{\rho_r^3}{T_r^3} \frac{1}{\left(1 + A_8 \rho_r^2\right) \cdot e^{-A_8 \rho_r^2}}\right] \tag{1.20}$$

[8] This can be calculated as the quotient between the gas molecular mass, approximately 18 g/mol and the air's molecular mass, which is 28.9647 g/mol.

With the pseudo-reduced density $\rho_r$ being equal to,

$$\rho_r = \frac{0.27 \cdot p_r}{Z_c \cdot T_r} \tag{1.21}$$

This renders the DPR method an implicit one, requiring several iterations to calculate the compressibility factor. The DPR method renders accurate results for the range of $1.05 < T_r < 3$ and $2 < p_r < 30$. The coefficients $A_i$ in the equations are equal to,

$A_1 = 0.31506237$ $A_2 = 1.04670990$ $A_3 = 0.57832729$ $A_4 = 0.53530771$
$A_5 = 0.61232032$ $A_6 = 0.10488813$ $A_7 = 0.68157001$ $A_8 = 0.68446549$

As mentioned before, several correlations have been developed over the years trying to express mathematically the calculation of the compressibility factor. The reader is kindly requested to consult relevant literature, for instance, Benedict (Benedict et al. 1940), Hall-Yarborough (Hall and Yarborough 1973), Dranchuk-Abu-Kassem (DAK) (Dranchuk and Abou-Kassem 1975), Azizi (Azizi et al. 2010) or Mahmoud (Mahmoud 2014).

Regarding the *three-parameter theory of corresponding states*, first the formal definition of the acentric factor should be introduced. It can be expressed mathematically as a function of the reduced pressure,

$$\omega = -\log_{10}\left(p_r^{sat}\right) - 1, \quad \text{at} \quad T_r = 0.7 \tag{1.22}$$

The functional form of the compressibility factor as a function of the acentric factor is,

$$Z_c = Z_c^0\left(T_r, p_r\right) + \omega Z_c^1\left(T_r, p_r\right) \tag{1.23}$$

Though Pitzer (1955, 1955) calculated the values of $Z_c^0$ and $Z_c^1$ for different materials as a function of the reduced pressure and temperature, they were improved at a later stage by Lee and Kesler (1975) using a modified equation of state (Benedict–Webb–Rubin), as shown in the following equation, (Tosun 2021)

$$\frac{p_r \hat{v}_r}{T_r} = 1 + \frac{B}{\hat{v}_r} + \frac{C}{\hat{v}_r^2} + \frac{D}{\hat{v}_r^5} + \frac{c_4}{T_r^3 \hat{v}_r^2}\left(\beta + \frac{\gamma}{\hat{v}_r^2}\right) e^{-\frac{\gamma}{\hat{v}_r^2}} \tag{1.24}$$

where the variables are calculated by,

$$B = b_1 - \frac{b_2}{T_r} - \frac{b_3}{T_r^2} - \frac{b_4}{T_r^3} \tag{1.25a}$$

$$C = c_1 - \frac{c_2}{T_r} + \frac{c_3}{T_r^3} \tag{1.25b}$$

**Table 1.1** Constant values used in the calculation of the compressibility factor by the three-parameter method

| Constant | Simple fluid | Reference fluid |
|---|---|---|
| $b_1$ | 0.1181193 | 0.2026579 |
| $b_2$ | 0.265728 | 0.331511 |
| $b_3$ | 0.154790 | 0.027655 |
| $b_4$ | 0.030323 | 0.203488 |
| $c_1$ | 0.0236744 | 0.0313385 |
| $c_2$ | 0.0186984 | 0.0503618 |
| $c_3$ | 0 | 0.016901 |
| $c_4$ | 0.042724 | 0.041577 |
| $d_1$ $(\times 10^4)$ | 0.155488 | 0.48736 |
| $d_2$ $(\times 10^4)$ | 0.623689 | 0.0740336 |
| $\beta$ | 0.65392 | 1.226 |
| $\gamma$ | 0.060167 | 0.03754 |

$$D = d1 + \frac{d_2}{T_r} \tag{1.25c}$$

All the constants appearing in Eqs. 1.24 and 1.25 were tabulated for two different fluids: a simple one, with an $\omega = 0$; and a reference one (*n-octane*), with an $\omega = 0.3978$. These are found in the Table 1.1.

Therefore, the compressibility factor using the three-parameter method is calculated according to the following steps (Table 1.2),

**Table 1.2** Critical values and acentric factors for the elements most commonly transported by pipelines

| Element | $T_c[K]$ | $p_c[MPa]$ | Acentric factor |
|---|---|---|---|
| Argon | 150.8 | 4.87 | 0.001 |
| Helium-4 | 588 | 10.3 | 0.108 |
| Hydrogen | 33 | 1.29 | –0.216 |
| Nitrogen | 126.2 | 3.39 | 0.039 |
| Oxygen | 154.6 | 5.04 | 0.025 |
| n-Butane | 425.2 | 3.80 | 0.199 |
| Ethane | 305.4 | 4.88 | 0.099 |
| Methane | 190.4 | 4.60 | 0.011 |
| Propane | 369.8 | 4.25 | 0.153 |
| Carbon Dioxide | 304.1 | 7.38 | 0.239 |
| Water | 647.3 | 22.12 | 0.344 |

1. Solve equation 1.24 with the parameter of *simple fluid* and obtain $\hat{\upsilon}_{r,sf}$.
2. Calculate $Z_c^0 = \frac{p_r \hat{\upsilon}_{r,sf}}{T_r}$
3. Solve equation 1.24 with the parameter of *reference fluid* and obtain $\hat{\upsilon}_{r,ref}$.
4. Calculate $Z_c^{ref} = \frac{p_r \hat{\upsilon}_{r,ref}}{T_r}$
5. Calculate $Z_c^1 = \frac{Z_c^{ref} - Z_c^0}{0.3978}$
6. Calculate the compressibility factor $Z_c = Z_c^0 + \omega Z_c^1$

There are different Equations of State (EoS) providing accurate results for the real behavior of gases using the *three-parameter method*, such as the Benedict-Webb-Rubin previously mentioned. Nonetheless, there is another type of EoS, the *cubic equations of state*, giving reasonable resuls to determine the behavior of real gases. These are called like this because they can be rewritten as cubic functions of the molar volume. Among these cubic EoS, one can mention Van der Waals (1873), Redlich-Kwong (1949), Soave-Redlich-Kwong (SRK) (1972), and Peng-Robinson (PR) (1976).

## Viscosity

It is observed from the values recorded for gases at different temperatures that, in contrast to liquids, the dynamic viscosity increases with temperature, and also increases with pressure. This is due to the molecular structure of gases, since their molecules have more freedom than those of solids and liquids. Therefore, when the pressure and/or temperature increase, these molecules have a greater kinetic energy, increasing the amount of collisions and thus decreasing their transfer of momentum between the fluid layers. As a general rule, it can be said that the viscosity of any (mixture of) gas(es) will depend on the temperature, pressure and its composition.

$$\mu = f(p, T, y_i) \vee \mu = f(\rho, T, y_i), \qquad i = \text{number of components} \tag{1.26}$$

The viscosity variation with the temperature can be approximated using the Sutherland's equation (Sutherland 1893),

$$\mu = \mu_0 \sqrt{\frac{T}{273.15}} \frac{1 + \frac{C}{273.15}}{1 + \frac{C}{T}} \tag{1.27}$$

where $\mu_0$ is the dynamic viscosity at 0 °C; $T$ is the absolute temperature at which the viscosity should be calculated; and $C$ is the Sutherland's constant (for instance, 113 for air; 240 for carbon dioxide; 72 for hydrogen; 169 for methane).

The pressure influences only when its value is very high, for example the viscosity of the air at 50 bar increases approximately 7% with respect to the standard condition (approx. 1 bar).

For natural gas, there are many polynomial expressions to estimate its viscosity, frequently used in the industry, such as Dempsey (Carr et al. 1954; Dempsey et al. 1965), LBC (Lohrenz et al. 1964), LGE (Lee et al. 1966), VW (Vesovic and Wakeham 1989; Vesovic 2001), SLP (Sanjari et al. 2011), JAH (Jarrahian et al. 2015) or Yang (Yang et al. 2017). For instance, the Lee-Gonzalez-Eakin (LGE) is expressed as,

$$\mu = 10^-4 K e^{X \rho_g^Y} \tag{1.28}$$

$$K = \frac{(9.379 + 0.016 M_w)(1.8T)^{1.5}}{209.2 + 19.26 M_w + 1.8T} \tag{1.29a}$$

$$X = 3.448 + \frac{986.4}{1.8T} + 0.01009 M_w \tag{1.29b}$$

$$Y = 2.447 - 0.2224X \tag{1.29c}$$

where $\rho_g$ is the density of the natural gas density in g/cm$^3$; $M_w$ is the gas molecular weight in g/mol; $T$ is the temperature in K; and finally, $\mu$ is the viscosity of the natural gas viscosity, expressed in mPa·s.

It is noteworthy to mention that one should previously check for non-hydrocarbon components (i.e. $CO_2$, $N_2$ and $H_2S$) and corrections to the critical properties and the viscosity should be made if these are present in concentrations greater than 5 mole percent.

Another empirical polynomial expression was developed by Sanjari et al. (2011). He presented the viscosity as a function of the reduced pressure and temperatures in the ranges of $1.01 \le T_{pr} \le 3.0$ and $0.01 \le p_{pr} \le 21$ in the following rational expression,

$$\mu = 10^{-4} \frac{A(p_r) + B(T_r)}{C(p_r + D(T_r)}, \qquad \mu = [mPa \cdot s] \tag{1.30}$$

where,

$$A = \alpha_1 + \alpha_2 p_{pr} + \alpha_3 p_{pr}^2 + \alpha_4 ln(p_{pr}) + \alpha_5 ln^2(p_{pr}) \tag{1.31a}$$

$$B = \frac{\alpha_6}{T_{pr}} + \alpha_7 ln^2(T_{pr}) \tag{1.31b}$$

$$C = 1 + \alpha_8 p_{pr}^2 \tag{1.31c}$$

$$D = \frac{\alpha_9}{T_{pr}} + \frac{\alpha_{10}}{T_{pr}^2} + \frac{\alpha_{11}}{T_{pr}^3} \tag{1.31d}$$

The constants $\alpha_i$ are the coefficients used to minimize the least-square errors and are presented in the following table (Table 1.3),

**Table 1.3** Constant values used in the calculation of the viscosity by the Sanjari correlation

| Constant | Value |
|---|---|
| $\alpha_1$ | –0.141645 |
| $\alpha_2$ | 0.018076 |
| $\alpha_3$ | 0.00214 |
| $\alpha_4$ | –0.004192 |
| $\alpha_5$ | –0.000386 |
| $\alpha_6$ | 0.187138 |
| $\alpha_7$ | 0.569211 |
| $\alpha_8$ | 0.000387 |
| $\alpha_9$ | –2.857176 |
| $\alpha_{10}$ | 2.925776 |
| $\alpha_{11}$ | –1.062425 |

**Table 1.4** Constant values to be applied to calculate viscosity and density of pure water

| Density | | Viscosity | |
|---|---|---|---|
| $\alpha_p\ 1/Pa$ | $\alpha_T\ [1/K]$ | $A_p\ 1/Pa$ | $A_T\ [1/K]$ |
| $4597 \times 10^{-10}$ | $2.234 \times 10^{-4}$ | $6.9322 \times 10^{-10}$ | $2.40 \times 10^{-2}$ |

## *1.3.4 Examples*

### Water

Hereunder the values with the approximate values of the viscosity and density coefficient for Eqs. 1.9 and 1.10 are presented to take into account the effects of pressure and temperature on the viscosity and density, respectively. These values are applicable for the range $4\ °C < T < 45\ °C$ and pressures between $1\ bar < p < 103\ bar$ (Table 1.4).

### Crude Oil and Natural Gas

The following tables show the different ways of measuring densities and viscosities in the oil industry, expressed in SI units as well as its equivalence in API gravity (°API). The examples correspond to the different types of crude oil transported in Argentina by the oil state company YPF, which are mainly extracted in the southern part of the country (see Tables 1.5 and 1.6). Finally, an estimate of the density and viscosity correction coefficients for the Eqs. 1.9 and 1.10 is presented (see Table 1.7).

$$°API = \frac{141.5}{SG} - 131.5 \tag{1.32}$$

**Table 1.5** Typical characterization of crude oil based on the API gravity

| Crude oil type | Density $kg/m^3$ | API gravity (°API) |
|---|---|---|
| Extra heavy | >1000 | <10 |
| Heavy | 1000–920 | 10–22.3 |
| Medium | 920–870 | 22.3–31.1 |
| Light | 870–830 | 31.1–39 |
| Superlight (condensate) | 830–650 | 39–80 |
| LNG/CNG | <650 | >80 |

**Table 1.6** Typical properties of Argentinian crude oils

| Procedence | Escalante | Cañadón seco | Neuquén/Río Negro | Hydra | María Inés | Tierra del Fuego |
|---|---|---|---|---|---|---|
| Kinematic viscosity @ 10 °C [cst] | 7124.1 | 3341.1 | 14.9 | 7.3 | 8.0 | N/A |
| Kinematic viscosity @ 20 °C [cst] | 1305.3 | 733.5 | 13.5 | 6.4 | 7.4 | N/A |
| Kinematic viscosity @ 30 °C [cst] | 483.7 | 302.2 | 12.1 | 5.9 | 6.8 | N/A |
| Density min. [$kg/m^3$] | 907.2 | 904.2 | 848.8 | 736.1 | 777.0 | >870.0 |
| Density max. [$kg/m^3$] | 909.7 | 905.4 | 861.7 | 766.4 | 782.8 | >870 |
| API gravity [°API] | 22.3-31.1 | 22.3-31.1 | 31.1 | > 31.1 | >31.1 | > 31.1 |
| Type | Medium | Medium | Medium | Light | Light | Light |

**Table 1.7** Density and viscosity coefficients calculated for a medium oil type

| Density base $kg/m^3$ | Density $\alpha_p$ $1/bar$ | Density $\alpha_T$ $1/K$ | Viscosity $A_p$ $1/bar$ | Viscosity $A_T$ $1/K$ |
|---|---|---|---|---|
| 990 | $1.4831 \times 10^{-5}$ | $6.28 \times 10^{-4}$ | $25.76 \times 10^{-5}$ | 0.05 |
| 934 | $1.6747 \times 10^{-5}$ | $7.05 \times 10^{-4}$ | $25.76 \times 10^{-5}$ | 0.05 |
| 876 | $1.8869 \times 10^{-5}$ | $8.01 \times 10^{-4}$ | $25.76 \times 10^{-5}$ | 0.05 |
| 825 | $2.0864 \times 10^{-5}$ | $9.03 \times 10^{-4}$ | $25.76 \times 10^{-5}$ | 0.05 |
| 769 | $2.3200 \times 10^{-5}$ | $10.40 \times 10^{-4}$ | $25.76 \times 10^{-5}$ | 0.05 |

Where $SG$ is the specific gravity of relative density of the crude oil with respect to water (Table 1.5).

**Example: Densities and viscosities of crude oil**

With the following laboratory measurements from a crude oil, you are asked to calculate the viscosity and the density at the pipeline conditions, using the Eqs. 1.9 and 1.10.

- Reference Temperature = 290 K
- Reference pressure = 1 bar
- Reference density = 825 kg/m$^3$
- Reference viscosity = 150 mPa·s
- Specific heat = 1800 J/kgK
- Thermal conductivity = 0.014 W/mK
- Pressure in the pipeline = 50 bar(g)
- Temperature in the pipeline = 293 K

With the information from Table 1.7, one can calculate the new values of viscosity and density.

$$\mu = \mu_0 \cdot e^{[A_p(p-p_0)-A_T(T-T_0)]} = 150\,\text{mPa} \cdot \text{s} \cdot e^{[25.76\times10^{-5}\cdot 49-0.05\cdot 3]} = 130.75\,\text{mPa} \cdot \text{s}$$

$$\rho = \rho_0 \cdot e^{[\alpha_p(p-p_0)-\alpha_T(T-T_0)]} = 825\,\frac{\text{kg}}{\text{m}^3} \cdot e^{[2.0864\times10^{-5}\cdot 49-9.03\times10^{-4}\cdot 3]} = 823.60\,\frac{\text{kg}}{\text{m}^3}$$

**Example: Densities and viscosities of natural gas**

Hereunder you can see two tables, corresponding to the variation of the dynamic viscosity with the temperature (at 1.033 bar—first table) and the pressure (at 288 K—second table). The fluid information is as follows,

- Molecular weight = 17.773 g/mol
- Relative density = 0.614
- Reference viscosity (at 0 °C) = 0.0105 mPa·s
- Pressure at the pipeline = 71 bar(a)
- Temperature at the pipeline = 27 °C ≈ 300 K
- Thermal conductivity = 0.087 W/mK
- Specific heat = 2600 J/kgK

Calculate with this information the compressibility coefficient, absolute density and viscosity of the natural gas at the pipeline conditions, applying the equations described in this chapter.

| Temperature [°C] | 0 | 20 | 50 | 100 |
|---|---|---|---|---|
| $\mu[mPa \cdot s]$ | 0.0102 | 0.0108 | 0.0117 | 0.0133 |

| Pressure [bar(g)] | 1 | 20 | 60 | 100 |
|---|---|---|---|---|
| $\mu[mPa \cdot s]$ | 0.0107 | 0.0112 | 0.0124 | 0.0143 |

The first step is the compressibility factor. Using the equation from Hankinson, Thomas, and Phillips,

$$p_c = 48.9252 - 40.4722 \cdot \gamma_g = 46.44, \qquad p = [bar]$$

$$T_c = 94.7222 + 170.7222 \cdot \gamma_g = 199.55, \qquad T = [K]$$

Applying the Sutton correlation, one verifies the values obtained before,

$$p_c = 52.1795 - 9.0321 \cdot \gamma_g - 0.2482 \cdot \gamma_g^2, \qquad p = [bar]$$

$$T_c = 94 + 194.1666 \cdot \gamma_g - 41.1111 \cdot \gamma_g^2, \qquad T = [K]$$

With these values, one obtains the pseudo-reduced pressure and temperatures,

$$p_{sr} = \frac{p}{p_c} = 1.53$$

$$T_{sr} = \frac{T}{T_c} = 1.50$$

Using these values in Fig. 1.12, an approximate compressibility factor of $Z_c = 0.855$ is obtained. Using the empirical equations presented in this chapter,

$$Z_c = 1 + \left(\frac{0.257}{p_c} - \frac{0.533}{p_c}\frac{T_c}{T}\right) p_m = 0.851$$

The next step consists in calculating the gas density. In order to do that, first the density of the air at standard conditions (288 K and 101325 Pa) is calculated,

$$\rho_{air} = \frac{p}{R_{p,air} T} = \frac{101325\text{Pa}}{287.053 \frac{\text{J}}{\text{kgK}} 288\text{K}} = 1.2256 \frac{\text{kg}}{\text{m}^3}$$

$$\rho_{gas,STD} = \gamma_g \cdot \rho_{air} = 0.7513\frac{\text{kg}}{\text{m}^3}$$

$$\rho_{gas} = \frac{p_{pipe}}{Z_c R_{p,gas} T_{pipe}} = \frac{70 \times 10^5 \text{Pa}}{0.851 \frac{R_{univ}}{M_w} 300\text{K}} = 59.44\frac{\text{kg}}{\text{m}^3}$$

Finally, one proceeds to calculate the gas viscosity at the pipeline conditions. Using firstly Sutherland's correlation, considering only the temperature,

$$\mu = \mu_0 \sqrt{\frac{300}{273.15}} \frac{1 + \frac{169}{273.15}}{1 + \frac{169}{300}} = 0.0105\,\text{mPa}\cdot\text{s}\cdot 1.0854 = 0.0114\,\text{mPa}\cdot\text{s}$$

### *1.3.5 Heterogeneous Fluids*

It has been previously mentioned that in order to make possible the economical and reliable transport of any solid substances, they are mixed in adequate proportions with a widely available carrier fluid (i.e. water or air), which is generically called heterogeneous fluid flow, although a distinction must be made between one another. This denomination of heterogeneous fluids is also given to the immiscible fluids that are transported together in the pipe (although more frequently they are called biphasic or multiphasic flows).

In the case of solids—which do not naturally have an adequate granulometry—they must be subjected to grinding to bring them to a powdery condition that allows them to be incorporated in a dosified way into the carrier fluid prior to entering the pipeline. In all cases, the substance to be transported must not be affected in its quality by the carrier fluid, and furthermore, at the end of the journey, it must be possible to separate the transported product from the carrier fluid, in a relatively simple manner.

In the case of liquids and gases, in general they are fluids that are naturally mixed (e.g. oil and natural gas or oil, natural gas and water) and are called biphasic or triphasic flows. This happens specially in the upstream industry, after the extraction from the reservoir, i.e., at the wellhead. Its joint transport is economically convenient to take it to the plant that will be responsible of the (primary) separation of the products.

Although less common in the industry, some association of solids-liquids with gases can be transported together, for reasons of convenience.

## 1.4 Transport Systems

### *1.4.1 Pipelines—Mechanical Perspective*

A fluid transport system is basically made up of a pipeline that physically joins, by means of a continuous line of pipes, the injection point of the product or products in the so-called *head* of the pipeline, with the point of delivery or supply of the consumption to be satisfied.

The product or products enter the header at a certain pressure, depending on the length to be transited, intermediate pumping/compression plants are incorporated that keep the pressure drops within certain limits so as to enable delivery at the supply point with a pressure that is equal to or greater than that required by the supplied system, depending on whether it is supply systems for online consumption, to tanks/storage, or to processing plants.

Pipelines for long-distance transportation of fluids are generally built by welding (rolled-sheet) steel pipes together. Synthetic or composite materials have not had a great application in the industry for this type of system up to date and are limited to low/medium pressure applications, short distances and special fluids (e.g. transportation of $H_2$ by means of non-metallic pipelines). This is because steel pipes have a resistance well determined by relatively simple tests, and also the homogeneity and isotropy of steel allow simplified calculations. Synthetic or composite materials are generally anisotropic, which makes the choice of safety coefficients or allowable stresses difficult.

For several decades now, ducts have been installed underground for safety and environmental reasons, which in the most usual case—those built with carbon steel pipes—requires external protection from corrosion (due to the aggressiveness of the land), through anticorrosive coatings of different types (modernly, 3-layered high-density polyethylene or epoxy paints). There are some exceptions to this general rule, for instance, the Alaska oil pipeline, mounted above ground over variable spring supports (see Fig. 1.13) due to the extreme temperature conditions found in the region.

In some particular cases in which the flow temperature of the fluid or the state of the surrounding terrain must be preserved (for example, those of the "permafrost" type[9]), a thermal insulating coating is superimposed on the anticorrosive coating. Although it is not so frequent, the interior of the pipes is also covered in the case of fluids that are transported with some proportion of corrosive components,[10] or in the case of erosive materials that have been fluidized for transport (e.g., mineral pipelines) (Figs. 1.14 and 1.15).

---

[9] In areas of the Arctic (Alaska, Siberia, etc.), where the pipes are installed buried in permanently frozen ground, this condition must be preserved to avoid the loss of stability (and eventual breakage) of the pipes that would occur if the ground thaws around the pipe.

[10] The regulations for long-distance transport of hydrocarbon compounds generally restrict the presence of elements that may be corrosive in the presence of water in the fluids to be transported, generally called "minimum quality of the product to be transported".

**Fig. 1.13** Alaska Pipeline, photo taken by Galuzzi (2005)

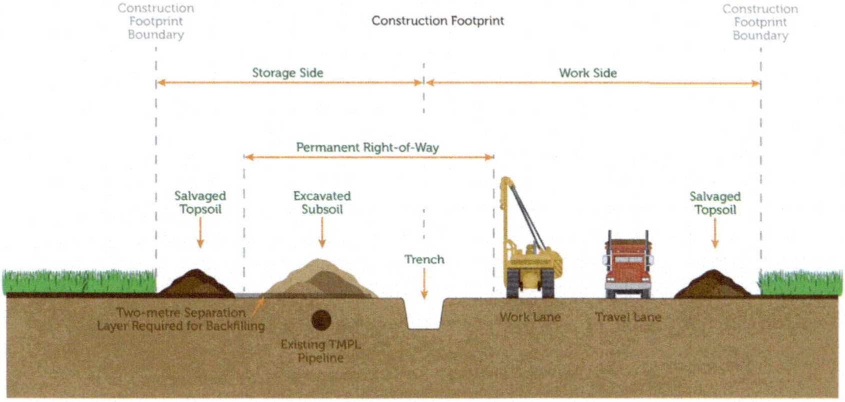


**Fig. 1.14** Schematic cut of the construction of a pipeline (Reproduced with permission of Trans Mountain Corporation)

## *1.4.2 Pipeline Efficiency*

All pipes are hydrostatically tested after construction, and consequently must be drained and dried. At this time, even a new pipe, which is not internally coated, will be covered with oxides and manufacturing scales and will contain all kinds of materials resulting from construction. The pipes must be protected against internal corrosion by eliminating all presence of liquid water before the commissioning of the line. If the line is dried prior to commissioning and the transported product is free of water, there is little chance of internal corrosion or hydrate formation.

**Fig. 1.15** Construction of the NEL gas pipeline near Hasenhäge, Germany (Niteshift 2011)

The *flow efficiency* of a pipeline is defined as the relationship between the measured flow and the flow calculated by the flow equation applied for the same conditions that prevail at the time. The delivery capacity of a line depends on the following physical parameters of the pipeline:

- Roughness of the pipe wall.
- Inside diameter.
- Factors that induce localized losses, such as curves, flanges, valves, circumferential welds, etc.

## Pipe Cleaning

A line that has dirt, rust, and scale deposits on its walls will be rougher, exert greater resistance to flow, and therefore have reduced ability to deliver product. At the same time, depending on the amount of material deposited on the walls, the inside diameter of the pipe will be reduced, reducing the volumetric capacity of the line. Both of these conditions—pipe roughness, which causes more friction, and reduced cross-sectional area—can be eliminated by cleaning the pipe. The localized resistances are part of the original design and will not be modified by the line cleaning operation. There are three main methods for cleaning installed pipes:

- Pigging.
- Sandblasted/shot blasted.
- Chemical cleaning.

In the case of gas pipelines, the exclusively used method,and as old as the gas pipelines themselves, is the passage of "scrapers, pigs or rabbits" plugs, there are different models and types. The length and size of the line that can be cleaned by this method depends on the gas flow and the pressure required to keep the pigs moving at an average speed of around 15 km/h ($\approx$ 4 m/s). The higher the speed of the pigs and the gas, the easier it will be to remove loose manufacturing scale, oxides and dirt. When dry air is used to clean large diameter pipes, large capacity compressors and dryers will be needed to ensure adequate plug velocities.

#### Cleaning Requirements

Although the degrees of cleanliness are difficult to define, there is agreement that a pipe is considered "clean" when the walls are free of oxides, scale and other deposits and the metal surface has the appearance of white metal. The absence of rust is a good barometer of the condition of the pipe, because if all the rust was removed, all other material on the metal can be considered removed.

#### Efficiency Improvements

The increase in friction losses due to roughness and the reduction in capacity due to the reduction of the cross section of the pipe, reduce the efficiency due to some of the following conditions:

- To maintain the same flow, an increase in differential pressure will be required, and in this case, more power. If there is additional power in the line and also there are no pressure limitations, compression costs will increase.
- If there is no additional power, the dirty line will cause a lower arrival pressure for the same flow (and in some systems it will cause problems or the need for additional compression installations).

A measure of the pipe's ability to deliver gas is known as the transmission factor, which is a dimensionless factor directly proportional to the ability to deliver volumes and is a function of frictional effects due to pipe diameter and pipe roughness wall. The enhancement of the effect of friction due to roughness is more pronounced at low roughness values. Therefore the efforts to clean the pipes are more effective in low roughness values and it is proven to be economically convenient if the selected cleaning method can reduce the roughness between approximately 0.0075 mm and 0.01 mm.

The absolute roughness of the bare pipes, leaving the factory, is in the range of 0.013 to 0.02 mm. Its exposure to normal atmospheric conditions for 6 months can increase it up to 0.030 mm, and after 12 months it can reach 0.038 mm. The rate of corrosion and deterioration decreases substantially after 12 months. The same pipe will easily increase its roughness during operation to values of 0.05 mm or more, especially if the line was not cleaned after construction and before construction.

The same pipe will easily increase its roughness during operation to values of 0.05 mm or more, especially if the line was not cleaned after construction and before commissioning. Normal service increases the roughness of the bare metal between 0.004 and 0.005 mm/year. By cleaning with a scraper (brushes or brush pigs) during the normal operating cycle, the effective normal roughness can be maintained at values less than 0.025 mm.

Because pipe efficiency is referred to at the same time as actual performance versus its theoretical performance and is also commonly used as a correction factor in flow calculations, care must be taken when using the term. Part of the confusion comes from the numerous empirical relationships that define the carrying capacity of pipelines and the variation of theoretical flows using the different equations.

### *1.4.3 Pumping and Compression Plants and Ancillary Facilities*

The ones mentioned as driving plants in the previous point are made up of either pump systems (in the case of transporting liquid substances through pipes) or compressors (in the case of gases). Pumping/compression equipment is normally arranged for operation in series or parallel arrangement, and rarely in flexible arrangements that allow operation in series or parallel at the operator's discretion (Mohitpour et al. 2007; Menon and Menon 2010; Hassan 2016).

Both pumps and compressors are machines that add energy to the fluid, although to be precise, a pump is a hydraulic machine, whilst a compressor is a thermal machine (Bentaleb 2015; Moran 2016). In both cases the objective of adding energy to the fluid results in an increase in its pressure and not necessarily in an increase in the speed of the fluid (i.e. kinetic energy) as it passes through the machine.

The pumps can be classified into (Fig. 1.16):

- Dynamics with radial (or centrifugal) type rotors, or with mixed type or axial type rotors.
- Alternative Positive Displacement (piston) or rotary (lobe, vane, gear or screw).

Compressors can be classified into:

- Dynamic with radial (or centrifugal) or axial type rotors.
- Alternatives to integral type pistons (motor-compressors) or separable type.
- Rotating lobes, vanes, etc.

In all cases, the fluid to be transported must arrive adequately clean of impurities to the pump (or the compressor) for which these plants always have filtering systems in

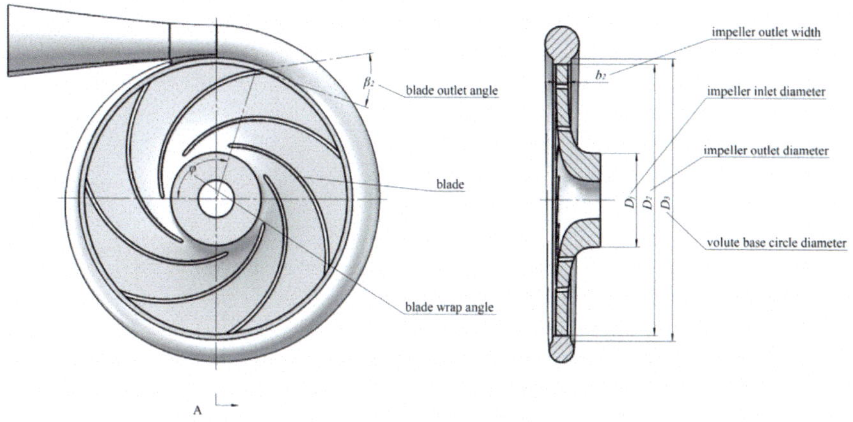


(a)

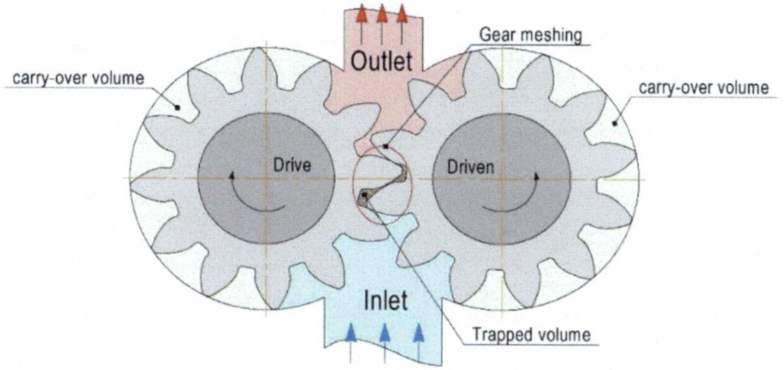


(b)

**Fig. 1.16** Schematic representation of an end-top centrifugal pump (**a**) (Yu-Qin and Ze-Wen 2022), and an external gear pump (**b**) (Romanenko et al. 2022)

the suction area. The inlet and outlet facilities, fuel supply systems that drive pumps and compressors, the generation of own electrical energy (i.e. base or emergency), fire-fighting systems, remote control systems, offices and maintenance workshops, in general, are part of the auxiliary facilities that any plant requires.

Liquid transportation systems generally work between tank yards that serve to store the product to be transported at the origin and to receive it at the destination site. In the oil industry the name "tanks" refers to vertical cylindrical metal containers for storing liquid products. These tanks are used to store crude oil, refined oil and chemical products; but depending on the type of service, the design and equipment of the tanks can vary considerably (see Fig. 1.17a).

**Fig. 1.17** View of the tank farm and pumping station of the company Axion-Pan American Energy (ex-OilTanking) in Brandsen, province of Buenos Aires, Argentina (**a**); detail of the pumping and metering station (**b**); and detail of floating rook tank 13 with a capacity of 45000 $m^3$

Tanks can be classified, according to their constructive form, in:

- Fixed roof tanks.
- External floating roof tanks (EFR).
- Fixed roof tanks with internal floating roof (IFR).
- Tanks with outer ring.
- Pressure tanks.

- Externally embanked tanks for Jet-fuel.
- Cryogenic tanks (LNG and LPG).

Added to these are horizontal cylindrical tanks for pressure storage (mainly for LPG), spheres, and low level tanks (Fig. 1.17).

## References

Ahmed TH. Hydrocarbon phase behavior. 1989.

Alomair O, Jumaa M, Alkoriem A, Hamed M. Heavy oil viscosity and density prediction at normal and elevated temperatures. J Petroleum Explorat Product Technol. 2016;6(2):253–63. ISSN 2190-0566. https://doi.org/10.1007/s13202-015-0184-8

Ayad Hassan AFS. Experimental and computational study of semi-open centrifugal pump. LAP LAMBERT Academic Publishing; 2016.

Azizi N, Behbahani R, Isazadeh MA. An efficient correlation for calculating compressibility factor of natural gases. J Nat Gas Chem. 2010;19(6):642–45. ISSN 1003-9953. https://doi.org/10.1016/S1003-9953(09)60081-5. https://www.sciencedirect.com/science/article/pii/S1003995309600815.

Babuk T, Mohitpour M, Jenkins A, Murray MA Pipeline cost reduction through effective project management and applied technology. 2005.

Benedict M, Webb GB, Rubin LC. An empirical equation for thermodynamic properties of light hydrocarbons and their mixtures i. methane, ethane, propane and n-butane. J Chem Phys. 1940;8(4):334–45.

Bentaleb T. Model-based control techniques for centrifugal compressors. PhD thesis. University of Siena; 2015.

Carr NL, Kobayashi R, Burrows DB. Viscosity of hydrocarbon gases under pressure. J Petroleum Technol. 1954;6(10):47–55.

Cookenboo L Jr. Costs of operating crude oil pipe lines. Rice Inst Pamphlet-Rice Univ Stud. 1954;41(1).

Dempsey JR, et al. Computer routine treats gas viscosity as a variable. Oil Gas J. 1965;63:141–3.

Dranchuk PM, Purvis RA, Robinson DB. Computer calculation of natural gas compressibility factors using the standing and katz correlation. In: PETSOC annual technical meeting. PETSOC. 1973. p. PETSOC–73.

Dranchuk PM, Abou-Kassem H. Calculation of z factors for natural gases using equations of state. J Canadian Petroleum Technol. 1975;14(03).

Druetta P, Tesi P, De Persis C, Picchioni F. Methods in oil recovery processes and reservoir simulation. Adv Chem Engin Sci. 2016;6:399–435. https://doi.org/10.4236/aces.2016.64039.

Druetta P. Numerical simulation of chemical EOR processes. PhD thesis, University of Groningen; 2018.

Dune KK, Oriji BN. An alternative correlation for the computation of pseudo-critical temperature and pressure. Glob J Engin Res. 2007;6(1):69–74.

Erdener BC, Sergi B, Guerra OJ, Chueca AL, Pambour K, Brancucci C, Hodge BM. A review of technical and regulatory limits for hydrogen blending in natural gas pipelines. Int J Hydrog Energy. 2023;48(14):5595–17. ISSN 0360-3199. https://doi.org/10.1016/j.ijhydene.2022.10.254. https://www.sciencedirect.com/science/article/pii/S0360319922050923.

Fimbres Weihs GA, Kumar K, Wiley DE. Understanding the economic feasibility of ship transport of co2 within the ccs chain. Energy Procedia. 2014;63:2630–37. ISSN 1876-6102. https://doi.org/10.1016/j.egypro.2014.11.285. URL https://www.sciencedirect.com/science/article/pii/S1876610214021006. 12th international conference on greenhouse gas control technologies, GHGT-12.

Galuzzi L. Alaska pipeline. 2005. https://upload.wikimedia.org/wikipedia/commons/8/8f/Trans-Alaska_Pipeline_System_Luca_Galuzzi_2005.jpg. distributed under a CC-BY-SA 2.5 license via Wikipedia Commons.

Global Data Energy. The us has the highest share in global oil and gas transmission pipeline network. 2022. https://www.offshore-technology.com/analyst-comment/us-oil-gas-pipeline-network/.

Global Energy Monitor. Tracker map - global gas infrastructure tracker. 2022a. https://globalenergymonitor.org/projects/global-gas-infrastructure-tracker/tracker/. [Accessed 15 Feb 2024].

Global Energy Monitor. Tracker map - global gas infrastructure tracker. 2022b. https://globalenergymonitor.org/projects/global-gas-infrastructure-tracker/tracker/. [Accessed 15 02 2024].

Guo B. Petroleum production engineering, a computer-assisted approach. Elsevier; 2011.

Hafner M, Luciani G. The Palgrave handbook of international energy economics. Springer International Publishing; 2022. ISBN 9783030868840. https://books.google.nl/books?id=XdxxEAAAQBAJ.

Hall KR, Yarborough L. A new equation of state for z-factor calculations. Oil Gas J. 1973;71(25):82–92.

Hansen ME, Dursteler E. Economic, environmental, and safety impacts of transporting oil and gas in the us. 2017.

Hopkins P. Pipelines: past, present, and future. In: The 5th Asian Pacific IIW international congress. Sydney, vol. 27. 2007.

Jarrahian A, Aghel B, Heidaryan E. On the viscosity of natural gas. Fuel. 2015;150:609–18.

Kareem LA, Iwalewa TM, Al-Marhoun M. New explicit correlation for the compressibility factor of natural gas: linearized z-factor isotherms. J Petroleum Explorat Product Technol. 2016;6:481–92.

Kay WB, et al. Density of hydrocarbon gases and vapors at high temperature and pressure. Ind Eng Chem. 1936;28(9):1014–9.

Kazi SR, Sundar K, Srinivasan S, Zlotnik A. Modeling and optimization of steady flow of natural gas and hydrogen mixtures in pipeline networks. Int J Hydr Energy. 2024;54:14–24. ISSN 0360-3199. https://doi.org/10.1016/j.ijhydene.2023.12.054. https://www.sciencedirect.com/science/article/pii/S0360319923063541.

Krieg D. Konzept und Kosten eines Pipelinesystems zur Versorgung des deutschen Straßenverkehrs mit Wasserstoff, vol. 144. Forschungszentrum Jülich; 2012.

Lee BI, Kesler MG. A generalized thermodynamic correlation based on three-parameter corresponding states. AIChE J. 1975;21(3):510–27.

Lee AL, Gonzalez MH, Eakin BE. The viscosity of natural gases. J Petroleum Technol. 1966;18(08):997–1000.

Lohrenz J, Bray BG, Clark CR. Calculating viscosities of reservoir fluids from their compositions. J Petroleum Technol. 1964;16(10):1171–6.

Mansour EM. Equation of state. In: Bhattacharya S, Ardekani MM, Biswas R, Mehta RC, editors. Inverse heat conduction and heat exchangers, chapter 8. Rijeka: IntechOpen; 2020. https://doi.org/10.5772/intechopen.89919. https://doi.org/10.5772/intechopen.89919.

Mahmoud M. Development of a new correlation of gas compressibility factor (z-factor) for high pressure gas reservoirs. J Energy Res Technol. 2014;136(1): 012903.

Mokhatab S, Mak JY, Valappil JV, Wood DA. editors. Chapter 1 - lng fundamentals. In: Handbook of liquefied natural gas. Boston; Gulf Professional Publishing; 2014. p. 1–106. ISBN 978-0-12-404585-9. https://doi.org/10.1016/B978-0-12-404585-9.00001-5. https://www.sciencedirect.com/science/article/pii/B9780124045859000015.

Menon ES, Menon PS. Working guide to pumps and pumping stations: calculations and simulations. Gulf Professional Publisher; 2010. ISBN 9781856178280. https://books.google.nl/books?id=YUyrPwAACAAJ.

Mohitpour M, Golshan H, Murray MA. Pipeline design and construction: a practical approach. ASME Press: Pipelines and Pressure Vessels; 2007. https://books.google.nl/books?id=_d1SAAAAMAAJ.

Mokhatab S, Purewal S. Letter to the editor: is lng a competitive source of natural gas? Petroleum Sci Technol. 2006;24(2):243–5.
Monge M, Romero Rojo MF, Gil-Alana LA. The impact of geopolitical risk on the behavior of oil prices and freight rates. Energy. 2023;269: 126779.
Moran S. Pump sizing: Bridging the gap between. The best of equipment series 3. 2016.
Nilsson PA, Apeland S, Dale HM, et al. The costs of co2 transport. Zero emissions platform. 2011.
Niteshift. Construction of nel gas pipeline, in the background the bundesautobahn 14, near hasenhäge, district ludwigslust, mecklenburg-vorpommern, Germany; 2011. https://upload.wikimedia.org/wikipedia/commons/3/3e/Hasenhäge_NEL-Erdgas-Pipeline_Bau_2011-05-15_009.JPG distributed under a CC-BY-SA 3.0 license via Wikipedia Commons.
Peng D-Y, Robinson DB. A new two-constant equation of state. Ind Engin Chem Fundamen. 1976;15(1):59–64.
Pitzer KS. The volumetric and thermodynamic properties of fluids. i. theoretical basis and virial coefficients1. J Am Chem Soc. 1955;77(13):3427–3433.
Pitzer KS, Lippmann DZ, Curl RF Jr, Huggins CM, Petersen DE. The volumetric and thermodynamic properties of fluids. ii. compressibility factor, vapor pressure and entropy of vaporization1. J Am Chem Soc. 1955;77(13):3433–40.
Poling B, Prausnitz J, Connell JO. The properties of gases and liquids. Chemical engineering series. Mcgraw-Hill; 2000. https://books.google.nl/books?id=E920LnqY2woC.
Redlich O, Kwong JNS. On the thermodynamics of solutions. v. An equation of state. Fugacities of gaseous solutions. Chem Rev. 1949;44(1):233–44.
Robinius M, Otto A, Syranidis K, Ryberg DS, Heuser P, Welder L, Grube T, Markewitz P, Tietze V, Stolten D. Linking the power and transport sectors-part 2: modelling a sector coupling scenario for germany. Energies. 2017;10(7). ISSN 1996-1073. https://doi.org/10.3390/en10070957. https://www.mdpi.com/1996-1073/10/7/957.
Romanenko I, Martseniuk Y, Bilohub O. Modeling the meshing procedure of the external gear fuel pump using a cfd tool. Computation. 2022;10(7):114.
Sanjari E, Lay EN, Peymani M. An accurate empirical correlation for predicting natural gas viscosity. J Nat Gas Chem. 2011;20(6):654–8.
Standing MB, Katz DL. Density of natural gases. Trans AIME. 1942;146(01):140–9, 12. ISSN 0081-1696. https://doi.org/10.2118/942140-G. https://doi.org/10.2118/942140-G.
Schaschke C, Fletcher I, Glen N. Density and viscosity measurement of diesel fuels at combined high pressure and elevated temperature. Processes. 2013;1(2):30–48. ISSN 2227-9717. https://doi.org/10.3390/pr1020030. https://www.mdpi.com/2227-9717/1/2/30.
Soave G. Equilibrium constants from a modified redlich-kwong equation of state. Chem Eng Sci. 1972;27(6):1197–203.
Sutherland W. Lii the viscosity of gases and molecular force. London, Edinburgh, Dublin Philos Mag J Sci. 1893;36(223):507–31.
Sutton RP. Compressibility factors for high-molecular-weight reservoir gases, volume All Days of SPE annual technical conference and exhibition, 09. 1985. https://doi.org/10.2118/14265-MS. https://doi.org/10.2118/14265-MS.
Tabora J. Aqua claudia. 2006. https://commons.wikimedia.org/wiki/File:Roma-parco_degli_acquedotti03.jpg. Distributed under a CC-BY 2.0 license via Wikipedia Commons.
Thomas LK, Hankinson RW, Phillips KA. Determination of acoustic velocities for natural gas. J Petrol Technol. 1970;22(07):889–95.
Tosun İ. The thermodynamics of phase and reaction equilibria. Elsevier; 2021.
Van der Waals JD. Over de Continuiteit van den Gas-en Vloeistoftoestand, vol. 1. Sijthoff; 1873.
Vesovic V. Predicting the viscosity of natural gas. Int J Thermophys. 2001;22:415–26.
Vesovic V, Wakeham WA. The prediction of the viscosity of dense gas mixtures. Int J Thermophys. 1989;10:125–32.
Yang X, Zhang S, Zhu W. A new model for the accurate calculation of natural gas viscosity. Nat Gas Ind B. 2017;4(2):100–5. ISSN 2352-8540. https://doi.org/10.1016/j.ngib.2017.07.011. https://www.sciencedirect.com/science/article/pii/S235285401730075X.
Yu-Qin W, Ze-Wen D. Optimization design of hump phenomenon of low specific speed centrifugal pump based on cfd and orthogonal test. Sci Rep. 2022;12(1):12121.

# Chapter 2
# Transport Phenomena and Science of Materials

> Aeroplanes are not designed by science, but by art in spite of some pretence and humbug to the contrary. I do not mean to suggest that engineering can do without science, on the contrary, it stands on scientific foundations, but there is a big gap between scientific research and the engineering product which has to be bridged by the art of the engineer
>
> *British Engineer to the Royal Aeronautical Society, 1922.*

Transport of mass, momentum and energy by pipelines is based on the principles of mechanical sciences, fundamentally those derived from the application of the developments in Mohitpour et al. (2007), Hopkins (2007), McAllister (2014):

- Fluid Mechanics;
- Thermodynamics; and
- Resistance of Materials.

Electrochemistry could also be mentioned if it is a question of considering the corrosion or erosion of materials, whether the pipe is expected to operate within the design parameters for many years. In this regard, it can be commented that it is a common practice in certain cases to add a *corrosion allowance* to the pipe thickness (determined for pressure and/or structural reasons), according to the estimated corrosion speed and/or the erosion determined by calculations or experimental data. This is a critical point to consider in the case of transport of what is known as "sour" service[1] (Bratland 2009; Menon 2014; Shashi Menon et al. 2013).

[1] See for instance the standards NACE MR0175/ISO 15156, which defines it as the exposure to certain oilfield environments or fluid that contain sufficient $H_2S$ to provoke cracking of materials.



P. Druetta et al., *Transport by Pipelines and Its Optimization*, Petroleum Engineering,
https://doi.org/10.1007/978-3-031-66566-0_2

## 2.1 Fluid Mechanics

### *2.1.1 Conservation Laws and the Bernoulli Equation*

*Continuum Mechanics* is the branch physics/engineering which studies the behavior of the materials modeled as a continuous mass and not as group of discrete particles.

In fluid dynamics there are two possible ways to study the flow of liquid and gases, always using the continuum approach (the reader is advised to see, for instance, the differences between Lagrange versus Euler on the internet and Fig. 2.1) (White and Xue 2021; Van der Waals 1873).

- The **Lagrangian** one, in which individual fluid particles or collections of fluid particles are followed moving through the flow.
- In the **Eulerian** description, a control volume is defined through which fluid flows in and out. One computes then the pressure field of the flow and not the pressure changes of a single particle.

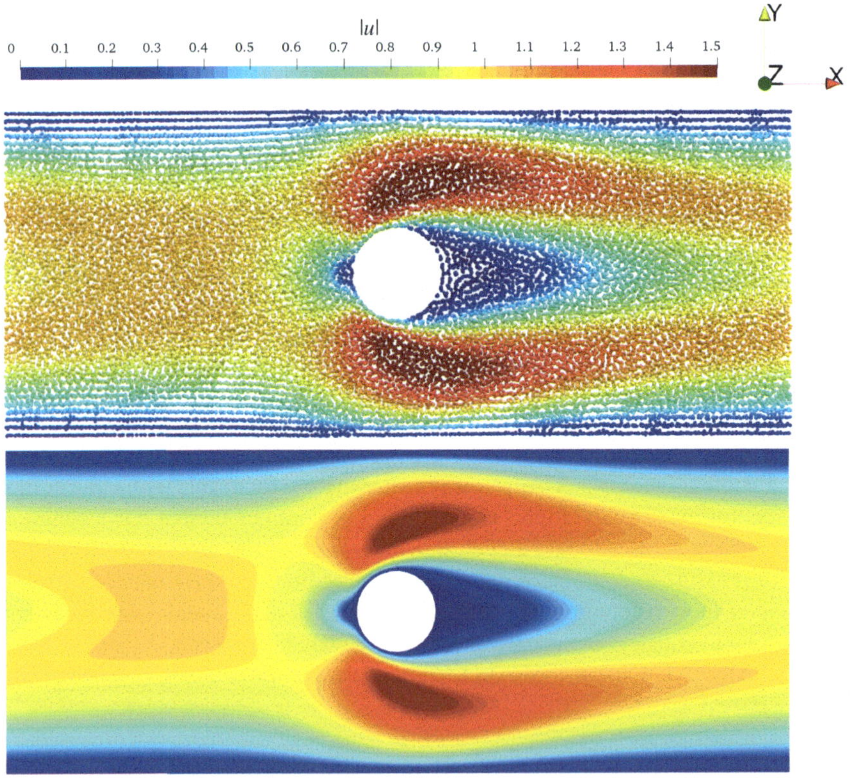


**Fig. 2.1** Different analysis in continuum mechanics, a Lagrangian (top) and Eulerian (bottom) images of a flow around a cylinder (Rakhsha et al. 2021)

For many of the problems faced in engineering, the Eulerian approach presents several advantages in order to study a system (and not each particle in particular) and therefore it is chosen to analyze and predict the behavior of a system.

### *2.1.2 Reynolds Transport Theorem*

The Reynolds Transport Theorem (also known as the Leibniz-Reynolds Transport Theorem or simply RTT) is a generalization of the Leibniz integral rule. You can see many different expressions and/or differential equations, but a simply "translation to English" would be:

***what was already there + what enters – what goes out = what is there now***

Quite simple to summarize, quite hard to derive (and integrate), the RTT is a very powerful tool in engineering, and many of the fundamental laws can be derived from it. Analyzing the system and a coinciding control volume with a control surface depicted in Fig. 2.2. The RTT states that the rate of change of an extensive property $B_{syst}$ is equal to rate of change with respect to the time of $B_{syst}$ within the control volume and the net rate of flux of $B_{syst}$ through the control surfaces (see Eq. 2.1).

$$\frac{d}{dt}\left(B_{syst}\right) = \int_{CV} \frac{\partial}{\partial t}(\beta\rho)\, dV + \int_{CS} \beta\rho\,(\mathbf{v}\cdot\mathbf{n})\, dA \tag{2.1}$$

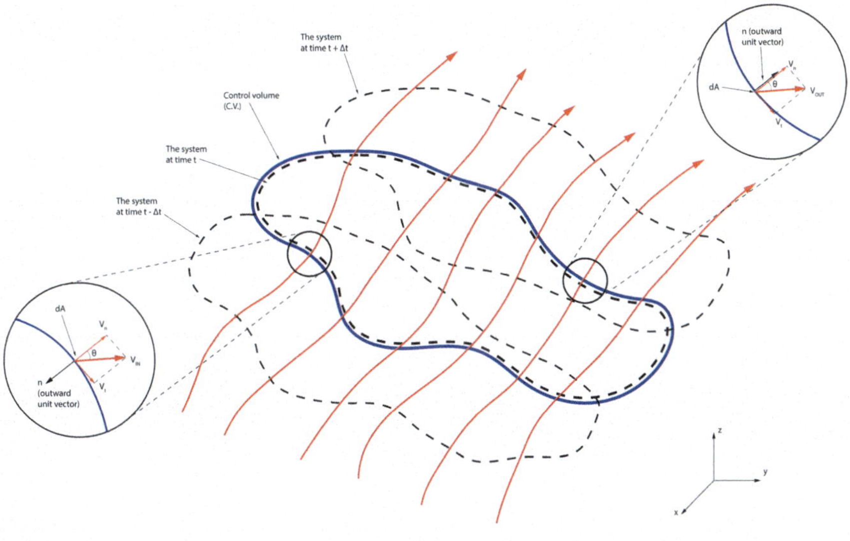


**Fig. 2.2** Scheme used to derive the Reynolds Transport Theorem

$$\beta = \frac{d}{dm}\left(B_{syst}\right) \tag{2.2}$$

The RTT can be applied in a fixed, moving or deformable $CV$ to study how the different properties change with the time. All the conservation laws can be derived from this.

#### Mass Conservation

$$B_{syst} = m \Rightarrow \beta = 1 \tag{2.3}$$

$$\left(\frac{dm}{dt}\right)_{syst} = 0 = \frac{d}{dt}\iiint_{CV} \rho dV + \iint_{CS} \rho\left(\mathbf{v}\cdot\mathbf{n}\right) dA \tag{2.4}$$

#### Momentum

$$B_{syst} = m\mathbf{v} \Rightarrow \beta = \mathbf{v} \tag{2.5}$$

$$\frac{d}{dt}\left(m\mathbf{v}\right)_{syst} = \sum \mathbf{F} = \frac{d}{dt}\iiint_{CV} \mathbf{v}\rho dV + \iint_{CS} \mathbf{v}\rho\left(\mathbf{v}\cdot\mathbf{n}\right) dA \tag{2.6}$$

In this case, this makes reference to an important set of equations in Fluid Mechanics: the *Navier-Stokes* equations. These are just an extension of Conservation of Momentum (or Newton's Second Law) in fluids, in order to describe their motion and deformation. The main difference with Newton's Law is that in this case it is considered that the stresses in the fluid are caused by a *pressure term* and by a *viscous term* which relates the deformation and the fluid viscosity. The so-called "Navier-Stokes existence and smoothness" is one of the Millennium Prize problems and it remains as one of the most important unsolved problems in engineering. The equation in vectorial form is (White and Xue 2021),

$$\frac{\partial}{\partial t}\left(\rho\mathbf{u}\right) + \nabla\cdot\left(\rho\mathbf{u}\otimes\mathbf{u}\right) = -\nabla p + \nabla\cdot\mathbf{T} + \rho\mathbf{g} \tag{2.7}$$

The left side describes the acceleration of the fluid, and it may be composed of time-dependent and convective components. The right side is composed by a summation of hydrostatic effects (i.e., pressure), the viscous stress and finally the body/interfacial forces (e.g., gravity). For Newtonian, incompressible fluids of viscosity $\mu$, the viscous stress is,

$$\mathbf{T} = \mu\left(\nabla\mathbf{u} + \nabla\mathbf{u}^T\right) \tag{2.8}$$

$$\frac{\partial}{\partial t}(\rho \mathbf{u}) + \nabla \cdot (\rho \mathbf{u} \otimes \mathbf{u}) = -\nabla p + \mu \nabla^2 \mathbf{u} + \rho \mathbf{g} \tag{2.9}$$

**Angular Momentum**

$$B_{syst} = \mathbf{H} = \int_{syst} (r \times \mathbf{v})\, dm \Rightarrow \beta = r \times \mathbf{v} \tag{2.10}$$

$$\left(\frac{d\mathbf{H}}{dt}\right)_{syst} = \sum \mathbf{M} = \frac{d}{dt}\iiint_{CV} (r \times \mathbf{v})\, \rho dV + \iint_{CS} (r \times \mathbf{v})\, \rho\, (\mathbf{v} \cdot \mathbf{n})\, dA \tag{2.11}$$

**Energy**

$$B_{syst} = \mathbf{E} \Rightarrow \beta = e \tag{2.12}$$

$$\left(\frac{d\mathbf{E}}{dt}\right)_{syst} = \frac{d\mathbf{Q}}{dt} - \frac{d\mathbf{W}}{dt} = \frac{d}{dt}\iiint_{CV} e\rho dV + \iint_{CS} \left(e + \frac{p}{\rho}\right) \rho\, (\mathbf{v} \cdot \mathbf{n})\, dA \tag{2.13}$$

$$e = u + \frac{1}{2}v^2 + \mathbf{g}z \tag{2.14}$$

**Entropy**

$$B_{syst} = \mathbf{S} \Rightarrow \beta = s \tag{2.15}$$

$$\left(\frac{d\mathbf{S}}{dt}\right)_{syst} = \iiint \left(\frac{\Upsilon + \Phi}{T}\right) dV = \frac{d}{dt}\iiint_{CV} s\rho dV + \iint_{CS} \left(\rho \mathbf{v} s - \frac{k_T}{T}\nabla T\right) \cdot \mathbf{n} dA \tag{2.16}$$

In this case, $\Upsilon$ represent the heat flux in the system and $\Phi$ is the viscous dissipation in the system.

### 2.1.3 The Bernoulli Equation

The *Bernoulli Equation*, also known as *Bernoulli's Principle*, can be considered a special case of the conservation of energy. Even though is applicability is quite limited in theory (valid for isentropic flows, i.e. when irreversible effects such as turbulence or friction and non-adiabatic processes can be neglected in the system), one can find in this equation a powerful tool to study the behavior of the systems in fluid mechanics. There are several forms of the Bernoulli equation, depending on the kind of flow one study (e.g. incompressible at low velocity , or compressible at high velocity—fluids, steady state or transient processes, etc.).

There are several ways to derive this equation, from the RTT, the first Law of Thermodynamics, or even Newton's second Law. The derivation of this equation is done analyzing the following system (Fig. 2.3) and using the mass and momentum conservation equations.

One of the most important assumption in this first stage is that the system is considered ideal, with no friction loses ($\tau_w = 0$). Later it will be discussed what happens with real fluids, when the viscosity cannot longer be neglected.

Thus, in order to get to the equation, the conservation of mass and momentum is applied to the system.

#### Conservation of Mass

$$\frac{d}{dt}\left(\int_{CV} \rho d\forall\right) + \dot{m}_{OUT} - \dot{m}_{IN} = 0 \approx \frac{\partial \rho}{\partial t} d\forall + d\dot{m} \tag{2.17}$$

$$-\frac{\partial \rho}{\partial t} A ds = d\left(\rho A V\right) \tag{2.18}$$

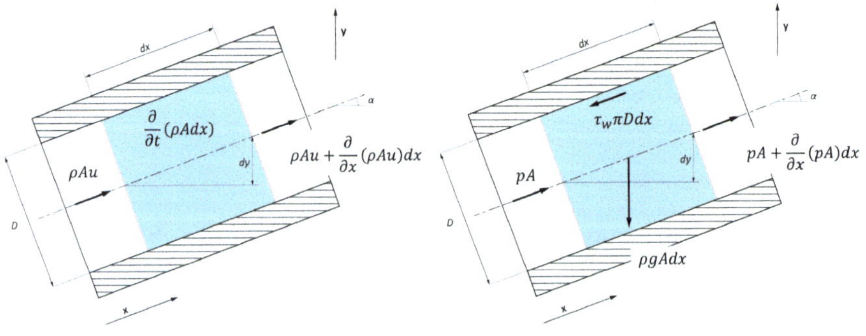


(a) Conservation of mass. (b) Conservation of momentum.

**Fig. 2.3** Control volume analysis for the derivation of the Bernoulli Equation

The conservation of mass, as expected, does not require the frictionless assumption in the analysis. The next step consists in the analysis of the momentum equation.

### Momentum

$$\sum dF_{syst} = \frac{d}{dt}\left(\int_{CV} V\rho d\forall\right) + (\dot{m}V)_{OUT} - (\dot{m}V)_{IN} \approx \frac{\partial}{\partial t}\rho V A ds + d(\dot{m}V) \tag{2.19}$$

One evaluates the forces acting on the system: primarily the action of the acceleration field, in this case only the gravity, and the pressure forces acting on the surface of the control volume.

$$dF_{s,grav} = -\rho \mathbf{g} A dz \quad \wedge \quad dF_{s,press} = -A dp \tag{2.20}$$

$$\frac{\partial V}{\partial t} ds + \frac{dp}{\rho} + V dV + \mathbf{g} dz = 0 \tag{2.21}$$

Integrating this differential equation,

$$\int \frac{\partial V}{\partial t} ds + \int \frac{dp}{\rho} + \int V dV + \int \mathbf{g} dz = \int 0 \tag{2.22}$$

$$\int \frac{\partial V}{\partial t} ds + \int \frac{dp}{\rho} + \frac{1}{2} V^2 + \mathbf{g} z = constant \tag{2.23}$$

For the special case when the flow is steady-state and the fluid is incompressible the Bernoulli Equation is reduced to,

$$\frac{p}{\rho} + \frac{1}{2} V^2 + \mathbf{g} z = constant \tag{2.24}$$

The total energy in a fluid is composed by several terms: pressure, velocity and potential energy. These may (and usually do) vary, but the total value **does not** (**Important: this is only valid for frictionless fluids**). These components can be transformed into other types of energies, but the total will remain unaltered, this means that the total **usable** energy will remain unaltered.

However, ideal fluids do no exist in real life, and the viscosity always plays an important role in the design of industrial facilities. Thus, how does one consider these 'real effects'? In this case, the Bernoulli Equation is slightly modified to consider these effects.

$$\left(\frac{p}{\rho} + \frac{1}{2} V^2 + \mathbf{g} z\right)_1 = \left(\frac{p}{\rho} + \frac{1}{2} V^2 + \mathbf{g} z\right)_2 + \Delta H_{1-2} \tag{2.25}$$

The head loss is a function of the velocity, the pipe diameter, the length of the pipe and a factor, called *friction factor* which relates the flow conditions, the fluid itself and the pipe (see Sect. 2.1.11).

$$\Delta H_{1-2} \propto \left(V^2; D^{-5}; L_{1-2}; f\right) \tag{2.26}$$

$$f \propto \left(V; \rho; D; \mu^{-1}; \epsilon\right) \tag{2.27}$$

In the last equation $\epsilon$ represents the absolute rugosity of the pipe [mm].

## 2.1.4 Steady-State Flows

It is common to use the term steady-state or steady flow to define that none of the flow characteristics (i.e., density, viscosity, pressure, velocity, etc.) at a given point changes over time. Mathematically, they are usually defined as any partial derivative with respect to time (for a property at a given point) equal to zero ($\frac{\partial\ldots}{\partial t} = 0$).

During a steady flow the properties of the fluid may change from point to point, within the pipe or equipment, but at any given point they remain constant over time. An important part of the engineering development of pipeline fluid transport systems is determining whether steady-state studies, in some cases averaging over time, are sufficient or whether a more detailed non-steady-state (i.e., transient) study is required.

## 2.1.5 Isothermal Flow of Incompressible Fluids

An ideal or inviscid fluid is one whose viscosity is considered zero. According to fluid mechanics, the principle of conservation of mass, or better known as the equation of continuity, in a steady state, results for the flow of an ideal liquid (i.e., considering—in addition to zero viscosity—a constant density):

$$0 = \frac{d}{dt}\int_{\forall} \rho d\forall + \int_{A} \rho(\mathbf{v} \cdot \hat{n}) dA \tag{2.28}$$

$$0 = \cancel{\frac{d}{dt}\int_{\forall} \rho d\forall} + \rho \int_{A} (\mathbf{v} \cdot \hat{n}) dA \rightarrow \mathbf{v}A = \mathbf{Q} = \mathcal{C} \tag{2.29}$$

Furthermore, the application of the Bernoulli principle, derived from the conservation of energy, yields for two different section of the pipeline the following relationship,

$$\left(\frac{p}{\rho} + \frac{v^2}{2} + gz\right)_1 = \left(\frac{p}{\rho} + \frac{v^2}{2} + gz\right)_2 \tag{2.30}$$

Simplifying it further for a simple pipeline considering that the diameter does not change, hence the velocity is constant. Thus,

$$\left(\frac{p}{\rho}+gz\right)_1=\left(\frac{p}{\rho}+gz\right)_2\rightarrow p_1=p_2+\rho g\Delta H \tag{2.31}$$

## 2.1.6 *Real Flows and the Importance of Temperature and Compressibility*

A real fluid is one whose viscosity is not considered to be negligible. In the case of real liquid flows through pipes, the problem is generally stated as: *given a liquid of a known density* $\rho$ *(assumed constant) and viscosity* $\mu \neq 0$ *(also assumed constant), it is desired to transport it through a distance L, following a certain topography*. In addition, being $m$ the mass to be transported within a certain time $t$, the mass flow $\dot{m}$ to be transported is defined as,

$$\dot{m}=\frac{m}{t}=\rho\mathbf{v}A \tag{2.32}$$

In the case of real fluids, the viscosity has an influence on the application of conservation of energy, that is, the so-called Bernoulli equation with the addition of a pressure loss term:

$$\left(\frac{p}{\rho}+\frac{v^2}{2}+gz\right)_1=\left(\frac{p}{\rho}+\frac{v^2}{2}+gz\right)_2+\frac{\Delta p_{1-2}}{\rho} \tag{2.33}$$

Being the extra-term the pressure loss due to viscous effects in a pipeline of length $L$ and internal diameter constant $D_{int}$, given by the following equation,

$$\frac{\Delta p_{1-2}}{\rho}=f\frac{v^2}{2}\frac{L}{D_{int}}=\frac{f}{2}\frac{L}{D_{int}}\frac{\dot{m}^2}{\rho^2\frac{\pi^2 D_{int}^4}{16}}=8f\frac{\dot{m}^2}{\rho^2\pi^2}\frac{L}{D_{int}^5} \tag{2.34}$$

Introducing in the pressure loss equation the friction factor $f$, which is an empirical/analytical value, function of the Reynolds[2] number and the relative rugosity. Follow the same approach as in the case of ideal fluids, the final equation renders,

$$p_1=p_2+\rho g\Delta H+8f\frac{\dot{m}^2}{\rho\pi^2}\frac{L}{D_{int}^5} \tag{2.35}$$

[2] Osborne Reynolds (Belfast, Northern Ireland, 1842—Watchet, England, 1912) was an engineer and physicist who made important contributions to hydrodynamics and fluid dynamics, the most notable being the dimensionless number that bears his name in 1853.

in which $p_1$ is the initial pressure in the pipe section which, for this book, it represents the outlet pressure of the pumping plant; $p_2$ is the final pressure in the pipe section or suction pressure in the pumping plant; and $D_{int}$ is the inner diameter of the pipe.

For these reasons, the mass flow rate $\dot{m}$ of a certain fluid that can be transported through a pipe of length $L$ is a function of the diameter $D_{int}$ and the pressure drop $p_1 - p_2$; this is to the extent that density and viscosity are assumed to be constant, which can be reasonably verified in the case of liquids.

Nevertheless, in reality the density and viscosity depend on the temperature of the fluid and the pressure to which it is subjected, and both vary from section to section due to the transfer of heat between the fluid and the medium, and due to the pressure drop produced in the direction of circulation. Considering these continuous variations in density and viscosity, section after section, can only be approximated by a finite difference calculation methodology.

### 2.1.7 *Flow of Ideal Gases—Compressibility Factor*

Gases are fluids, i.e., matter that can undergo significant deformation under the action of relatively weak forces. They do not have their own shape and this is a property that distinguishes them from liquids: filling up all the space that is available to them.

So far, the analysis has been limited to flows for which density variations and compressibility effects are negligible. However, when fluids involving important changes in density are considered, they are treated as compressible flows that combine fluid mechanics and thermodynamics. Here the relationships associated with compressible fluids for ideal gases with constant specific heats will briefly discussed and then proceed in the following chapters to discuss also real aspects of different gases.

An important parameter in the study of compressible flows is the speed of sound, which is the speed at which a sound (or pressure) wave travels through an elastic medium, in this case the gas in question. For an ideal gas, i.e., one that satisfies the equation of state Eq. 1.11, the speed of sound can be found with the expression:

$$c^2 = \left(\frac{dp}{d\rho}\right)_{s=constant} \tag{2.36}$$

If the evolution is adiabatic, then one can write,

$$\frac{p}{\rho^k} = constant \Rightarrow d\left(\frac{p}{\rho^k}\right) = 0, \quad \Rightarrow \quad \rho^{-k}dp - kp\rho^{-(k+1)}d\rho = 0 \Rightarrow$$

$$\frac{dp}{d\rho} = c^2 = k\frac{p}{\rho} = kRT \tag{2.37}$$

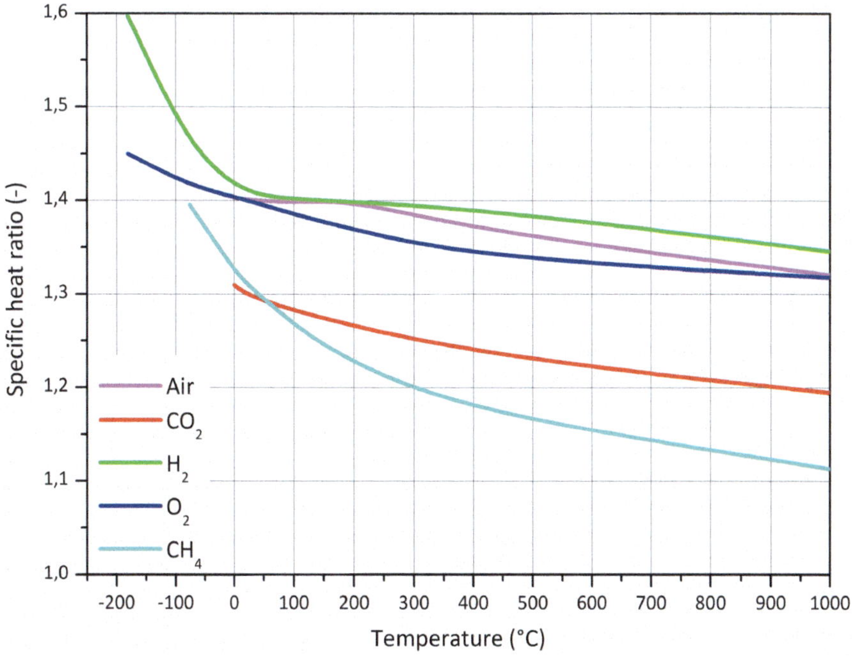


**Fig. 2.4** Specific heat ratio for several gases as a function of the temperature

The constant $R$ has a fixed value for a particular ideal gas and that the ratio of specific heats $k = c_p/c_v$ of an ideal gas is a function of temperature, hence the speed of sound in a given ideal gas is only a function of the absolute temperature (see Fig. 2.4).

A second important parameter in compressible flow is the Mach[3] number, which is the relationship between the actual speed of the fluid and the speed of sound in the fluid in the same state.

$$Ma = \frac{v}{c} = \frac{v}{\sqrt{kRT}} \tag{2.38}$$

The flow regime of a compressible fluid is often described in terms of the Mach number. Indeed a flow is called subsonic flow when $Ma < 1$; sonic flow when $Ma = 1$ (or transonic if $Ma \approx 1$); supersonic flow when $Ma > 1$ and hypersonic flow when $Ma \gg 1$.

For real gases, the equation of state can be expressed as in Eq. 1.12, where $Z_c$ is the compressibility factor, indicating the deviation between the "ideal gas" model and the real behavior of the gas. It was already seen in the previous chapter that the compressibility factor is a variable and depends on the pressure, the temperature and the composition of the gas.

[3] Ernst Mach (February 18, 1838–February 19, 1916) was an Austrian scientist and philosopher noted for his studies of shock waves.

### *2.1.8 Flow of Gases by Pipelines—Types of Flows*

A gas flowing through a pipe without exchanging heat with the medium (i.e., an adiabatic process) and considering negligible the influence of its viscosity, hence no friction (reversible process), sets up a flow where the entropy can be considered constant, this is, a reversible or isentropic adiabatic flow. These assumptions can be linked in real life to the flow taking place in devices such as nozzles, diffusers and the passages between the compressor blades, where the gas passes very rapidly, with no possibility of heat exchange whatsoever, and the sections are short (i.e. viscosity losses can be neglected). In all these cases, the flow can be approximated with enough accuracy as a unidirectional isentropic flow.

Determining the pressure loss of a compressible fluid flowing through a pipe requires knowledge of the relationship between pressure and density, which is not easy to determine for each particular problem. The extreme cases normally considered are adiabatic and isothermal flow. Adiabatic flow is assumed to occur in short, well-insulated pipes. This is because no heat is transferred to or from the pipe, except for a small amount of frictional heat that is added to the flow. Isothermal flow is considered to occur very often, because it is closer to reality than what happens in long pipes. Some authors show that the flow of gas in isolated pipes is very close to the isothermal flow for high pressures.

The stationary unidirectional flow of a gas that circulates through a pipe of constant area, where a heat transfer process with negligible friction takes place is known as Rayleigh flow.[4] When sections of pipe of constant cross-sectional area are studied with significant wall friction effects, but negligible heat transfer (especially for small sections) the flow is referred to as Fanno Flow.[5]

In general, for large diameter lines (significant heat transfer) and great length (significant friction), the flow model that is closest to the process taking place in a large part of the duct is at constant temperature, that is, Isothermal (See Fig. 2.5).

It is possible to relate the stationary unidirectional Fanno and Rayleigh flows with the condition of assuming that the temperature remains unchanged (T = constant) and obtain a theoretical model that has served as a basis for many authors—engineers involved in the natural gas industry—for propose relationships for calculating gas pipelines (Weymouth, Panhandle, Modified Panhandle, General, Barbato, etc.) that were used during the 20th century for the design of large gas transportation systems.

#### Complex Systems—Looping

The equations seen for flow in simple pipe lines can be extended in their application to parallel circuits, series circuits of different sizes, combined parallel and series

[4] Lord Rayleigh (1842–1919) English scientist named John William Strutt, discoverer of the Argon gas for which he was awarded the Nobel Prize in 1904.

[5] Gino Fanno (1882–1962) was an Italian mechanical engineer who developed the viscous resistance flow model.

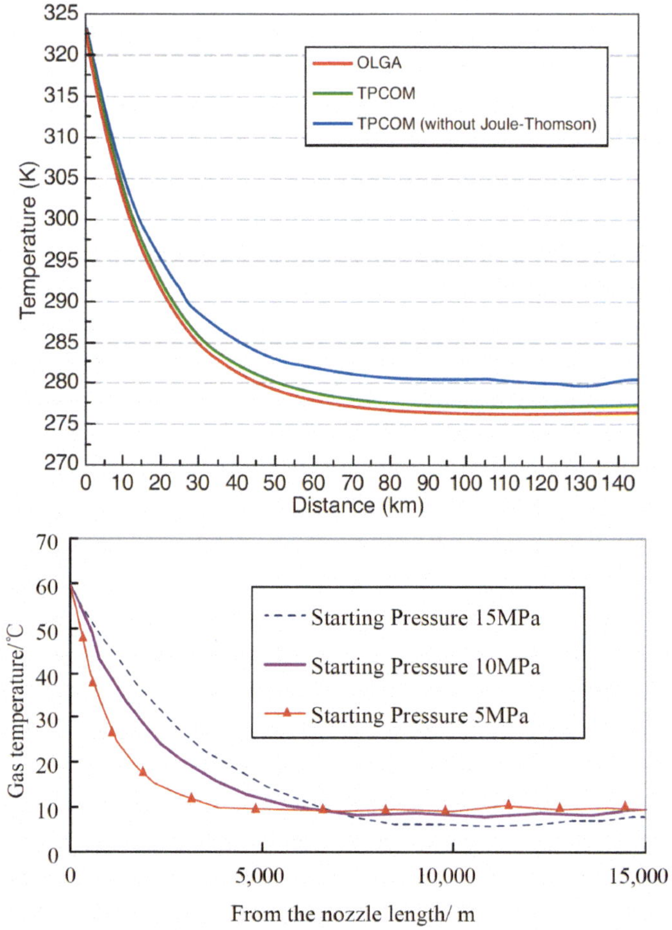


**Fig. 2.5** Temperature profile in gas pipelines as a function of the pipeline length (Duan et al. 2013; Xiuwei 2014)

circuits, and distribution networks. In order to simplify the calculations, it is often convenient to ignore the fact that the friction factor $f$ is a function of Reynolds and the diameter of the pipe, assuming that $f$ is the same for all diameters. This is not strictly correct, but it avoids involving the diameter in the expressions which greatly facilitates the solution.

It is generally accepted that the carrying capacity of a pipeline varies with the fifth power of the diameter, which is not strictly accurate in certain cases. For water at normal atmospheric temperatures, the fifth power relationship gives correct results. The equivalent diameter $D_e$ for a section that has a parallel loop in its total length, in the case of natural gas is:

$$D_e = \left(D_1^{\frac{8}{3}} + D_2^{\frac{8}{3}} + D_3^{\frac{8}{3}} + \cdots + D_n^{\frac{8}{3}}\right)^{\frac{3}{8}} \tag{2.39}$$

As previously mentioned, it is considered that the transport capacity varies with the power 8/3 of the diameter, as proposed by Weymouth.

### *2.1.9 Real Gases with Flow at Variable Entropy and Non-isothermal*

The isothermal hypothesis of the gas flow is approximate, since in fact in a gas pipeline the temperature of the fluid varies along the line (since the hot gas is discharged from the compressors due to the effect of compression and it cools down in its circulation), affecting the rest of the variables.

When it comes to analyzing the behavior of real gases in a line of constant section under a steady state flow, one should consider the variable entropy, i.e., taking into account the effects of viscous friction and non-isothermal evolution, this is considering the heat transmission that will occur naturally between the fluid and the surrounding medium. The solution of the system of differential equations describing the flow properties allow, if the *Mach number* at a certain inlet section is known, to calculate the pressure, density, speed and temperature (and consequently the outlet *Mach number*) in each section.

In the case of real flows, being $dt \neq 0$, $\partial Q \neq 0$ and $\mu \neq 0$, there is no possible analytical solution of the solving system of differential equations (conservation equations using Reynolds Transport Theorem or the Lorenzelli's unified method—see Chap. 6) by mathematical integration, and one must try to find the solution by numerical methods. It is then necessary to appeal to the resolution by an approximate methodology by application of finite differences, i.e. sections of the pipeline of a certain length in which certain properties are assumed constant. In the case of large natural gas transportation pipelines, e.g. DN 350 (14") $\leq$ Diameter (NPS—Nominal Pipe Size) $\leq$ DN 1100 (42"), analyzing sections of roughly 1,000 to 2,000 m in length is a solution according to the precision required.

The resolution by the finite difference method implies defining previously for these sections which variables will remain constant in the segment: one can consider keeping the density constant, it may be logical for low pressures, which is not the case in trunk lines; or keeping the temperature constant, i.e., application of the isothermal flow section by section; or, with less deviation from reality, keeping the *Mach number* constant in the section, by knowing the speed and temperature, the *Mach number* can be obtained with it.

It is worth mentioning that in the case of real gases in flow at variable entropy and temperature, corrections must be incorporated for altimetry, since the pipes always pass through non-flat terrain, and for the effect of the viscosity itself, which when generating heat due to friction, a part of it is reincorporated into the flow, modifying the viscosity and density, i.e., the viscous dissipation term in the energy conservation equation. With the calculation of the pressure at the end of a section, one can calculate the heat transferred to or from the outside to the flow and find the temperature at the end of the section and with both correct the density and viscosity, and thus continue with the next section until completing the desired length.

### *2.1.10 Non-steady Flows*

The terms non-stationary or non-permanent/impermanent and transient are often used interchangeably. The terms non-steady or non-permanent apply to any flow that is not steady, but transient is commonly applied to developing flows. Traditionally, long distance fluid transport piping systems have been designed using steady state models which are suitable for optimizing design when supply and demand are relatively stable. However, there are situations where the load factors are important, the oscillations in the flows supplied or received and the entrance and exit of impulsion installations, etc. require more than conventional stationary analysis. There are many -and very important- factors that must be considered before a pipeline is designed, although the most important of all are the flow requirements. Permanent regime hydraulic analyzes cannot be applied too much to abnormal operating conditions such as abrupt changes in user consumption profiles, internal cleaning ("pigging"), line breaks, loss of compression or pumping units, shocks battering ram, among others. *Line-pack* is the term used to describe the effect of gas storage in pipelines. Overpressure or surge (or water-hammer) is the term used to describe a relatively rapid process that occurs with nearly incompressible fluids. Steady regime analyzes do not allow modeling either line pack fluctuations or the water-hammer phenomenon.

#### Liquid Pipelines

Dynamic simulations for the analysis of transient flows in the design of transportation systems for liquids such as water and crude oil tankers have been commonly used in recent decades, particularly for the following purposes:

- Economy optimization: use of overpressure control systems to avoid increasing the thickness of the pipes (basically, determining a minimum valve closing times).
- Water hammer control: location and sizing of vacuum and pressure relief valves in aqueducts.
- Design and operation of pumping plants: control systems to protect pipelines from overpressure, effects of rapid fluctuations in demand, effects of the sudden closing of plant and line valves, effects of putting a pump online, design of pressure relief systems, calibration of line break systems, detection of losses or failures, etc.

To study a particular transient, the so-called water hammer, it will be assumed to be a simple system composed of a tank, a pipe and a valve (it is transient if the tank is emptied), but it is very slow and it is not the movement that matters, which could be resolved as successive equilibrium states, without taking into account the acceleration of the masses.

**Gas Pipelines**

Contrary to liquid systems, there is very limited information available on the application of transient analysis to pipeline design. The available information mainly refers to loss control systems for detection of line breaks, venting in gas pipelines, line-pack determination and flow pulsations. In Chap. 9, the mathematical analysis of transient systems is introduced in both systems and the influence of the phenomena previously mentioned in the pipeline design is studied.

## *2.1.11 Friction Factor*

Considering a fully developed, laminar, incompressible flow in a pipe, as shown in Fig. 2.6, one can apply the Reynolds Transport Theorem and analyze the transport of momentum in a differential volume element. This force balance in the pipe axial direction is,

$$2\pi r dr p - 2\pi r dr \left(p + \frac{\partial p}{\partial x}dx\right) + 2\pi r dx \tau - 2\pi(r + dr)dx\left(\tau + \frac{\partial \tau}{\partial r}dr\right) = 0 \tag{2.40}$$

Simplifying and dividing by $2\pi dx dr$, it yields,

$$-r\frac{\partial p}{\partial x} - \tau - r\frac{\partial \tau}{\partial r} = 0 \Rightarrow r\frac{\partial p}{\partial x} = -\frac{\partial}{\partial r}(r\tau) \tag{2.41}$$

Limiting the analysis to Newtonian fluids, then the relationship between stresses and deformations is given by,

$$\tau = \mu\frac{\partial u}{\partial r}$$

In case of different type of fluids, such as non-elastic, non-Newtonian fluids or slurries and viscoelastic solutions, the appropriate fluid model should be used. Then, replacing in the previous equation,

$$\frac{\partial p}{\partial x} = -\frac{1}{r}\frac{\partial}{\partial r}\left[r\mu\frac{\partial u}{\partial r}\right] \tag{2.42}$$

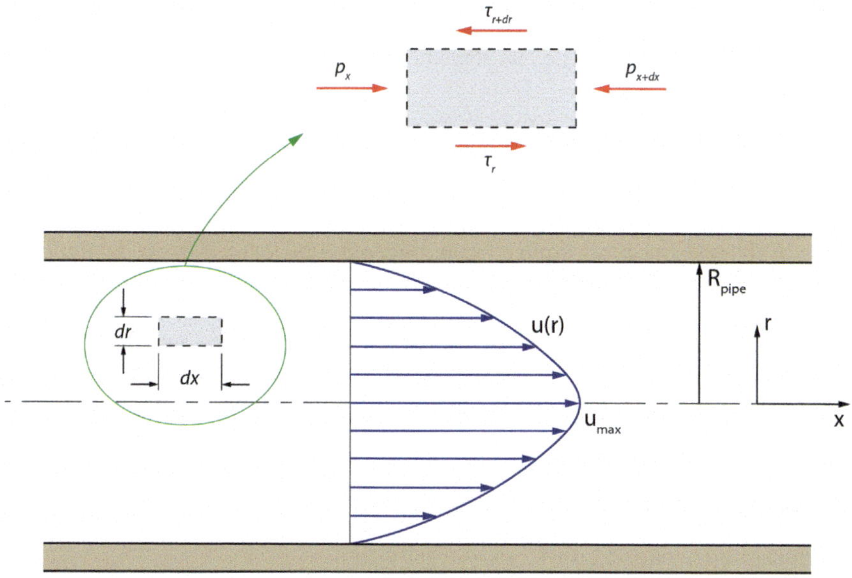


**Fig. 2.6** Scheme of the laminar flow in a pipe, showing the velocity profile and the differential volume control and stresses acting on it

Equation 2.42 shows an interesting conclusion: the left-hand side is a function of the variable $x$, and the right-hand side is solely a function of the radius $r$, yielding an equality of the form $f(r) = g(r)$. The only way that this equality can be kept is that both functions are equal to the same constant value. Integrating this differential equation and applying appropriate boundary conditions, one obtains the well-known parabolic velocity profile for Newtonian fluids (Fig. 2.6),

$$u(r) = -\frac{R^2}{4\mu}\left(\frac{dp}{dx}\right)\left(1 - \frac{r^2}{R^2}\right) \tag{2.43}$$

Even though the previous analysis was based on laminar fluids, the same approach can be used to make a more general model, comprising as well turbulent flows. In this case, the final differential equation is of the form,

$$\frac{\partial p}{\partial x} = -\frac{4}{D}\tau \tag{2.44}$$

The shear stress can be written also as a function of the parameters involved in the momentum-transport process,

$$\tau = f(\rho, \mu, \bar{u}, D) \Rightarrow \frac{\tau}{\frac{\rho\bar{u}^2}{2}} = f\left(\frac{\rho\bar{u}D}{\mu}\right) = f(Re) \tag{2.45}$$

Being $\bar{u}$ the average velocity in the pipe. In reality, the shear-stress is also function of the pipe's characteristics, i.e., its absolute rugosity $\varepsilon$,

$$\frac{\tau}{\frac{\rho\bar{u}^2}{2}} = f\left(Re, \frac{\varepsilon}{D}\right) \Rightarrow \frac{dp}{dx} = -2f\rho\frac{\bar{u}^2}{D} \tag{2.46}$$

The Moody diagram (Figs. 2.7 and 2.8) is a dimensionless coordinate graph that relates the Darcy-Weisbach (from now on, Darcy friction factor) friction factor to the Reynolds number and relative roughness for fully developed flow in circular pipes, as a graphical way to determine the friction factor. In 1944 Lewis Ferry Moody plotted the Darcy friction factor in what is known as the Moody Diagram. The Fanning factor, used in the so-called General Equation and countries employing the Imperial System of Units, is one-fourth of the Darcy factor and therefore the pressure drop equation must incorporate a compensating factor of four (Moody 1944).

The diagram shows three distinct regions (see also Fig. 2.9, showing the onset of turbulent flow as a function of the Reynolds number):

- Laminar region (Re < 2100);
- Transition region (2100 < Re < 4000);
- Turbulent region (Re > 4000)

In the case of new steel pipes (or those with good maintenance and cleanliness), manufactured under American (API 5L) or European standards (EN 14161), used in the construction of pipelines of a certain importance, an absolute roughness

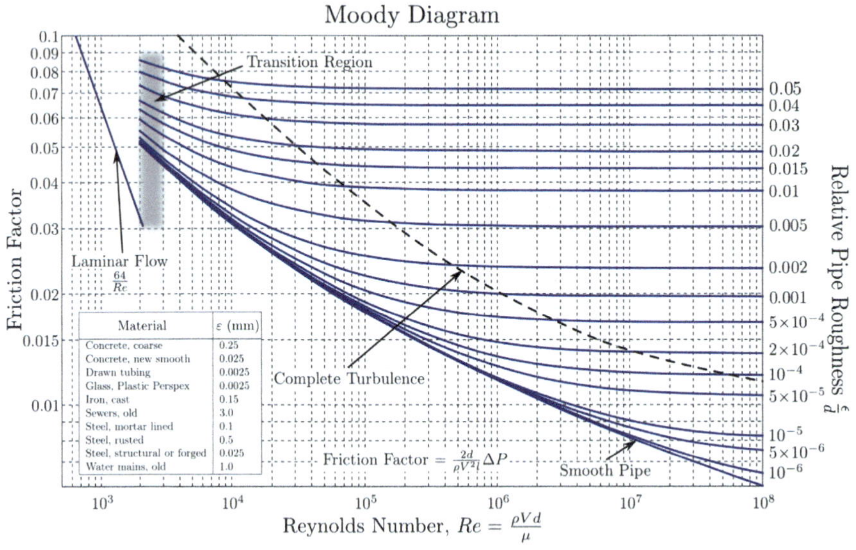


**Fig. 2.7** Moody chart to obtain graphically the Darcy friction factor (Derumaux 2016)

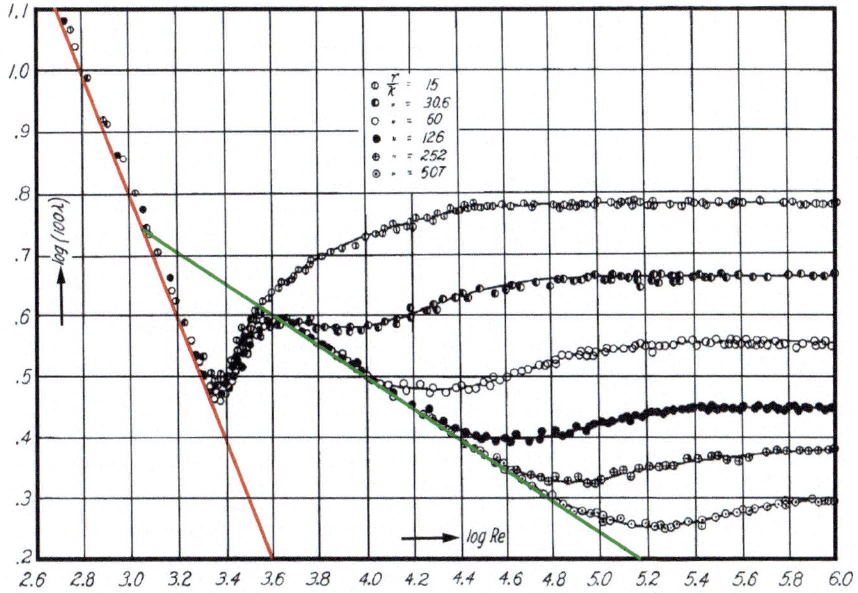


**Fig. 2.8** Detail of the transition region in the Moody chart showing Nikuradse's experimental values for different relative rugosities, from the laminar regime to the early onset of turbulence. The red line is the laminar correlation and the green line is Blasius' correlation for smooth pipes (adapted from Nikuradse et al. 1950)

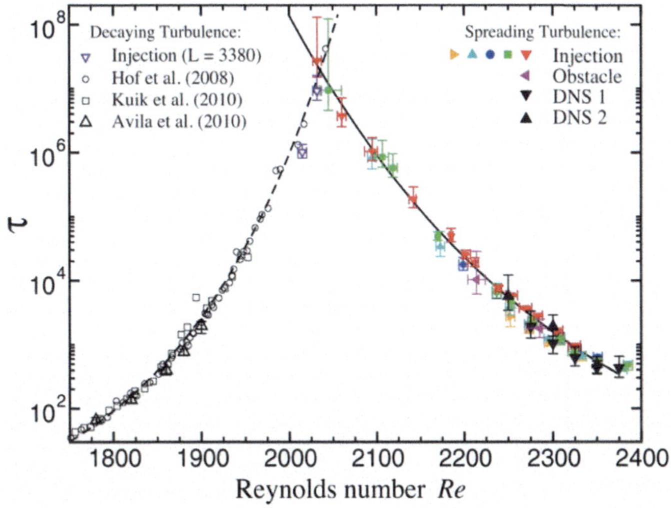


**Fig. 2.9** Plot showing times required for eddies to spread and decay. At Reynolds = 2040, the eddies spread faster than they decay, generating a turbulent flow (Avila et al. 2011)

$\varepsilon = 0.0254$ mm is generally adopted.[6] Given the complexity of using a diagram, there are many mathematical correlations, implicit as well as explicit, developed by researchers over time to determine the friction factor, which can be used to calculate pressure loss in pipes. In general, they are specific expressions for certain fluids and flow regimes. See Table 2.1 listing the different analytical or empirical correlations and their range of applicability.

In the graph it can be appreciated three different areas, based on the dependence of the friction factor on the Reynolds number and the relative rugosity, namely: the laminar regime, in which the friction factor is entirely dependent on the Reynolds number only; the transition and the first part of the turbulent regimes, in which the friction factor is a function of both the Reynolds number and the relative rugosity; and finally the fully developed turbulent flow, in which the friction factor becomes only a function of the relative rugosity. The explanation to this can be observed in the following figures, depicting schematically the phenomena taking place in regions next to the pipe's wall.

The whole concept lays on the idea that within the boundary layer, some sub-regions may be observed, namely, the *laminar sub-layer*. In the laminar flow, this layer cores completely the pipe's rugosity and thus, the skin friction therefore is not affected by the system operational behavior; the second area is the transition and the starting of turbulent flow regime, where the rugosity is of the same order of the viscous sub-layer, then both Reynolds and the relative rugosity affects the skin friction factor; and finally, the fully developed turbulent flow, in which the rugosity becomes larger than the viscous sub-layer, becoming the main factor in determining the friction within the pipe, which is represented by the relative rugosity (Figs. 2.10 and 2.11).

The equations in the Table 2.1 are just a brief summary of those that can be found in the literature. In case the reader is interested in other correlations, please see Plascencia et al. (2019, 2020), Zeghadnia et al. (2019) and Tkachenko et al. (2020).

## 2.2 Resistance of Materials in Cylindrical Shells

### *2.2.1 General State of Stresses*

When a cylindrical envelope, such as a pipeline, is subjected to an internal or external pressure, stresses appear in the walls causing a state of deformations, and a 3D stress state is present,[7] with a longitudinal, a hoop and a radial stress acting at any point (See Figs. 2.12 and 2.13). Depending on the stress conditions, the material can be in different mechanical states, namely: elastic state under less intense forces; in an elasto-plastic state under the presence of more important forces; and rupture if the stresses are greater and cracks appear.

---

[6] In general, this value ranges between 0.017 mm and 0.030 mm; in the case of well-maintained pipelines in operation, the value previously indicated is usually adopted

[7] Sign convention for strength of materials: tensile + (positive) and compression – (negative).

**Table 2.1** Different implicit and explicit correlations to calculate the friction factor

| Name | Equation | Range |
|---|---|---|
| Laminar | $\frac{64}{Re}$ | $Re < 2000$ |
| Blasius (2024) | $0.3164Re^{-\frac{1}{4}}, \varepsilon = 0$ | $2100 < Re < 10^5$ |
| Nikuradse (1933) | $\frac{1}{\sqrt{f}} = 1.74 + 2\log(\frac{D}{2\varepsilon})$ | $10^5 < Re < 3 \times 10^6$ |
| Colebrook-White (1939) | $\frac{1}{\sqrt{f}} = -2\log(\frac{\varepsilon}{3.7D} + \frac{2.51}{Re\sqrt{f}})$ | $Re > 4000$ |
| Churchill (1977) | $8\left[\left(\frac{8}{Re}\right)^{12} + \frac{1}{(A+B)^{1.5}}\right]^{\frac{1}{12}}$<br>$A = \left\{-2.457\ln\left[0.27\frac{\varepsilon}{D} + \left(\frac{7}{Re}\right)^{0.9}\right]\right\}^{16}$<br>$B = \left(\frac{37530}{Re}\right)^{16}$ | All regimes and rugosities |
| Cheng (2008) | $\frac{1}{f} = \left(\frac{Re}{64}\right)^{a}\left(1.8\log\frac{Re}{6.8}\right)^{2(1-a)b}\left(2\log\frac{3.7D}{\varepsilon}\right)^{2(1-a)(1-b)}$<br>$a = \frac{1}{1+\left(\frac{Re}{2720}\right)^{9}}$<br>$b = \frac{1}{1+\left(\frac{Re}{160\frac{D}{\varepsilon}}\right)^{2}}$ | All regimes and rugosities |
| Bellos et al. (2020) | $\frac{1}{f} = \left(\frac{Re}{64}\right)^{a}\left(0.75\ln\frac{Re}{5.37}\right)^{2(1-a)b}\left(0.88\ln\frac{3.41D}{\varepsilon}\right)^{2(1-a)(1-b)}$<br>$a = \frac{1}{1+\left(\frac{Re}{2712}\right)^{8.4}}$<br>$b = \frac{1}{1+\left(\frac{Re}{150\frac{D}{\varepsilon}}\right)^{1.8}}$ | All regimes and rugosities |

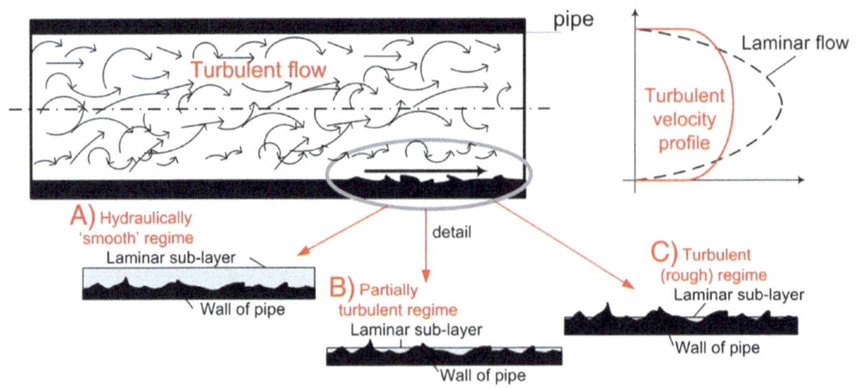


**Fig. 2.10** Cut away of a pipeline showing the turbulent flow and a detail next to the wall, comparing the laminar sub-layer for different regimes (Brkić 2011)

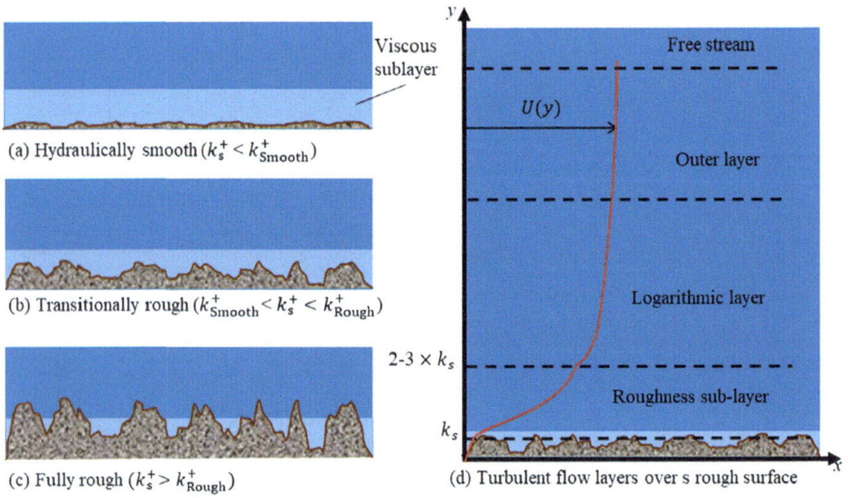


**Fig. 2.11** Cut details showing the turbulent flow over different rough surfaces, **a** a hydraulically smooth regime within the viscous sub-layer; **b** a transition rough regime with partially affected viscous sub-layer; **c** a fully rough regime with completely affected viscous sub-layer; and **d** a schematic of velocity profile and thickness compared to the indication of roughness, showing the different velocity layers (Kadivar et al. 2021)

In the study of the resistance of materials, the mechanical states are defined by the stresses that originate in the material. The stress states in which the transition from one mechanical state to another occurs are called *limit states*. If a material that is available can be subjected to a limited number of tests and it is necessary to predict the conditions of its transition from one state to another in all classes of stress states, the theory that allows us to solve this problem is the *theory of limit states*. Limit state

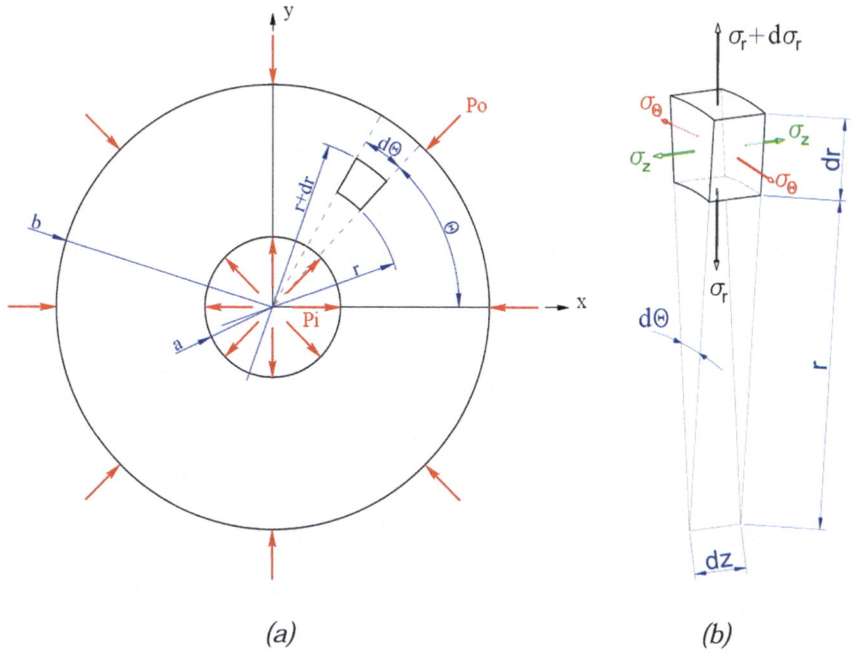


**Fig. 2.12** 3D scheme showing a thick-cylindrical shell under a state of pressure, showing the balance of forces and stresses in a differential control volume (Partovi and Shamili 2012)

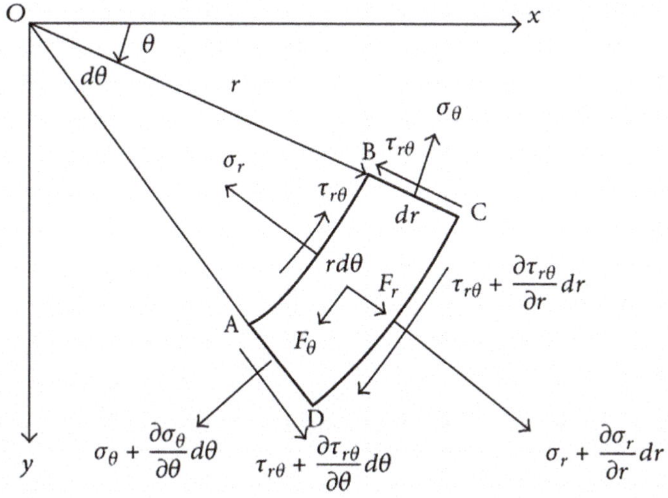


**Fig. 2.13** Detail showing the differential element in polar coordinate system and the stresses in a 2D plane (Nagler 2016)

theory is related to the study of material properties and the stress state of the entire solid is analyzed only to the extent necessary to locate the most dangerous point.

The failure theories of *maximum shear stresses* or *shape change energy (distortion)* have been applied for the calculation and provide appreciable advantages in solving problems in the case of ductile materials such as low alloy carbon steels. The most widespread form of the stress equation in this case is,

$$\sigma_i = \frac{\sqrt{2}}{2}\sqrt{(\sigma_x - \sigma_y)^2 + (\sigma_x - \sigma_z)^2 + (\sigma_y - \sigma_z)^2 + 6(\tau_{xy}^2 + \tau_{xz}^2 + \tau_{yz}^2)} \quad (2.47)$$

The magnitude $\sigma_i$ determines, according to the theory of distortion, the passage from the elastic to the plastic state. Within the elastic zone according to Hooke's law, the so-called "stress intensity" $\sigma_i$ can be defined,

$$\begin{bmatrix} \sigma_{xx} \\ \sigma_{yy} \\ \sigma_{zz} \\ \tau_{yz} \\ \tau_{xz} \\ \tau_{xy} \end{bmatrix} = \frac{E}{(1+\nu)(1-2\nu)} \begin{bmatrix} 1-\nu & \nu & \nu & 0 & 0 & 0 \\ \nu & 1-\nu & \nu & 0 & 0 & 0 \\ \nu & \nu & 1-\nu & 0 & 0 & 0 \\ 0 & 0 & 0 & \frac{1-2\nu}{2} & 0 & 0 \\ 0 & 0 & 0 & 0 & \frac{1-2\nu}{2} & 0 \\ 0 & 0 & 0 & 0 & 0 & \frac{1-2\nu}{2} \end{bmatrix} \begin{bmatrix} \epsilon_{xx} \\ \epsilon_{yy} \\ \epsilon_{zz} \\ \gamma_{yz} \\ \gamma_{xz} \\ \gamma_{xy} \end{bmatrix} \quad (2.48)$$

$$\begin{bmatrix} \sigma_{xx} & \tau_{xy} & \tau_{xz} \\ \tau_{yx} & \sigma_{yy} & \tau_{yz} \\ \tau_{zx} & \tau_{yz} & \sigma_{zz} \end{bmatrix} = 2\mu_L \begin{bmatrix} \epsilon_{xx} & \gamma_{xy} & \gamma_{xz} \\ \gamma_{yx} & \epsilon_{yy} & \gamma_{yz} \\ \gamma_{zx} & \gamma_{yz} & \epsilon_{zz} \end{bmatrix} + \lambda_L \mathbf{I}(\epsilon_{xx} + \epsilon_{yy} + \epsilon_{zz}) \quad (2.49)$$

where $\mathbf{I}$ is the identity tensor, $\lambda_L$ and $\mu_L$ are the Lamé's constants, $\epsilon_{ij}$ and $\gamma_{ij}$ and the longitudinal and rotational deformations, respectively. Thus, the "deformation intensity" is,

$$\epsilon_i = \frac{\sqrt{2}}{3}\sqrt{(\epsilon_x - \epsilon_y)^2 + (\epsilon_x - \epsilon_z)^2 + (\epsilon_y - \epsilon_z)^2 + \frac{2}{3}(\gamma_{xy}^2 + \gamma_{xz}^2 + \gamma_{yz}^2)} \quad (2.50)$$

Beyond the limits of elasticity, it is necessary to pass to the Theory of Plasticity. If there is a symmetry between the loads and the envelope around the axis, there will be no tendency for any circumference to change its circular shape unless buckling occurs.

### 2.2.2 *Thin Cylindrical Shells*

If the cylindrical shell walls are relatively thin (usually less than 1/10 of the radius) and do not have abrupt changes in thickness, inclination, or curvature, and the load is uniformly distributed or varies smoothly and is axial-symmetric, a "thin-wall" approach can be followed. The longitudinal and hoop stresses are the only ones with a significant presence, whilst the radial and those of bending that occur are negligible (Bazán and Beck 2016).

### 2.2.3 Thick Cylindrical Shells

If the wall thickness of the shell is greater than 1/10 of the radius, the longitudinal and hoop stresses cannot be considered uniform through the wall thickness, and the radial stress cannot be neglected (see Figs. 2.14 and 2.15).

Analyzing the equations, it will be possible to observe that the hoop (also known as girth or tangential) stress on the inner surface of the cylinder approaches the inner pressure, when the relationship between the outer and inner radius approaches infinity. It is apparent, however, that if the stresses are limited to a specific value $\sigma$ the pressure $p$ will never exceed it ($p = \sigma$) no matter how thick the pipe is made.

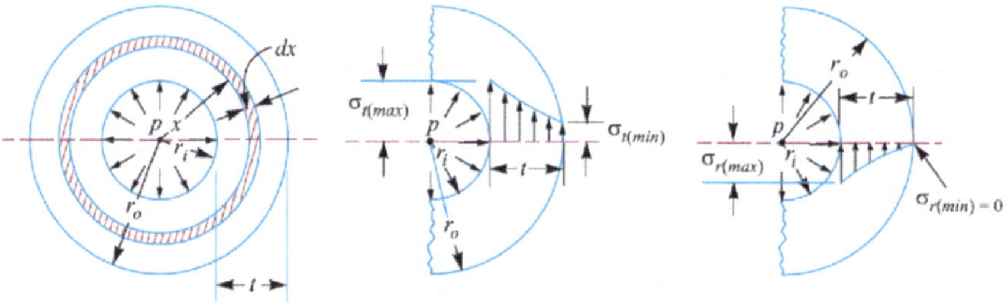


**Fig. 2.14** Stress distribution in a thick-wall cylindrical object (Wibawa et al. 2020)

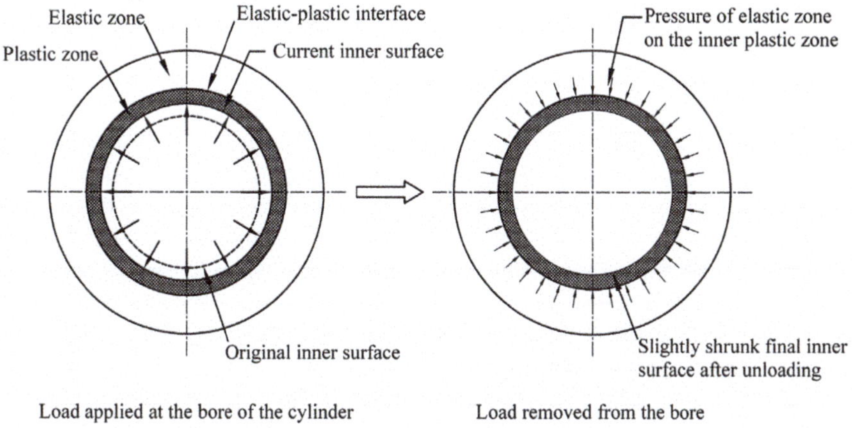


**Fig. 2.15** Plot showing the areas in the pipe after an autofrettage process (Dixit and Shufen 2020)

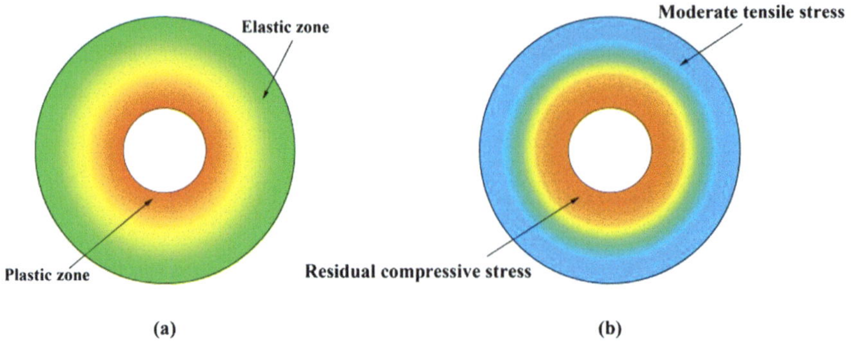


**Fig. 2.16** Cut detail showing elastic and plastic zones in a thick cylinder pipe during the autofrettage (**a**), and the final residual stresses after the process (**b**) (Partovi and Shamili 2012)

To overcome this limitation, the material in the vicinity of the inner surface must be subjected to a state of initial compression. This can be done by hooping one or more shells together or by subjecting the shell to high internal pressures so as to cause the interior fibers to creep, which then when depressurized leaves residual compression on the interior surface and residual tension on the outside part. This last procedure is called autofrettage (self-hooping) (Fig. 2.16).

### *2.2.4 Homogeneous Materials—Steels—Thick and Thin Walled Shells*

In the case of pipes made of a homogeneous material (e.g. steel), it is practically verified that the mechanical tests on specimens taken in the longitudinal or transverse direction to the direction of lamination give a very similar resistance in the tests and for this reason they are called isotropic materials.

Steel tubes and pipes are built with strips, hot-formed, with a butt-welded longitudinal joint (with or without filler). Seamless pipes are made by rolling and drawing and have no longitudinal joint. Steel pipes have been meticulously standardized in terms of wall thickness and the pressure they can tolerate for safety. In the American pipe system, the nominal diameter of "*standard* weight pipes up to 323.85 mm (12.75"—DN 300) is approximately equal to the inside; from 356 mm (14"—DN 350) and larger, the nominal diameter is the outside diameter. The additional metal in thicker pipes (e.g. extra strong—XS, and double extra strong—XXS) is added to the inside part, leaving the outside diameters equal.

In the case of pipes made of composite materials, for example, fiberglass-reinforced plastic, there are very important differences depending on the direction of the reinforcement fabric (longitudinal, at 45°, etc.) and the type of test tube established for the test. Their use in conducting fluids under pressure and at great distances is limited and they are not so common to find in projects. Woven blankets are often developed to *order* and the resistance of the resulting pipes has to do with their quality and direction.

As the mechanical resistance results from specific tests (therefore plastic composite pipes are difficult to predict in their behavior), the chapter will be focused on the developments applicable to those that are manufactured with homogeneous materials, particularly steel. Summing up:

- Pipes made of homogeneous materials are generally isotropic and the specimens taken in different directions give practically the same resistance in the tests.
- Pipes made of fiber-reinforced synthetic materials give important differences depending on the direction of the fiber weave. Woven cloaks are sometimes developed to order and strength has to do with the quality of the cloaks and the direction of the cloaks.

In dealing with thin-walled cylinders, it will be assumed that the stress is evenly distributed through the entire thickness of the wall. This assumption is approximately accurate for very thin walls only. In the case of an enclosure with internal pressure, the stress varies from a maximum on the internal surface to a minimum on the external surface.

In the more general case of thick-walled tubes, there will be pressure applied simultaneously to the inside and outside of the cylinder. The axial stress can be considered as uniform through the wall thickness. However, the most common case is that of thick casings subjected to internal pressure only. The analysis of the equations indicates that the radial and tangential stresses increase as the radius decreases, therefore the maximum stresses are on the inner surface.

The maximum tangential principal stress in the shell wall has been substituted for those of the radial and tangential stresses in the general expression, and since the theory of maximum tangential stress postulates that the tangential stress is equal to half the tension at the moment of failure, an equation applicable to the calculation of thicknesses for ductile materials is obtained, but it is not considered as exact as the Clavarino (closed cylinders) and Birnie (open cylinders) equations (Eqs. 2.51 and 2.52, respectively), since it is not taken into account in it the effect of transverse strain (see following section). In these cases, the stress $\sigma_t$ is usually taken equal to $0.8 \times \sigma_{yield}$.

$$t = \frac{D_{int}}{2}\left[\sqrt{\frac{\sigma_t + (1 - 2\mu)p_{int}}{\sigma_t - (1 + \mu)p_{int}}} - 1\right] \tag{2.51}$$

$$t = \frac{D_{int}}{2}\left[\sqrt{\frac{\sigma_t + (1 - \mu)p_{int}}{\sigma_t - (1 + \mu)p_{int}}} - 1\right] \tag{2.52}$$

### 2.2.5 *Thick Wall Pipes*

Lamé's[8] classical theory of elasticity, without considering plastic deformation, provides the following results for the tri-axial stress state, ignoring tangential deformations (Silvano 2010; Gupta 2020; Gupta et al. 1995).

**Hoop Stress**

$$\sigma_t = \frac{p_a a^2 - p_b b^2}{b^2 - a^2} + \frac{a^2 b^2}{r^2} \frac{p_a - p_b}{b^2 - a^2} \tag{2.53}$$

**Radial Stress**

$$\sigma_t = \frac{p_a a^2 - p_b b^2}{b^2 - a^2} - \frac{a^2 b^2}{r^2} \frac{p_a - p_b}{b^2 - a^2} \tag{2.54}$$

**Longitudinal Stress**

$$\sigma_x = \frac{p_a a^2 - p_b b^2}{b^2 - a^2} \tag{2.55}$$

Combining the three stresses, with a theory of failure[9] appropriate to the type of material and request, the highest stresses are found. In the case of ductile steels, and using the Theory of Maximum Distortion or Maximum Octahedral Stress, the von Mises yield criterion, which indicates:

$$\tau = \frac{1}{3}\sqrt{(\sigma_t - \sigma_r)^2 + (\sigma_r - \sigma_x)^2 + (\sigma_x - \sigma_t)^2} \tag{2.56}$$

where $\tau$ is the resulting tangential stress. Since $\sigma_t$ and $\sigma_r$ are functions of the radius, it is necessary to look for the radius to which the highest stresses correspond, calculate them and combine them. Then, the calculated $\sigma_{max}$ should be compared with the results of the simple tensile test (that is, considering only $\sigma_t$ and null $\sigma_r$ and $\sigma_x$) from which the yield stress of the material is obtained (see Fig. 2.17),

$$\tau = \frac{1}{3}\sqrt{(\sigma_t - 0)^2 + (0 - 0)^2 + (0 - \sigma_t)^2} = \frac{1}{3}\sqrt{2\sigma_t^2} = \frac{\sqrt{2}}{3}\sigma_t \tag{2.57}$$

Replacing $\sigma_t$ by the specified minimum yield stress (SMYS) obtained from the simple tensile test,

$$\tau_{max} = \frac{\sqrt{2}}{3}\sigma_{SMYS} \tag{2.58}$$

[8] Gabriel Léon Jean Baptiste Lamé (1795–1870) was a French mathematician who contributed to the theory of partial differential equations through the use of curvilinear coordinates and the mathematical theory of elasticity.

[9] Being ductile materials (steels with low and medium carbon content and other alloys) the following theories are applicable: maximum shear stress (Tresca criterion); maximum strain energy theory; or maximum distortion energy.

**Fig. 2.17** Scheme and picture of a commonly used tensile test in order to determine the steel properties. **a** Example of a stress-strain curve for a carbon steel used in pipelines (Pratheesh Kumar et al. 2021). **b** Tensile test showing the necking occurring in the sample, (Swiss Proficiency 2024)

It is important then to apply an appropriate safety factor (FOS) (generally variable between 0.8 and 0.4 in gas transportation systems, depending on the standards used for the design) so that the expression results,

$$\tau < \tau_{adm} = \frac{\sqrt{2}}{3}\sigma_{SMYS} \cdot FOS = \frac{\sqrt{2}}{3}\sigma_{adm} \tag{2.59}$$

### Maximum Stresses

If the external pressure can be neglected ($p_b = 0$), the maximum circumferential and radial stresses correspond to the internal radius $r = a$, the longitudinal stress being constant. The equations are then simplified as follows,

**Hoop Stress**

$$\sigma_{t,r=a} = \frac{p_a(a^2 + b^2)}{b^2 - a^2} \tag{2.60}$$

**Radial Stress**

$$\sigma_t = -p_a \tag{2.61}$$

**Longitudinal Stress**

$$\sigma_x = \frac{p_a a^2}{b^2 - a^2} \tag{2.62}$$

From the simplified hoop stress equation, one can calculate the thickness,

$$t = \frac{D}{2}\left[\sqrt{\frac{\sigma_t + p_a}{\sigma_t - p_a}} - 1\right] \tag{2.63}$$

However, when $p_b$ is not null and, in an limiting case, the internal pressure is non-existent $p_a = 0$, the maximum stress is located in the outer radius, $r = b$ (compression), and here it is necessary to elucidate where the maximum is. If the pipe is of the submarine type ("offshore") sometimes the external pressure $p_b$ predominates and since the thicknesses are generally important, it is always convenient to see in which radius the maximum stress is located.

## 2.2.6 *Thin Wall Pipes*

For pipes considered to be thin-walled, i.e. pipelines with t/D ratios between 1/50 to 1/100,[10] the one that predominates is the hoop stress. If the external pressure $p_b \neq 0$, the maximum hoop stress occurs for $r = a$.

[10] ASME standards set $t/r_{int} \leq 0.2$ as a limit for pipes.

$$\sigma_{t,r=a} = p_a \frac{(a^2 + b^2)}{b^2 - a^2} \tag{2.64}$$

Simplifying and considering $a \approx b$ (to be on the safe side),

$$\sigma_{t,r=a} \approx p_a \frac{(b^2 + b^2)}{(b + b)(b - a)} = p_a \frac{(b^2 + b^2)}{(b + b)t} = \frac{p_a b}{t} = \frac{p_a D_{ext}}{2t} \tag{2.65}$$

For simple calculations, this last slightly modified equation is applied (see Barlow's equation, Eq. 2.68), affected by a corresponding safety factor $F$, but then once the pipe has been selected and more details of the project are known, the stresses must be verified with the complete expressions (equations for thick-walled pipes).

In the case of longitudinal tension, simplifying in an analogous way, it results:

Longitudinal Stress

$$\sigma_x = \frac{p_a b}{2t} = \frac{p_a D_{ext}}{4t} \tag{2.66}$$

In other words, the longitudinal stress is 50% of the hoop stress. If the longitudinal stress results from traction (+) and the tangential stress also gives traction (+), the stress combination equation, the term $(\sigma_x - \sigma_t)^2$ helps to make the result of the tangential stress $\tau$ smaller.

The question would be, why does the longitudinal tension appear if the pipe is not "covered"? And the answer is: since the pipe is long, there is friction between the fluid and the interior of the pipe that acts as if there were a virtual "cover", which causes longitudinal stresses.

However, if the pipeline is buried, there will be friction between the ground and the outer surface of the pipeline (Fig. 2.18). However, the external friction with the ground will act as not infinitely rigid restraints and then the longitudinal stress decreases—or disappears[11] in the central part far from the ends—, but near the exit to the surface (or entrance) and in the aerial sections (swan necks), or scraper traps, the longitudinal tension appears and it has an important influence.

In effect, in general in the case of a buried pipeline, when the pipe emerges, its longitudinal movement is restricted with—for example—the installation of surrounding concrete blocks/anchors that immobilize it, and thus the entire duct can be designed with pipes of a single thickness, but outside the underground parts (already on the surface, i.e., the aerial sections) thermal stresses appear due to solar radiation, which are added to those resulting from pressure, which can be important and complex calculations, *strain-stress analysis*, must be performed.

Hooke's law expresses that for there to be (mechanical) tension, there must necessarily be a deformation. When the pipe is buried, away from the aerial ends, it is hypothesized that it does not deform longitudinally, and therefore the $\sigma_x$ is not taken into account.

[11] Remember that with regard to mechanical stresses without deformation there is no stress.

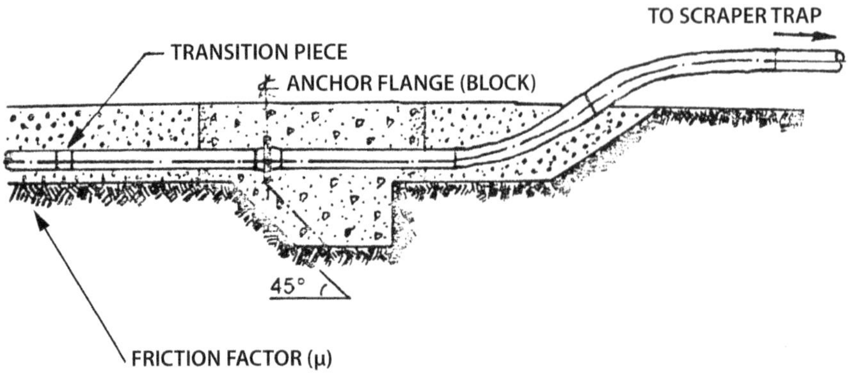


**Fig. 2.18** Sectional 2D detail for the anchor block design of a pipeline connecting to a scraper trap (McAllister 2014)

If the pipe is buried, the question to ask would be: are there charges due to the underground? If it were water or a dense fluid (mud) it would be easy to predict, but with solid soils it is difficult to know. This is because when filling the trench after the pipes have been lowered into them, the replacement of the fill soil is compacted very little (normally it is not carefully compacted) and then it only affects a part of the weight or sometimes it is considered as the load of the natural slope of the land. The uncertainty or ignorance is very large and its influence on the stresses must be treated with safety coefficients.

*Warning: depending on how the fill is compacted, the pipe can become oval on any side of the circumference. The solution would be to fill the part of the trench where the pipe is located with a fluid material (for example: lean concrete or foamed cement), but in general this technique is not applied because of its cost (except in cities and for a matter of of protection against the eventual action of third parties).*

## 2.2.7 *External Critical Pressure*

The stability of a structure is its ability to preserve its state and it is said that it is "stable", but the question arises as to how to carry out the stability analysis. The solution of the problem is reduced to determining the possible forms of equilibrium.

For example, in a compressed ring it is assumed that its shape is perfectly circular and that the load is evenly distributed throughout the circumference. The loss of stability is identified with the appearance of conditions that determine the existence of new forms of equilibrium. The loads under which these conditions are met are called critical loads. In the calculation of engineering works, the critical load is admitted as limit load where the stability margin is assigned.

When there is a very predominant external pressure or almost negligible internal one, e.g. submarine pipelines during its installation, the negative sign predominates in the combination of stresses and the pipe fails outside the theory of elasticity and does so by "buckling", hence it does not fail due to stresses but due to Young elasticity modulus, E.[12] The critical pressure and the critical failure stress are obtained from the following expressions, being the Poisson's[13] modulus for steel, $\mu = 0.3$, appears if longitudinal stress is taken into account:

$$p_{crit} = \frac{2E}{1-\mu^2}\left(\frac{t}{D_{int}}\right)^3 \Rightarrow \sigma_{crit} = \frac{E}{1-\mu^2}\left(\frac{t}{D_{int}}\right)^2 \tag{2.67}$$

Internal pressure on a thin-walled shell that is not truly cylindrical tends to make it cylindrical. However, if the material is quite rigid, it will resist this action and bending stresses of unknown magnitude will occur in the wall.

The acting loads due to the effect of pressure (internal, external, or both acting simultaneously) depend fundamentally on said pressures and on the shape of the pipe, its out-of-roundness, but also on the clamping situation with which the pipe is restricted or not to deform.

It is known that the acting loads due to internal pressure derive from the calculation of the mechanical stresses in three orthogonal axes (and that the absence of deformation in a certain direction denotes the absence of stress in that axis). In the case of external pressure, it is more complex because, as will be seen, the acting loads have more to do with an inelastic situation than with a composition of elastic stresses that allow the acting loads to be calculated.

However, the expressions are correct only for the case of a request by forces constantly oriented according to the normal to the curve of the ring, i.e., a uniform external pressure. If the behavior of the forces is different, the critical load will also be different.

### 2.2.8 Barlow's Equation

Considering that $D_{ext} - 2 \cdot t = D_{int}$ in the Lamé equation for pipelines subjected to internal pressure only,[14] it is found that, if the thickness is small with respect to the internal diameter, it results in the Barlow's equation,[15]

[12] The modulus of elasticity does not vary by structure, but is determined by the bond forces between atoms.

[13] Siméon Denis Poisson (Pithiviers, France, June 21, 1781—Sceaux, ididem, April 25, 1840) was a French physicist and mathematician who is known for his different works in the field of electricity and for his publications on differential geometry and probability theory.

[14] Simplifying, setting $r = r_{ext}$ to be on the safe side.

[15] Equation named after Peter Barlow (1776–1862), English mathematician, who made numerous contributions to the theory of strength of materials. Some of his work was collected after his death by his sons Peter and William (both also prominent 19th century civil engineers).

$$\sigma_t = \frac{p_{int} D_{ext}}{2t} \tag{2.68}$$

It is slightly more exact and is commonly used in the industry to calculate the thickness of pipes for high pressure oil and gas conduction, with the application of safety factors depending on the installation site.

### 2.2.9 Simplified Equations

It was shown during this chapter that for thin-walled (steel) pipes, hoop stress predominates, and the Barlow's equation is used to determine the circumferential stress for a specified internal pressure and thickness.

This equation is the one that many standards (e.g. Argentine, American, European) use for the design and operation of oil and gas pipelines, and it is generally taken as a basis for (pre-)calculating the thickness of the pipes.

In all cases, they are affected by certain design coefficients, namely: $F$, according to the location and distance of the duct with respect to the adjacent population, its density and the type of construction; $S$, weld coefficients according to the type of pipe manufacturing, mainly the longitudinal joint welding method; and $T$, thermal coefficients, according to the temperature range in which the fluid will be transported. For example, the thickness calculation equation for a gas pipeline is:

$$t = \frac{p_{int} D_{ext}}{2\sigma_t FST} \tag{2.69}$$

For instance, in the normative cases of Argentina, the diameter to be included in the equation is the outer diameter, the pressure is the relative design pressure (or the maximum operating pressure), and the stress is the yield stress (SMYS), and the design factor $F$ corresponds to the class of layout and type of construction according to the area and location that the pipe crosses. In other countries (see Table 2.2), the applicable regulations use the average diameter, minimum or average thickness, other design factors, etc.[16]

### 2.2.10 Pipe Diameter Variation Due to Pressure and Temperature

During this section, how the diameter of a pipe (i.e., a thin-walled pipeline) varies due to the simultaneous effect of internal pressure and temperature is discussed. The

[16] F = design factor; E = weld joint factor; T = temperature factor; $M_f$ = material performance factor; $Y$ = variation of ductility with temperature.

**Table 2.2** Different pressure design codes using equations similar to the simplified Barlow's equation with extra factor related to the safety design

| Standard | Scope | Country | Equation | Remarks |
|---|---|---|---|---|
| NAG-100 | Gas | Argentina | $t = \frac{p_{int} D_{ext}}{2\sigma_{SMYS} FET}$ | $F = 0.72/0.6/0.5/0.4$; $E$ and $T$ |
| ASME B31.4 | Liquid | U.S. | $t = \frac{p_{int} D_{ext}}{2\sigma_{SMYS} FE}$ | $F = 0.72$; $E$ |
| ASME B31.8 | Gas | U.S. | $t = \frac{p_{int} D_{ext}}{2\sigma_{SMYS} FET}$ | $F = 0.8/0.72/0.6/$... ...$0.5/0.4$; $E$ and $T$ |
| ASME B31.11 | Slurry | U.S. | $t = \frac{p_{int} D_{ext}}{2\sigma_{SMYS} FE}$ | $F = 0.80$; $E$ |
| ASME B31.12 | Hydrogen | U.S. | $t = \frac{p_{int} D_{ext}}{2(\sigma_{adm} EM_f + p_{int} Y)}$ | $\sigma_{adm}$ allowable stress; $E$ mat. quality factor; $M_f$ and $Y$ |
| EN 13480-3 | Both | Europe | $t = \frac{p_{int} D_{ext}}{2fz + p_{int}}$ | $f$ allowable stress; $z$ mat. quality factor |
| CSA 662 | Both | Canada | $t = \frac{p_{int} D_{ext}}{2\sigma_{SMYS} FE}$ | $F = 0.8/0.72/0.56/0.44$; $E$ and $T$ |

variation of the initial diameter $D_0$ of the steel pipe due to thermal expansion from $T_0$ to $T_1$, is expressed by the equation,

$$\Delta D = \alpha D_0 (T_1 - T_0) \quad (2.70)$$

where $\alpha$ is the coefficient of linear expansion of the material. For carbon steel, the coefficient is $\alpha = 11/12 \times 10^{-6}$ 1/K; $D_0$ the outside diameter of the pipe at temperature $T_0$; $T_1$ the temperature at which it is carried to the pipe. Writing the expression in terms of the final diameter, expanded by temperature increase,

$$D_1 = D_0 \left[1 + \alpha(T_1 - T_0)\right] \quad (2.71)$$

In general, $D_0$ is the diameter measured at room temperature $T_0$.

Due to the effect of internal pressure, and since steel is an elastic material with an elasticity modulus E (until yield occurs),

$$\sigma = E\epsilon = \frac{pD}{2t} \rightarrow E\frac{\Delta(\pi D)}{\pi D} = E\frac{\Delta D}{D} = \frac{pD}{2t} \quad (2.72)$$

$$\Delta D = \frac{pD^2}{2Et} \quad (2.73)$$

If the nominal diameter is taken as a basis, the initial temperature and pressure must also be taken as a basis (i.e., the ambient temperature $T_0 = T_{amb}$ and the absence of internal relative pressure $p_{int} = p_0 = p_{atm}$), where $\alpha$ is the coefficient of linear expansion of the material. Considering the above mentioned effects taking place simultaneously,

$$\Delta D = \alpha D_0(T_1 - T_0) + \frac{(p_1 - p_{atm})D_0^2}{2Et} \tag{2.74}$$

$$D_1 = D_0\left[1 + \alpha(T_1 - T_0) + \frac{(p_1 - p_{atm})D_0}{2Et}\right] \tag{2.75}$$

### Stresses Produced by the Temperature Variation

Structures subjected to the action of temperature behave in a similar way to how they would behave if they were supporting external loads. The stresses due to temperature variations originate as a consequence of the thermal deformations of the solid. Its value is a function of temperature, fixing conditions and material properties.

In the field of elastic deformations, thermal stresses are the product of the Young modulus $E$, the coefficient of linear expansion $\alpha$ and the temperature variation $\Delta T$.

When a structure is simultaneously subjected to the action of heat and external forces, it is not possible to separate thermal stresses from those of forces. Both are gathered in a single stress group and since their correlations are not linear, the principle of superposition does not apply.

For example, the constructive properties of carbon steel are preserved up to approximately 600 °C. Its thermal elongation at that temperature will be,

$$\epsilon = \alpha\Delta T = 12 \times 10^{-6}\frac{1}{K} \cdot 600\ K \approx 0.7\% \tag{2.76}$$

This value significantly exceeds the deformation corresponding to the elastic limit and therefore high temperatures can give rise to plastic deformations in the steel, however an elongation of the order of 0.7% is much lower than the elongation at break, where it is difficult, if not impossible, to imagine the breakage of a piece of steel due to the thermal effect.

According to what has been exposed—and as a simple rules of thumb—for ductile materials, thermal stresses in themselves do not represent a danger and can be ignored in the calculation of a structure. The effect of temperature has to be taken into account only to the extent that the mechanical characteristics vary; on the contrary, in brittle materials, thermal stresses represent an appreciable danger and must be taken into account together with mechanical stresses, in resistance calculations.

### 2.2.11 *Pipe Thickness and Manufacturing Standards*

Pipe manufacturers normally limit the manufacture of tubes to certain minimum thicknesses for practical reasons (e.g. rigidity, ovalization, handling, welding, manufacturing method). Likewise, the construction of pipelines with pipes of much smaller thicknesses in relation to their diameter entails construction difficulties. e.g., dents due to support at the bottom of an irregular trench, handling difficulties, and security against third parties (possibility of impacts from adjacent works cause ruptures with leakages causing a potential environmental damage).

For this reason, it is technical practice to adopt an adequate manufacturing relationship between thickness and diameter. The inverse relationship $D/t$ is commonly established, with values in the order of 100 to 120, although manufacturers' tables generally do not go beyond $D/t \approx 80/90$.[17]

**Example: Thickness verification**

Verification of a Submarine Pipe laid in a sea depth of 70 m by the "S" curve method, also known as the "lay-barge method", by the limit states method, the thickness of the pipe and the diameter variation being known due to internal pressure.

- Pipe = NPS 24 (DN 600).
- Thickness = 15.9 mm (5/8")
- Material grade and standard = Steel API 5L X65 (L450)
- External diameter = 610 mm
- Internal diameter = 578.2 mm
- External pressure = 0.704 MPa (70 m depth in sea water)
- Internal pressure = 7 MPa
- SMYS = 450 MPa
- Design factor = 0.5

Considering no external pressure $p_2 = 0$ for $r = R_1$, the maxima of $\sigma_t$ and $\sigma_r$, with $\sigma_x = const$ can be calculated, with the following results obtained by applying the equations for thick shells (i.e. Lamé's equations):

$\sigma_t = +130.87$ MPa;

$\sigma_r = -7.00$ MPa;

$\sigma_x = +61.93$ MPa.

[17] The ASME B31.4 standard, applicable to lines for the transportation of liquid hydrocarbons, states in its article 403.2.5 "Use of High $D/t$ Ratio" that the designer must take additional precautions if the ratio $D/t$ is greater than 100.

Assuming thin-walled pipes, simplifying the formulations $D_{int} \approx D_{ext}$ and calculating with $D_{ext}$, it renders:

$\sigma_t = +134.27$ MPa, this overestimates the stresses acting on the pipe's wall;

$\sigma_r = -7.00$ MPa;

$\sigma_x = +67.14$ MPa.

Knowing the combination the stresses to obtain the highest values, and using the theory of maximum octahedral stresses, it renders,

With Lamé's equations, $\tau = 56.28$ MPa. From the simple tensile test $\sigma_t = \sigma f$ and $\sigma_x = \sigma_r = 0$ for the maximum stress (beginning of yield):

$$\tau = \frac{\sqrt{2}}{3}\sigma_f$$

Looking for the $\tau_{max}$ (admissible), one must incorporate a safety factor (denominated generally as $F$, the design factor, see Table 2.2),

$$\tau_{max} = \frac{\sqrt{2}}{3}\sigma_f F \Rightarrow \tau_{max} = 106.06\,\text{MPa}$$

This value is higher than the one obtained previously, verifying the design conditions with the material selected.

If one calculates the same with the stresses obtained using the thin-walled approach, it renders: $\tau = 57.70$ MPa. This value is also lower than the maximum allowed $\tau_{max} = 106.06$ MPa.

In order to find the maximum allowable operating pressure (MAPO), one must obtain $p_{int}$ from the expression of $\tau$ using the calculated one $\tau_{max}$. Solving $p_{int}$ as a function of and the radii $r_{int}$ and $r_{ext}$, obtaining $p_{int} = 14.13$ MPa.

Calculating with the simplified equations, in order to see what difference can be noticed when compared to taking the stresses with the Lamé's equations, one would obtain $p_{int} = 12.87$ MPa.

Finally, for non-zero external pressure $p_{ext} \neq 0$ and zero internal pressure $p_{int} = 0$, the critical buckling pressure is $p_{crit} = 8.173$ MPa, for the ratio $D/t = 38.36$.

Decreasing the thickness by half to 10.31 mm (13/32"), one would obtain a $p_{crit} = 1.11$ MPa and the result would be equal to 208.50 MPa, close to the admissible stress of 225 MPa (450 MPa × 0.5). Nevertheless, with 10.31 mm there would be also installation stress problems.

**Example: Diameter variation**

Calculate the variation of diameter due to internal pressure and operating temperature, knowing the reference pressure and temperature and the operating pressure and temperature, for a steel pipe in the following conditions.

- Pipe = NPS 30 (DN 750).
- Thickness = 6.35 mm (1/4")
- Material grade and standard = Steel API 5L X60 (L415)
- Reference temperature = 15 °C
- Operating temperature = 65 °C
- External pressure = 0 bar(g)
- Internal pressure = 49.5 bar(g)
- Coefficient of linear expansion ($\alpha$) = 11.6 $\times 10^{-6}$ 1/°C
- Young modulus = 2.1 $\times$ $10^{11}$ Pa

Applying Eq. 2.75 in order to find the new value of the pipe diameter,

$$D_1 = D_0\left[1 + \alpha(T_1 - T_0) + \frac{(p_1 - p_{atm})D_0}{2Et}\right]$$

$$D_1 = 0.762\,m\left[1 + 11.6 \times 10^{-6}\frac{1}{°C} \times 50\,°C + \frac{49.5 \times 10^5\,Pa \times 0.762\,m}{2 \times 2.1 \times 10^{11}\,Pa \times 0.00635\,m}\right] = 0.7635\,m$$

## 2.2.12 *Stresses Produced by Variable Loads Over Time—Pulsations*

Loads of this type must be given special consideration due to the added stress that can be incorporated by the rate of application of the motivating influence and the fact that the yield point can be appreciably elevated by very fast stresses. Localized points of stress concentration can be exacerbated under such conditions and fracture much more easily initiated. The significant dynamic loads to consider in a piping design are indicated below:

**Earthquake**: Accelerations associated with earthquakes are generally on the order of 0.3 to 2.5 $\mathrm{m/s^2}$. These values represent 3 to 25% of the acceleration due to gravity ($g = 9.80665\,\mathrm{m/s^2}$. For this reason, anti-seismic designs commonly apply a horizontal force at the structure's center of gravity of the order of 10 to 20% of its weight. These charges are not normally assumed unless specifically required by the location.

**Water hammer or effects of flow fluctuations**: the pressure wave due to the sudden stop of a flow of liquid is a function of its speed, the closing time and the elasticity and support of the pipe. The effects of pressure fluctuations are present when reciprocating pumps or compressors are applied. The apparent mechanical vibrations may in some circumstances be sufficient to result in material fatigue failure, if not promptly corrected.

### Fundamental Considerations in the Vibration of Pipes

It must be clear about the following:

- Vibration period T [s]: designates the time of a complete oscillation that is repeated each time.
- Vibration frequency [cycles/s = Hz]: it is the reciprocal of the period. The angular frequency [radians/s] is equal to $2\pi/T$.
- The Number of degrees of freedom is equal to the number of independent quantities defining the position of the system. Thus a simply supported pipe has an infinite number of degrees of freedom due to the flexibility of each element relative to the adjacent elements, requiring an infinite number of deflections to fully describe the position.
- The main mode of vibration is the free vibration of the system at a defined frequency. The frequencies of the main oscillation modes are called natural frequencies. The lowest natural frequency is called the *fundamental frequency* and it corresponds to the fundamental oscillation mode.
- Damping can be defined as the reduction of the vibratory amplitude by frictional forces. It is a solution for resonant vibrations, but the non-resonant ones are little reduced by these devices.
- From a practical point of view, the biggest problem related to vibrations is the phenomenon of *resonance*. This occurs when a system, mechanical or acoustic, is periodically excited with a frequency similar, very close, or equal to the natural frequency of vibration.

With respect to the vibration types, the following types of situations can be defined:

- Free vibrations take place when the system is excited by an external transient impulse of short duration, and the system vibrates without external forces.
- Forced vibration is when the system oscillates under an external excitation from an external disturbing force.
- Self-excitation is a complex phenomenon since the system vibrates without the action of periodic external forces and the vibration persists due to internal energy sources, even in the presence of damping. In pipes they have been found associated with unstable flows, compressor pumping, vortices due to fluctuating winds, etc.

### Vibration Relief

One will refer later to the relief of vibrations in pipes, particularly due to the action of reciprocating compressors in gas pipelines. In the case of liquids in cases of Water Hammer, it may be found that if the pressure increase derived from a valve closing is excessive and cannot be tolerated, several measures can be decided to reduce said effect. The simplest is obviously to increase the closing time of the valves. If for some reason it is not feasible to increase the closing time of the valve, accumulator tanks or relief valves can be used.

### Troubleshooting Procedure

When faced with a specific case of vibrations in pipes, the first step is to define the need for immediate action, and if the case is serious enough, one can decide to order the "shutdown" of the installation, to take corrective measures that limit the amplitude. of the vibrations. On the other hand, it is often possible and also completely adequate to reduce vibrations, provide supports and temporary restraints while the fundamental solution of the problem is analyzed.

Regarding the derived stresses, when the total stress (based on the double amplitude, obtained from a simple beam calculation) exceeds by 2 the yield stress of the material in the area of greatest stress, there will be a danger of imminent damage, while that minor stresses can be compared against the curve of stresses versus the number of cycles of the material concerned, with an indication of the probable life in the absence of harmful defects. It must be emphasized that fatigue failure is related to peak stresses so that flanges, etc. where local stresses are above the average, they are the focus of possible failures. Finally, it should be noted that the interpretation of vibration data, e.g. measurements, is a difficult task, since the records obtained in the field frequently show a diversity of small oscillations or "noises" superimposed on systematic oscillations of greater amplitude.

As a rule "noise" type vibrations are so small that they are not dangerous. In general only well-developed sinusoidal vibrations should be reduced in amplitude for safe operation. Pulsations in a plumbing system can be described as a complex pressure wave having frequency components that can increase or decrease from one point in the system to another. In the case of a gas transmission system, the maximum peak-to-peak amplitudes are often more than 10% of the static pressure.

The successive intake and exhaust of gas in the cylinders of reciprocating compressors create pressure waves that can cause pipes to vibrate, influencing the delivered flow. It is therefore very important to eliminate them in order to obtain a gas flow that is as uniform and regular as possible. The causes of pulsations lie within the system as a whole and are not governed by any one component to the exclusion of others. Among the preponderant factors that cause pulsations are the configuration of the reciprocating compressor cylinders, including harmful clearances, number of cylinders in a given unit, and their ratio of crankshaft angles and components such as nozzles, cylinders, lateral lines, and buried pipes.

Probably the best known and most easily recognizable effect of flow pulsations is pipe vibration which leads to pipe and support failure. To combat support failures, it is common for the hold down clamps to have been increased in size, but in many cases this results in the foundation bolts breaking. If then bolts are strengthened, then cracking of foundations can turn out. More serious from a system economics standpoint are the effects of pulsation on reciprocating compressor load, compression capacity, and overall efficiency. Power losses of 4% to 20% have been verified in these circumstances. The Southwest Research Institute[18] developed the following measures to prevent pulsation in gas systems that apply to reciprocating compressors:

- An electrical analogy of alternative compression installations.
- Refinement of physical principles in acoustics, compressors, piping design through the use of electrical analogy.
- Effective use of the developments in the correction of problems in existing stations.
- Effective use of developments in piping and equipment design at new stations to ensure: compressors operating at their maximum capacities and efficiency; effectively damped or controlled pulsations with minimal pressure losses; forces resulting from pulsations effectively controlled to prevent vibration damage.

A plumbing system could be designed to be comparatively vibration free. This could be accomplished by avoiding "resonant" lengths of piping, providing adequate passages for fluid passage through the compressor, and arranging for secure piping anchorage. In this sense, there are some rules related to the design of the facilities that should be respected:

- Locate the anti-pulsation bottles as close as possible to the cylinder flanges.
- Avoid symmetries that can facilitate phase matching and overlapping of pulsation waves.
- Avoid short radius elbows where the forces at play are concentrated.
- Reduce as much as possible the changes of direction of the pipes.
- Do not distribute the supports of the pipes regularly.
- Solidly fix the equipment, allowing the necessary expansions.

## 2.3 Internal Corrosion

In some cases, the "battle" against internal corrosion in carbon steel ducts is focused on a double objective: the adequate conservation of the thickness of the pipes and the reduction of the coefficient of friction.

The internal corrosion of the pipes can be due to the corrosive characteristic of the transported product, e.g., sulfurous crudes or, more frequently, traces of water that

[18] The Southwest Research Institute, headquartered in San Antonio, Texas (USA), is an independent and nonprofit applied research and development (R&D) organization.

always accompany the transported fluids and that accumulate on the lower intrados of the oil pipeline/multi-product, or condense on the inside wall of the pipeline.

It should be noted that in the case of natural gas, the presence of $CO_2$ or sulfur compounds in its composition and water increase the corrosive effect on low alloy steels. Hence, it is so important to limit as much as possible the possibility of dragging water in both crude oil and natural gas, and even more so if it is brackish formation water. In all cases, a dehydration treatment is recommended, and it could be said to be essential, before introducing the liquid or gaseous products into the transport lines built in carbon steel pipes.

In gas pipelines, there was a tendency in the 80s to coat the pipes internally with an epoxy paint that improves internal roughness and protects against corrosion, with the idea that its cost was offset by a lower application of compression energy; but the system has not proliferated too much because with the passage of time the epoxy layer is coming off. In the case of corrosive crudes, it is more common to use corrosion inhibitor products and, in some cases, fluids that improve flow, especially for extremely viscous crudes; or use pipes made of composite materials, e.g., reinforced plastics.

## 2.4 Heat Transfer

The heat transfer between the system, i.e. fluid, and the environment, i.e. ground or surrounding atmosphere, plays a major role in the transport process since it affect the state parameters of the fluid system. At this point it is necessary to appreciate the fundamental differences between thermodynamics and heat transfer. Although thermodynamics is concerned with the interaction of heat, and its importance in the First (quantitative energy transfer) and Second (qualitative energy transfer) Laws, it does not provide the mechanisms of heat transfer nor the methods for computing that phenomenon. Thermodynamics concerns equilibrium states, while heat transfer is an intrinsically unbalanced phenomenon or process, since for it to exist there must be a temperature gradient, and therefore a thermodynamic imbalance.

### *2.4.1 Heat Transfer Between the System and the Environment*

The phenomenon of heat transfer plays a very important role in numerous technical and environmental problems. For example, in the area of electric energy production and conversion where all applications involve conduction, convection and radiation processes for the development of the design of boilers, condensers, turbines and cooling towers. There are three well differentiated phenomena as heat transfer mechanisms: *conduction, convection* and *radiation.*

Although this subject is going to be developed in detail, an overview of the phenomenon of heat transfer must be taken. However, before going into detail, one must have an intuition to determine what is happening physically.

**Conduction**

Conduction can be seen as the transfer of energy from the more energized to the less energized particles of a substance, due to the interaction between the particles. In a solid, conduction can be attributed to atomic activity in the form of lattice waves. Examples of conduction are well known, for example the end of a metal spoon immersed in a hot liquid (e.g. in a cup of coffee) will be heated by the conduction of energy through the metal.

**Convection**

The convection mode of heat transfer comprises two mechanisms: in addition to the energy transfer due to molecular random motion, i.e. diffusion, there is also energy transferred by the macroscopic motion of the fluid.

It is of special interest to this book the transfer by convection that occurs between a fluid in motion and the limiting surfaces when both are at different temperatures. As a consequence of the fluid-surface interaction, the kinematic boundary layer develops, this is, the region where the relative velocity goes from being zero on the surface to taking the velocity of the flow, and also a thermal boundary layer, which can be greater, less than or equal than to the kinematic one.

Convection can be classified according to the nature of the flow. *Forced convection* is called when the flow is caused by an external medium (pump, compressor or wind) and *natural convection* when the flow is induced by buoyant forces due to density differences that appear due to temperature differences in the air. fluent.

Typically the energy that is transferred is the sensible energy or thermal internal energy of the fluid; however, there are convection processes in which there is also latent heat exchange. This latent heat is associated with a phase change between the liquid and vapor states of the fluid (Table 2.3).

**Radiation**

Thermal radiation is energy emitted by matter that is at a certain temperature. Although the primary focus is radiation from solid surfaces, emissions can occur from liquids and gases. The emissions can be attributed to changes in the electronic configurations of the constituent atoms or molecules. The energy of the radiation field is carried by electromagnetic waves (or photons).

**Table 2.3** Typical convection coefficients based on the process in [W/m²K]

| Process | Coefficient $h$ |
|---|---|
| Natural convection in liquids | 50–1000 |
| Natural convection in gases | 2–25 |
| Forced convection in liquids | 50–20000 |
| Forced convection in gases | 25–250 |
| Convection with phase change | 2500–100000 |

While conduction or convection transfer requires the presence of a material medium, radiation does not. In fact the phenomenon of radiation is more efficient in a vacuum. The heat flux emitted by a real surface is less than that defined as a black body and the Stefan-Boltzmann equation,[19] with the so-known Stefan-Boltzmann constant ($5.6704 \times 10^{-8}$ W/m²K) is affected by the radiant property of the surface called emissivity or absorption factor $\epsilon$ (Table 2.4).

$$q\left[\frac{W}{m^2}\right] = \epsilon\sigma_{SB}T^4 \tag{2.77}$$

**Table 2.4** Different heat transfer mechanisms

| Mode | Description | Equation | Coefficient |
|---|---|---|---|
| Conduction | Diffusion of energy by the random movement of molecules | $q[W/m^2] = -k\nabla T$ | $k = [\frac{W}{mK}]$ |
| Convection | Diffusion of energy by random motion of molecules + energy transfer due to "bulk" or fluid's macroscopic motion | $q[W/m^2] = h\left(T_\infty - T_f\right)$ | $h = [\frac{W}{m^2K}]$ |
| Radiation | Transfer of energy by electromagnetic waves | $q[W/m^2] = \epsilon\sigma\left(T_{sup}^4 - T_f^4\right)$ | $\epsilon = [\frac{W}{m^2K}]$ |

[19] This Law was derived in 1884 by Jozef Stefan (1835–1893) in 1879 on the basis of experimental measurements made by John Tyndal, and was derived from theoretical considerations applying the thermodynamic studies of Ludwig Boltzmann (1844–1906).

### 2.4.2 Fluid State Parameters Variations Due to Heat Transfer

The transmission of heat affects the system since it modifies its response from basically two points of view: the mechanical, i.e., stresses in pipes and on equipment to which they are coupled, and that of the flow itself (pressures involved).

Surface installations are highly affected by solar radiation (air pipes exposed on days of high solar radiation can take very high temperatures, 70 or 80 °C), which are transmitted to the fluid that circulates inside them.

Particularly in oil transportation and treatment plants, it is practical to install reliefs for thermal overpressures, in sections of aerial pipes sectioned by blocking valves that can remain closed and that the internal product has no way to expand, except at through the relief valve whose discharge flows into a tank.

In buried pipes, the total transmission coefficient, which considers the surrounding terrain and the meteorological/atmospheric situation on the cover of the duct, will allow knowing for each segment the heat transmitted and how this affects the flow, by making the density, the viscosity, Reynolds and/or Mach number to vary, particularly the loss of pressure as the flow progresses through the pipeline.

The drive machines are also affected in their operation, since at higher ambient temperatures the delivered power decreases and vice versa. In addition, in some cases, the higher temperature of the pipes can affect the alignment between pumps and/or compressors and their motors, causing vibrations.

### 2.4.3 Heat Transfer in Underground Pipelines

In principle, it should be considered that the ground temperature is fairly constant for a given season of the year. Indeed, if it is a question of transporting oil through the pipeline in summer and in winter, the flows will be different for the same power of pumping power involved (Chung et al. 1999). It will be the other way around if it is a question of transporting natural gas, due to the increase in its viscosity with higher temperatures (Katsura et al. 2020; Xie et al. 2022). To ensure a minimum flow, the worst condition must be assumed during the design: if it is oil for the ground, the ground temperature in winter will be adopted, and if it is natural gas, the ground temperature in summer.[20]

In lands without the presence of rock strata, the ducts are normally buried with a minimum cover of 0.8 to 1 meter[21] for the pipeline right-of-way, and it is known that the ambient temperature on the ground surface has large oscillations between day and night, in different times of the year. Thus, at a depth of 1 meter, the line is in the zone of relatively wide thermal oscillations (see Fig. 2.19).

---

[20] This is due to the different trend in the variation of viscosity with temperature in gases and liquids.

[21] This depth may vary, going up to 2.0 m in highway and railroad crossings. It also depends on the soil conditions: lower depths in rocky areas and higher depths in unstable soil areas.

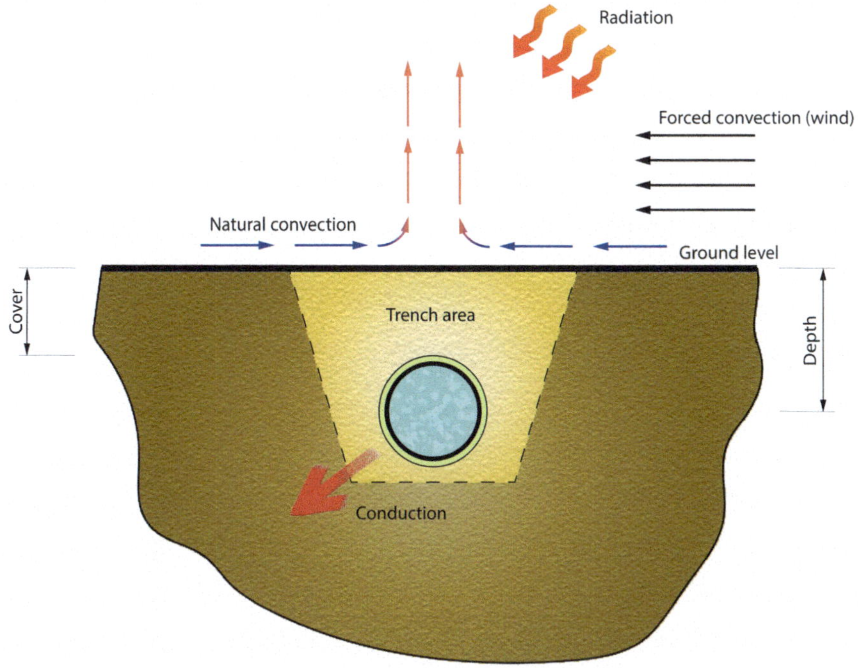


**Fig. 2.19** Scheme of a buried pipeline, showing the main heat transfer mechanisms present. It is also important to mention that the thermal properties of the soil used to fill the trench may differ from those found in other parts of the domain

Another condition to consider is that the soil—in the first layers—is not very compact, but deeper down it becomes more compact and the heat transfer coefficient changes a lot with compaction. In this book, for the simplicity of the calculations, one will define a single transmission coefficient of heat. In addition, the degree of soil moisture also greatly affects said coefficient.

For the optimization calculation, a single heat transmission coefficient is defined with the data from the so-called average soil or preponderant to the depth of the duct and it is assumed that the ground surrounding the pipe is the one that has the greatest influence on the transmission. The ground will be assumed to be a source of infinite heat capacity, and a radius will be set from which it is assumed that the temperature of the ground surrounding the duct remains unchanged (see Fig. 2.20).

It is possible to solve the issue at a steady-state, but when the flow begins, especially if the pipe is of small diameter, it can take many days, or even weeks, to heat the mass of land to bring it to the regular temperature, will require a long transient. If it is a liquid, its viscosity will vary significantly with temperature and it will take time to get the ground around the duct to heat up, so that once it is "hot" the duct starts to work properly.

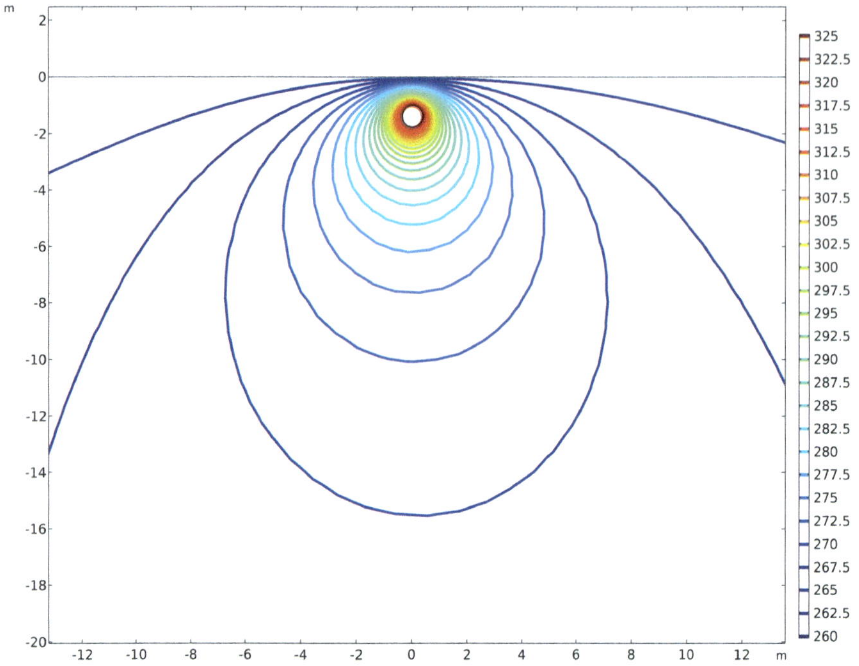


**Fig. 2.20** Numerical simulation of the temperature profile (in K) that can be usually encountered in a buried pipeline heat transfer process

## *2.4.4 Simplified Calculation of the Global Transmission Factor in Buried Pipes*

As mentioned in the previous sections, the temperature and the energy transfer play a major role in the operation of (underground) pipelines, and therefore it is of the utmost importance to consider its their effects during the economical studies and the engineering design of the project. It is well-known the example of the steady-state heat transfer of a pipeline, commonly found in Fluid Mechanics or Transport Phenomena books and university courses. The temperature profile as the function of the length for this case is expressed by the following equation,

$$T(x) - T_w = (T_i - T_w)\, e^{-\frac{4U}{\rho v D c_p} L} \tag{2.78}$$

where $T_w$ is the temperature of the pipe's wall; $T_i$ is the fluid's inlet temperature; $\rho$ is the density of the fluid; $v$ is the velocity; $D$ is the diameter of the pipe; $c_p$ is the fluid specific heat; $L$ is the length of the pipe; and finally, $U$ is what is known as the *overall heat transfer coefficient*, which is a function of the materials involved in the whole heat transfer process.

The total heat transferred in a system per unit of time can then be calculated using the overall heat transfer coefficient,

$$\dot{Q} = U\pi DL\left(T_{ext} - T_f\right) \tag{2.79}$$

Conventions establish that the positive sign (+) is adopted for the transfer of heat from the medium (i.e. ground/atmosphere) to the system (i.e. fluid). Being $U$ the total (or global) transmission coefficient; $D$ the inside diameter of the pipe; $L$ the length of the section; $T_{ext}$ the temperature of the medium; and $T_f$ the temperature of the fluid. In a buried pipe, especially in gas pipelines after leaving the compressor station, this heat will have a negative sign, an indication that the fluid is giving heat to the ground and therefore is cooling down as it circulates.

For approximate and computational calculations, a certain thickness of land is taken, after which the soil temperature remains unchanged. Commonly, the most difficult problem is precisely to predict the behavior of the terrain in terms of its thermal conduction properties.

Indeed, the transmission of heat from the fluid inside the pipe to the insulating coating can be calculated theoretically, then if one wants to be rigorous it should be done by finite elements until reaching the ground temperature $T_{ext}$, at the distance that that is, generally between 3 and 7 m from the pipe, where it could be set to reach 99% of the temperature $T_{ext}$, the same approach used, for instance, to calculate the thickness of boundary layers. The true insulator is the soil layer (almost certainly), except for thermal insulation used in the pipeline (e.g. expanded polystyrene, perlite).

However, adequate approximations can be made, where the sum of the effects of each heat transfer mechanism is like adding, as far as the coefficients are concerned, as a "parallel" circuit in electricity applications:

$$\frac{1}{U} = \frac{1}{h_f} + \frac{t_p}{k_p} + \frac{t_c}{k_c} + \frac{t_i}{k_i} + \frac{D_i \ln \frac{D_{exts}}{D_{ints}}}{2k_s} + \frac{1}{h_{ext}} \tag{2.80}$$

Being $f_f$ the fluid convection coefficient; $t_p$ and $k_p$ the thickness and thermal conductivity of the pipe sheet; $t_c$ and $k_c$ the thickness and thermal conductivity of the anti-corrosive layer; $t_i$ and $k_i$ the thickness and thermal conductivity of the thermal insulator; $k_s$ the thermal conductivity of the soil; and $h_{ext}$ is the air convection coefficient.

The conduction coefficient in a material such as steel is generally not a problem, being the coefficients for carbon steel materials quite known; the problem lies in the soil, that is "porous", with parts filled with air and being as well not an isotropic medium, rendering this an approximate thermal conductivity coefficient. The approximate values of the coefficients are shown in the following table (Fricke et al. 1992; Rerak 2017) (Table 2.5):

**Table 2.5** Typical heat conduction/convection coefficient for underground pipelines

| Mechanism | Constant | Value |
|---|---|---|
| Natural gas convection (LP) | $h_f$ [W/m$^2$K] | 12–60 |
| Natural gas convection (HP) | $h_f$ [W/m$^2$K] | 60–80 |
| Liquids | $h_f$ [W/m$^2$K] | 600–2400 |
| Carbon steel pipeline | $k_p$ [W/m$^2$K] | 40–60 |
| Polypropylene (PP) pipeline | $k_p$ [W/m$^2$K] | 0.11 |
| High-density polyethylene (HDPE) pipeline | $k_p$ [W/mK] | 0.44 |
| Anti-corrosive layer | $k_c$ [W/mK] | 0.2 |
| Thermal insulation layer | $k_i$ [W/mK] | 0.01–0.05 |
| Soil | $k_s$ [W/mK] | 0.9–2.5 |

### Reference Values for Air Convection Coefficient

The atmospheric convection coefficient in [W/m$^2$K] can be obtained from the expression:

$$h_{air} = 9.42 + 3.68V \tag{2.81}$$

Where $V$ is the average wind speed measured in [m/s] that is estimated to be on the area of the pipe cover. For this, it is necessary to collect statistical data from the meteorological records of the installation area. There are different expressions to estimate the turbulent convection coefficient for the air as a function of the wind speed, and most of these follow one of the following approaches, where $a$, $b$ and $n$ are empirical constants (Palyvos 2008).

$$h_{air} = a + bV \tag{2.82a}$$

$$h_{air} = a + bV^n \tag{2.82b}$$

### Reference Values for Soil Conduction Coefficient

In general, the conduction coefficient for the terrain varies between 0.9 and 2.5 [W/m$^2$K], it strongly depends on the type, humidity and final compaction. More precisely, the conduction coefficient $k_s$ for a terrain as a function of its degree of humidity, can be estimated with the expression:

$$S(\%) = C_1[\sinh(C_2 k_s + C_3) - \sinh C_4] \tag{2.83}$$

Where $S$ is percentage degree of soil moisture; $C_i$ are coefficients defined according

**Table 2.6** Soil parameter to calculate the heat transfer coefficient

| Type of soil | $C_1$ | $C_2$ | $C_3$ | $C_4$ |
|---|---|---|---|---|
| Clay | 27 | 0.265 | –1.5 | –0.97 |
| Gravel | 6.5 | 0.380 | –3.0 | –1.48 |
| Peat | 28 | 0.865 | –1.9 | –1.875 |
| Sand | 6.8 | 0.400 | –2.9 | –1.5 |
| Silt | 17 | 0.400 | –2.6 | –1.6 |

to the type of terrain, defined in Table 2.6; and $k_s$ is the soil's conduction coefficient. For a unfrozen silt-clay soil, one of the most commonly used correlations is the Kersten model, based on the data available for five different soils, with a moisture validity range up to 7%.

Kersten correlation for fine-grained soils (silt, clay, etc.)

$$k = 0.1442\left[0.9\log S(\%) - 0.2\right]\cdot 10^{\frac{0.6243\rho_{soil}}{1000}} \tag{2.84}$$

Kersten correlation for coarse-grained soils (sand)

$$k = 0.1442\left[0.7\log S(\%) + 0.4\right]\cdot 10^{\frac{0.6243\rho_{soil}}{1000}} \tag{2.85}$$

Once the degree of humidity of the installation ground is known (in percentage of saturation) and its type (by taking samples of the ground on the pipe trace and its subsequent test), the coefficient is calculated with the previous equations.

Where $k_s$ is measured in [BTU.inch/feet$^2$.h.K] and must be converted to SI units for application in equations. The common soil conduction coefficient for type of soil and materials can be found in the following Table 2.7.

### 2.4.5 *Calculation of Total Heat Exchanged in Pipes*

**Differential Energy Balance**

Using the Reynolds Transport Theorem studying the energy balance in a differential control volume, one can find the well-known energy equation, expressed as a function of the temperature, considering all the possible transfer mechanisms (see Fig. 2.21) (Duan et al. 2013),

$$\nabla\cdot\left(\mathbf{k}\nabla T - \mathbf{q''}_{\mathbf{rad}}\right) + \mathbf{T} : \nabla\mathbf{v} + \dot{E}_{other} = \rho c_p \frac{\partial T}{\partial t} + \nabla\cdot\left(\rho c_p T\mathbf{v}\right) \tag{2.86}$$

**Table 2.7** Different materials and their thermal conductivities

| Material | k $[W/mK]$ | Material | k $[W/mK]$ |
|---|---|---|---|
| Air | 0.026 | Peat (dry) | 0.17 |
| Ice | 2.2 | Peat (wet) | 0.54 |
| HDPE | 0.35 | Peat (icy) | 1.89 |
| Propylene | 0.22 | Sandy soil (dry) | 0.43/0.69 |
| Sandstone | 1.63/2.08 | Sandy soil (moist) | 0.87/1.04 |
| Snow | 0.34 | Sandy soil (soaked) | 1.90/2.42 |
| Soil (clay 10% moisture) | 0.4/1.3 | Clay (dry) | 0.35/0.52 |
| Permafrost | 1.4 | Clay (moist) | 0.69/0.87 |
| Carbon Steel | 50 | Clay (wet) | 1.04/1.56 |
| Stainless Steel | 21 | Clay (frozen) | 2.51 |
| Water | 0.6 | Gravel | 0.90/1.25 |
| Limestone | 1.3 | Gravel (sandy) | 2.51 |
| Seawater | 0.65 | Deep-water soil | 0.7/1.3 |

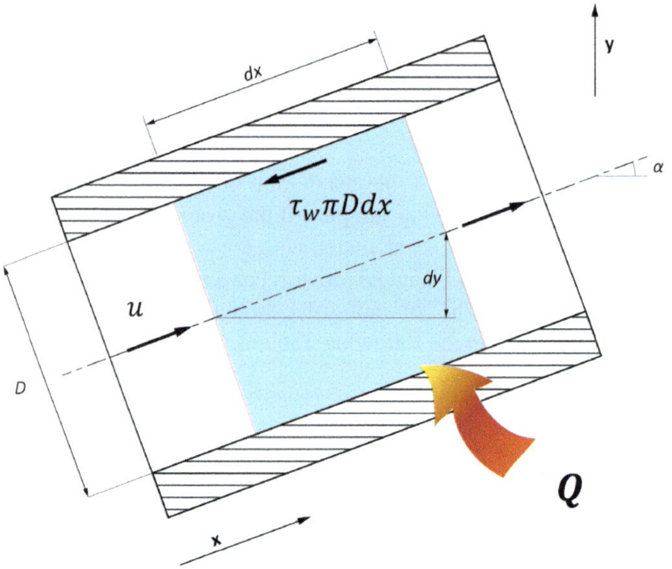


**Fig. 2.21** Control volume analysis of the energy transmission in a pipeline, using the RTT

With the radiation term being expressed as,

$$-\nabla \mathbf{q}_{\text{rad}''} = \int_0^\infty \alpha(\lambda, T)\left[\int_0^{4\pi} I(\lambda, \Omega)d\Omega\right]d\lambda - 4\pi\int_0^\infty \epsilon(\lambda, T)I(\lambda)d\lambda \tag{2.87}$$

with $\Omega$ being the solid angle, expressed in steradian [sr]. The conduction term is expressed using Fourier's law of conduction,

$$\mathbf{q''_{cd}} = -\mathbf{k}\nabla T \tag{2.88}$$

The convection term is generally calculated using Newton's law of convection,

$$q''_{cv} = -h\left(T_f - T_\infty\right) \tag{2.89}$$

Where $h$ is the convection coefficient, which can be calculated using a dimensionless approach, with the Reynolds, Prandtl and Nusselt numbers. These last two are defined as,

$$Pr = \frac{c_p \mu}{k} \tag{2.90}$$

$$Nu = \frac{hD}{k} \tag{2.91}$$

Generally, once the velocity and the system main dimensions are known, i.e. pipe's diameter, both Reynolds and Prandtl can be straightforwardly calculated. Nusselt number is usually expressed as a function of these two previous groups, and several correlations have been developed, depending on the type of flow and boundary conditions (Table 2.8).

The modern standards that regulate the installation of pipelines for long-distance transport of fluids (ASME, NAG, EN, etc.) require that the pipes be installed buried at variable depths, depending on the type of terrain and population density and surrounding, between 0.45 and 1.00 m.[22] This determines that atmospheric phenomena (temperature, snow, solar radiation, winds) influence the sub-surface portion of the ground in which the pipe has been installed and therefore the heat transmission between the fluid that circulates through the pipe and the middle.

Applying a global coefficient $U$ like the one developed in this chapter, omits precisely the fact that the upper part of the pipe is highly exposed precisely to atmospheric phenomena due to the reduced "clogging", while another part is in a different situation because the surrounding terrain has a much greater thermal capacity and inertia.

For this reason, two completely different circular sectors are defined below, Sector I for which the heat transmitted depends on a variable "path" depending on the angle with respect to the vertical up to an angle $\alpha_{max}$, i.e., the transmission coefficient is variable, while Sector II can be treated with a global transmission coefficient for the whole sector.

[22] The distance between the top of the pipe and the natural ground level is referred to in industry jargon as 'depth of cover'. It is only allowed to reduce the cover from 0.8 to 0.45 m in rocky terrain or with shallow subsurface rock strata due to the difficulty of trenching.

**Table 2.8** Different Nusselt correlations as a function of the type of flow encountered

| Condition | $Nu$ | Remarks |
|---|---|---|
| Laminar | 3.66 | Constant wall temperature |
| Laminar | 4.36 | Constant wall heat flux |
| Thermal laminar entry | $3.66 + \frac{0.0668 \frac{D}{L} Re Pr}{1+0.04\left[\frac{D}{L} Re Pr\right]^{2/3}}$ | – |
| Combined entry | $1.86\left(\frac{Re Pr}{\frac{L}{D}}\right)^{1/3}\left(\frac{\mu}{\mu_w}\right)^{0.14}$ | $T_s = constant, 0.48 < Pr < 16700$, $0.0044 < \frac{\mu}{\mu_w} < 9.75$ |
| Dittus-Boelter | $0.023 Re^{4/5} Pr^n$ | $Re > 10^4, 0.7 < Pr < 160$, $\frac{L}{D} > 10, \mu_w \approx \mu, n = 0.3$ (cooling) and $= 0.4$ (heating) |
| Siedler and Tate | $0.027 Re^{4/5} Pr^{1/3}\left(\frac{\mu}{\mu_w}\right)^{0.14}$ | $Re > 10^4, 0.7 < Pr < 16700$, $\frac{L}{D} > 10$ |
| Gnielinski | $\frac{\left(\frac{f}{8}\right)(Re-1000)Pr}{1+12.7\left(\frac{f}{8}\right)^{1/2}(Pr^{2/3}-1)}$ | $3000 < Re < 5 \times 10^6$, $0.5 < Pr < 2000$ |
| Petukhov | $\frac{\left(\frac{f}{8}\right)Re Pr}{1.07+12.7\left(\frac{f}{8}\right)^{1/2}(Pr^{2/3}-1)}$ | $10^4 < Re < 2 \times 10^7$, $0.5 < Pr < 2000$ |

In short, the sectors into which the pipe segment will be divided under analysis to calculate the light exchanged between fluid and earth are (see Fig. 2.22):

- Sector I: Its size depends on the depth of cover and the path of the heat flow from the fluid to the environment is assumed to be mainly in the radial direction, i.e., $\frac{\partial T}{\partial \theta} \approx 0$ for each angle (between 0°and $\alpha_{max}$).
- Sector II: The heat transfer is assumed to be constant and independent of the angle until the normal temperature of the terrain is reached at a specific radius $r_5$.

With respect to the heat transferred in the Sector I, one needs to carry out a more detailed analysis because the global transmission coefficient changes with the angle, since the distance to the surface is variable. Establishing the differential heat transferred,

$$dq = U(\alpha)dA(T_{amb} - T_f) = \frac{1}{\frac{1}{h_f} + \frac{t_p}{k_p} + \frac{t_i}{k_i} + \frac{r_1 \ln \frac{r_4}{r_3}}{k_s} + \frac{r_1}{r_4 h_a}} r_1 L_p d\alpha(T_{amb} - T_f) \tag{2.92}$$

Where $L_p$ is the total length of the pipeline considered in the study. Moreover, $r_4$ is a function of the angle, the cover of depth and the thicknesses and can be expressed as,

$$\cos\alpha = \frac{c_d + t_i + t_p + \frac{D_i}{2}}{r_4} \tag{2.93}$$

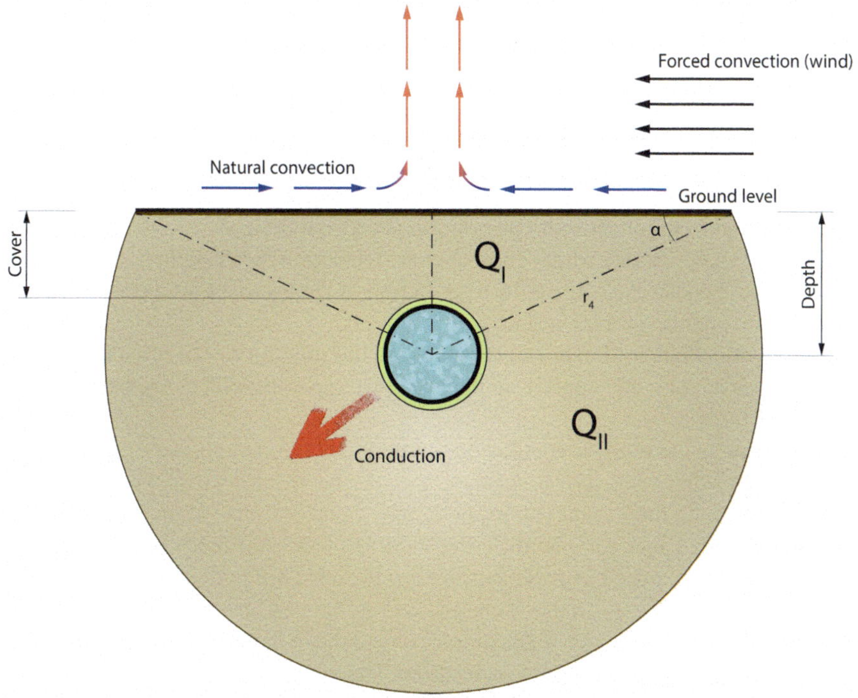


**Fig. 2.22** Scheme of a buried pipeline, showing the main heat transfer mechanisms present and the sectors used to the analysis of the heat transfer

Calling,

$$C_1 = \frac{1}{\frac{1}{h_f} + \frac{t_p}{k_p} + \frac{t_i}{k_i}}$$

$$C_2 = \frac{r_1}{k_s}$$

$$C_3 = \frac{r_1}{c_d + t_i + t_p + \frac{D_i}{2}}$$

It renders,

$$Q_I = \int dq = 2\int_0^{\alpha_{max}} \frac{1}{C_1 + C_2 \ln \frac{\frac{c_d + t_i + t_p + \frac{D_i}{2}}{\cos\alpha}}{r_3} + C_3 \frac{\cos\alpha}{h_a}} r_1 L_p (T_{amb} - T_f) d\alpha \tag{2.95}$$

Rewriting the integral leaving only terms function of the angle,

$$Q_I = 2r_1 L_p (T_{amb} - T_f) \int_0^{\alpha_{max}} \frac{d\alpha}{C_1 + C_2 \ln \frac{\frac{c_d + t_i + t_p + \frac{D_t}{2}}{\cos\alpha}}{r_3} + C_3 \frac{\cos\alpha}{h_a}} \tag{2.96}$$

The outer radius complicates the integration because it depends on the angle $\alpha$ and varies from 0, i.e. vertical, which corresponds to the cover, to an angle $\alpha_{max}$, positive and negative, which results from the intersection of the invariant temperature radius of the terrain and the ground surface.

Mathematical integration is very complex, so the solution can be found by means of an approximation by numerical integration (e.g. trapezoidal, Simpson's rule, Boole's rule, etc.).

In general, as the fluid enters the conduit with a higher temperature than the soil at the depth of the cover, $Q_I$ is negative, i.e., the heat transmission goes from the system, that is, the fluid to the atmospheric environment. Finally, this approach has the advantage that one can use different values for the soil thermal conductivity, since usually the soil used to cover the pipeline presents different mechanical and/or thermal properties than the native soil, away from the pipeline trench.

The calculations for sector II are simpler due to the independence of the variables involved with respect to the angles in question. The total transmission coefficient can be obtained as a constant, once the following expression has been integrated,

$$dq = UdA(T_{soil} - T_f) = \frac{1}{\frac{1}{h_f} + \frac{t_p}{k_p} + \frac{t_i}{k_i} + \frac{r_1 \ln \frac{r_4}{r_3}}{k_s}} r_1 L_p d\alpha (T_{soil} - T_f) \tag{2.97}$$

Integrating between the maximum angles of the section, one obtains the total heat transferred to the ground,

$$Q_{II} = \int dq = \int_{-\alpha_{max}}^{+\alpha_{max}} U(T_{soil} - T_f)dA = \frac{2}{\frac{1}{h_f} + \frac{t_p}{k_p} + \frac{t_i}{k_i} + \frac{r_1 \ln \frac{r_4}{r_3}}{k_s}} r_1 L_p (\pi - \alpha)(T_{soil} - T_f) \tag{2.98}$$

The term in parentheses represents the global transmission coefficient which is constant for Sector II ($U = constant$). In general, as the fluid enters the conduit with a higher temperature than that of the soil at the depth of the cover, the $Q_{II}$ is also negative, i.e., the heat transmission goes from the system, the fluid, to the surrounding medium, the land.

The total heat exchanged for a pipe segment is made up of terms which, as was seen, are generally negative, consequently the temperature of the fluid at the end of the segment defined between two sections, section $n$ and the following section $n+1$ is obtained from:

$$Q_T = Q_I + Q_{II} = \dot{m}c_p(T_{n+1} - T_n) \Rightarrow T_{n+1} = T_n + \frac{Q_I + Q_{II}}{\dot{m}c_p} \tag{2.99}$$

Hitherto, it has been considered a temperature balance based on the heat transferred from the fluid in the pipeline to the medium, without considering any effects that the flow conditions may have on the temperature. This, in other words, is knows as the *Joule-Thomson* effect,[23] which describes the change of temperature of a (real) fluid going under a pressure change, e.g., flow through a valve or a gradually decrease in the pressure, such as in pipelines (Li et al. 2021). The Joule-Thomson coefficient can be defined as,

$$\mu_{JT} = \left(\frac{\partial T}{\partial p}\right)_H = \frac{V}{c_p}(\alpha_T T - 1) = \frac{RT^2}{pc_p}\left(\frac{\partial Z_c}{\partial T}\right)_p \tag{2.100}$$

There are several approaches to calculate analytically this factor, from empirical equations to more complicated correlations, involving the EoS for real gases and the acentric factor. The empirical equation mentioned in this book has been tested for the normal range of operation of natural gas/pipelines and it yielded relatively good results between measurements and predictions (see Fig. 2.23 for values of the $\mu_{JT}$ in different blendings of hydrogen and natural gas.).

$$\mu_{JT} = \frac{4.1868T_c}{p_c c_p}(2.343T_r^{-2.04} - 0.071p_r + 0.0568) \tag{2.101}$$

Where the pressures are expressed in [MPa], the temperatures in [K] and $c_p$ is the molar specific heat of the gas in $[J/mol \cdot K]$. Finally, combining these effects, one can calculate the temperature drop in the line, considering also the Joule-Thomson effect.

$$\mu_{JT} \approx \frac{\Delta T}{\Delta p} \Rightarrow \Delta T = \mu_{JT}\Delta p \tag{2.102}$$

$$T_{n+1} = T_n + \frac{Q_I + Q_{II}}{\dot{m}c_p} + \mu_{JT}(p_{n+1} - p_n) \tag{2.103}$$

### Soil Temperature Model

In the previous model and analysis, it was considered that the soil temperature was constant and independent of the depth. However, substantial research has proven that the temperature is actually a function of many factors, namely: depth, time (hour/day of the year) and soil's thermal conductivity. Many models haven been developed

[23] This process is named after James Joule and William Thomson, 1st Baron Kelvin, who discovered it in 1852. It was based on earlier work by J. Joule on Joule expansion.

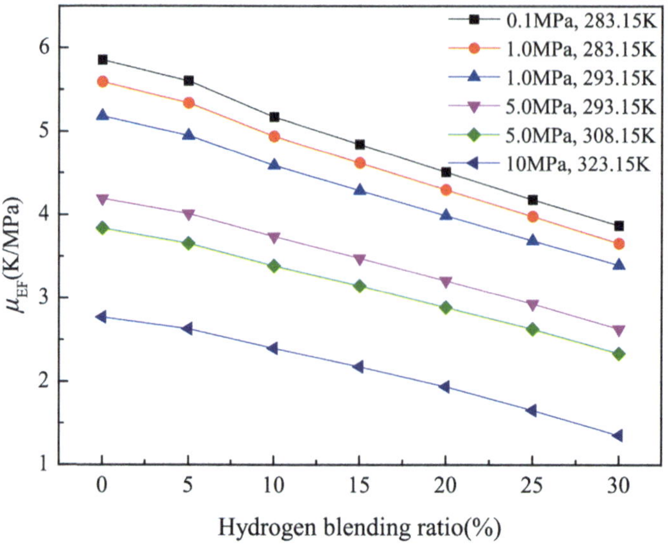


**Fig. 2.23** Joule-Thomson coefficient at different pressures and temperatures for a blending of hydrogen and natural gas, as a function of the percentage of blending with hydrogen (Li et al. 2021)

over the years, and this book presents hereunder one of the most commonly used to describe this temperature.

$$T(z, t) = T_m + Amp \cdot e^{-y/d} \sin\left(\frac{2\pi}{P}t + \phi - \frac{y}{d}\right) \tag{2.104}$$

Where, $y$ is the depth [meters]; $t$ is the day of year; $T_m$ is the annual average temperature at the surface; $Amp$ is the thermal amplitude, the average between minimum and maximum temperatures; $P$ is the total time frame considered, in this case equal to one year; $\phi$ is the phase constant, expressed as a function of a arbitrary starting point; $d$ is the damping depth [m], expressed as $\sqrt{\frac{2D}{\omega}}$, where $\omega$ is the angular frequency and $D$ is the soil's thermal diffusivity. This can be defined as $\frac{k_s}{C}$, where $k_s$ is the thermal conductivity and $C$ is the soil's volumetric heat capacity.

From the equation, it can be seen that as the depth is increased, the thermal amplitude decreases considerably, and moreover the timing between maximum and minimum temperatures exhibit a delay. This is obvious since it takes time a certain time for the energy to be transmitted from the surface to the deeper soil layers. Furthermore, the thermal fluctuations are more evident near the surface due to the radiation and air convection effects (see Fig. 2.24).

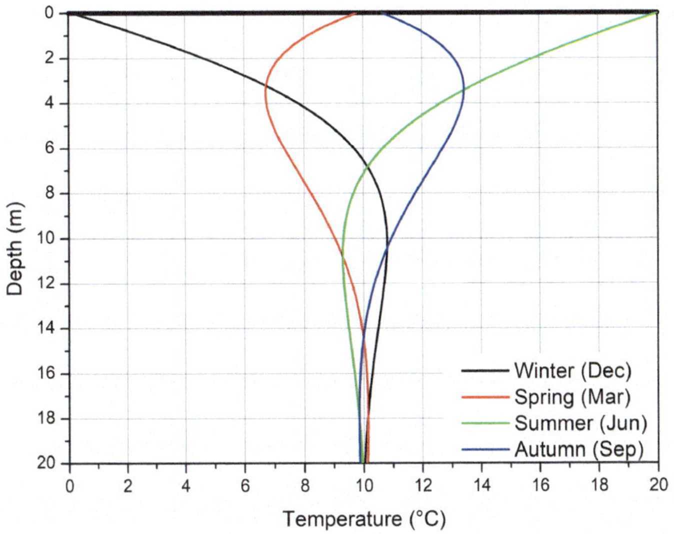


**Fig. 2.24** Soil temperature as a function of the soil depth for different times of the year

## Analysis by Means of Shape Factors

In a limited number of cases and in simplified geometries, a different approach can be followed to calculate in a swift way the heat transfer in the system. This is done by means of what it is known as the conduction shape factor ($S$), which is a tool allowing us to translate analytical solutions of multidimensional conduction problems into the form of solutions in a one-dimensional field. The heat transfer using this approach can be calculate using the following equation,

$$q[W] = k_s S \Delta T \tag{2.105}$$

As it can be seen, the shape factor has units of length. In the case of a horizontal, isothermal cylinder of length $L$, buried in a semi-infinite medium, such as the soil, the shape factor can be estimated as,

$$S = \frac{2\pi L}{\cosh \frac{2y}{D}} \tag{2.106}$$

Where $L$ is the length of the pipe, $y$ is the pipe's depth to the centerline and $D$ is the pipe's diameter. This equation is valid only if $L \gg D$.

### 2.4.6 Temperature Variation in Gas Pipelines

With respect to the seasonal variation of temperature of underground pipes, excellent field work was done in the last century by the US Bureau of Mines (BoM). The temperature record was made at the end of 1930 on a NPS 22" duct in Amarillo–Texas, with instruments from the Bureau of Mines itself. Observations were taken over 4 years (Deaton and Frost 1941; Deaton 1946).

For measurements of the gas temperature in the pipeline, bulbs inserted 150 mm into the gas vein were installed by the BoM. For pipe temperature measurements, the BoM thermocouple junctions were bonded with metal-based cement and insulated with a fleece lining.

Soil temperature measurements were made with mercury thermometers pushed to the bottom of perforations at the set depth. Normally the stabilization time was achieved in 5 min., then they were quickly withdrawn and read. Measurement accuracy: 0.5 °C.

Measurements were established in the 547 km of the 22" diameter gas pipeline from the Bivins Compression Plant in the Panhandle Field (Texas) to Denver. To establish when the gas temperature equals that of the ground ($T_{gas} = T_{soil}$), temperature measurements were installed in each of the blocking valves of the gas pipeline. Data collected by BoM showed that both temperatures were equal approximately 50 km measured from the compression plant (see Fig. 2.25).

During the testing period, the gas flow through the 22" line averaged 1.7 $\times 10^6$ Nm$^3$/day. In smaller diameter lines transporting smaller volumes, the distance

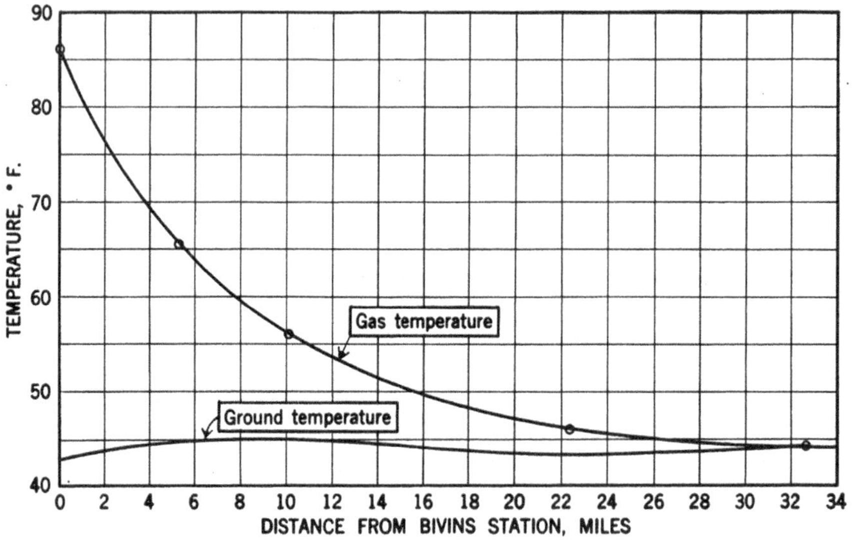


**Fig. 2.25** Temperature profile after the Bivins compressor station (Texas, USA) (Deaton and Frost 1941)

where the temperature equalize, i.e., the $T_g$ with that of the ground, is shortened. Several thermocouples were installed to determine how far the ground was affected by the gas temperature. The results of the measurements are shown in the Fig. 2.26.

The temperature of the gas in the pipeline, the average atmospheric temperature and the temperature of the ground at a depth of 15 and 45 cm in the Black Forest area (Colorado), were recorded over a long period of time (measurements were made over 4 years), shown in Figs. 2.27 and 2.28. The study determined that the gas temperature "lags" 3 to 5 days behind atmospheric temperature changes and that daily fluctuations in the gas were completely absent. The lag in gas temperature changes was typical for all observations at other points on the line. It was also observed that the surface temperature of the land was higher than that of the atmosphere.

The BoM concluded that the underground temperature is affected more by the surface temperature than by the atmospheric temperature. The type of soil, and its moisture, affects heat transfer and influences the change in underground temperature, as shown in the following Figs. 2.29 and 2.30.

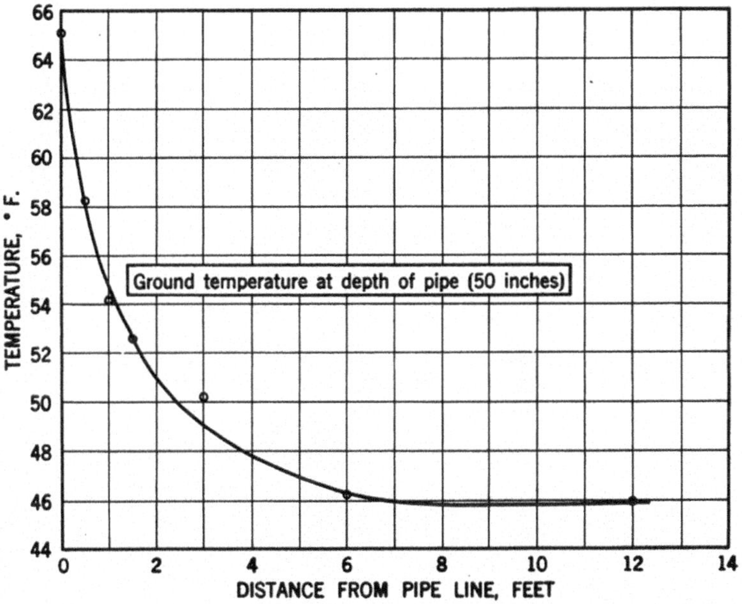


**Fig. 2.26** Distribution of underground temperature as a function of the distance to the pipeline, at 8 km from the Bivins compressor station (Texas, USA) (Deaton and Frost 1941)

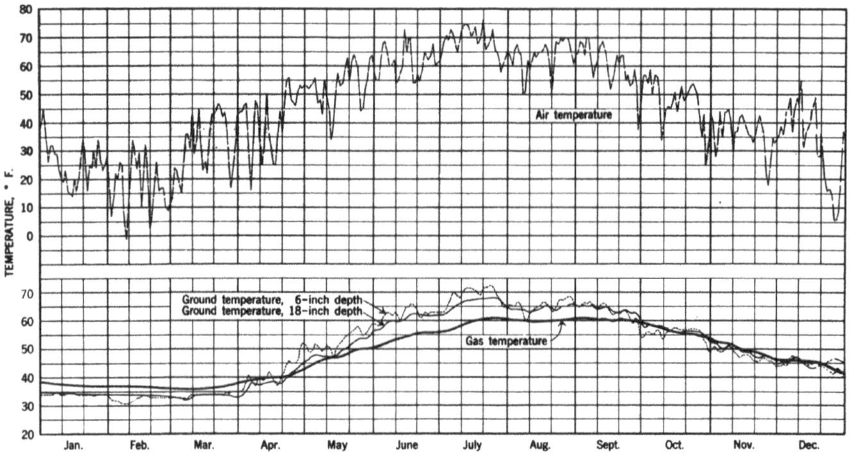


**Fig. 2.27** Temperature at the Black Forest station (Colorado, USA) as a function of the month in 1939 (Deaton and Frost 1941)

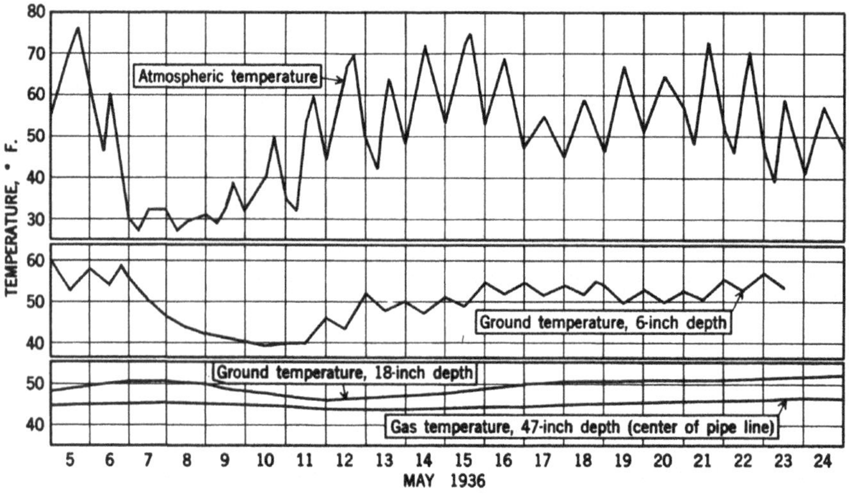


**Fig. 2.28** Plot showing the atmospheric, the ground and natural gas temperatures for the Black Forest station in 1939 (Deaton and Frost 1941)

### *2.4.7 Variations of the State Parameters of the Fluid and the External Environment—Transitional Regime Using Numerical Methods*

It is necessary to study heat transfer in underground fluid conduction pipes with greater precision than in the past, mainly in the oil industry due to the very frequent exploitation of oil fields with higher viscosity and very sensitive to temperature

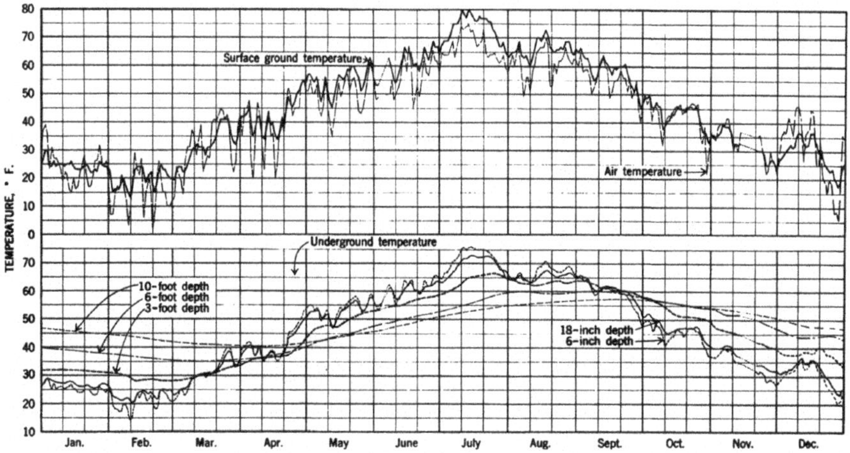


**Fig. 2.29** Plot comparing the atmospheric and the ground temperatures in Amarillo (Texas, USA) on the 20th of July, 1939 (Deaton and Frost 1941)

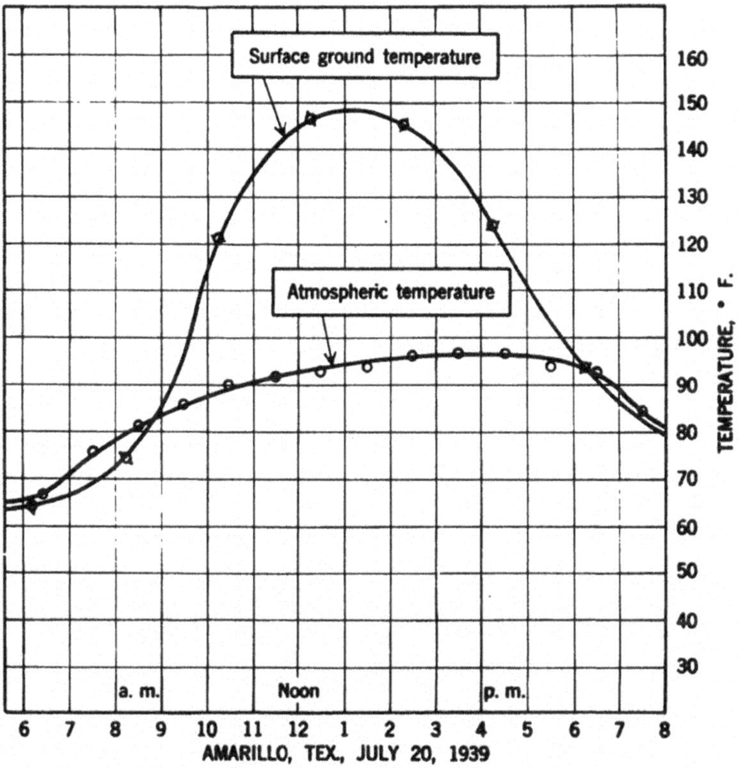


**Fig. 2.30** Atmospheric, ground and natural gas temperatures in the Black Forest station in 1936 (Deaton and Frost 1941)

changes. As said temperature is variable over time, an "always transient" fluid flow model is obtained (Chaczykowski 2010; Zhao 2020). The problem must then be solved with much more precision, fundamentally conduction with the ground, using for this advanced numerical techniques, e.g., Finite Difference, Finite Element, or similar. These techniques take as starting characteristics: the type of soil and its statistical temperature based on the geographical coordinates, the time and fluid state parameters. In order to obtain as a fundamental result the parameters of the state of the fluid and the temperature of the ground near the pipe as a function of time and space along the pipeline.

A study of this nature makes it possible to know the temperature of the ground around the pipeline, as a function of time, and consequently the flow. The transitory states thus obtained, from the practical point of view, can last days or months until reaching the stationary regime; Or perhaps it cannot even be started with the initial conditions imposed. Because of this and other factors, heat transfer in underground ducts and its influence on the flow of fluids inside them are extremely important. If a transversal cut is made to the axis of the underground pipe that will transport crude corresponding to a certain place and a certain time of the year for the following cases: pipe with product not circulating at ground temperature or without product; pipe with circulating product at a certain temperature. The following figure shows the temperature field obtained for a section and considered time. The calculations must be repeated for all the sections that make up the length of the pipe at each instant of time to obtain complete information on the fluid and soil (See Figs. 2.31 and 2.32).

From the figures, it can be seen that the ground heats up as oil circulates, at a higher temperature than the ground, through the pipeline. In such a way, that the crude oil gives up heat to the ground and in turn the ground gives up heat to the atmosphere; being the environmental characteristics of the latter (atmospheric temperature, wind speed, etc.) variables. Reason for which the parameters of the flow through the pipe, i.e., flow, pressure, density, temperature and viscosity of the crude oil are also variable; thus giving rise to a transport regime "always transitory".

As previously stated, this calculation methodology allows projects to be elaborated with greater precision. If said calculation methodology cannot be executed, the classic solutions of the projects must be oversized, by means of empirical criteria; which, of course, results in higher costs. In this way it is observed that the technical-economic contribution of the aforementioned calculation methodology is to be able to arrive at the optimal project; that is, in the "Minimum Cost" project.

The previous paragraph regarding the technical-economic contributions is an important conclusion regarding the use of this calculation methodology. But it is not all, since it also allows us to focus and solve the following problems:

- Commissioning of an oil pipeline (for the first time): Allows predicting how to carry out the Commissioning, if it can be started or not with the same product, or resort to another. Determine the time of the transitional start-up, which can result in hours, days or months depending on how the start-up has been planned.
- Prolonged Out of Service: It allows to know the time necessary to re-establish the permanent regime in the re-start.

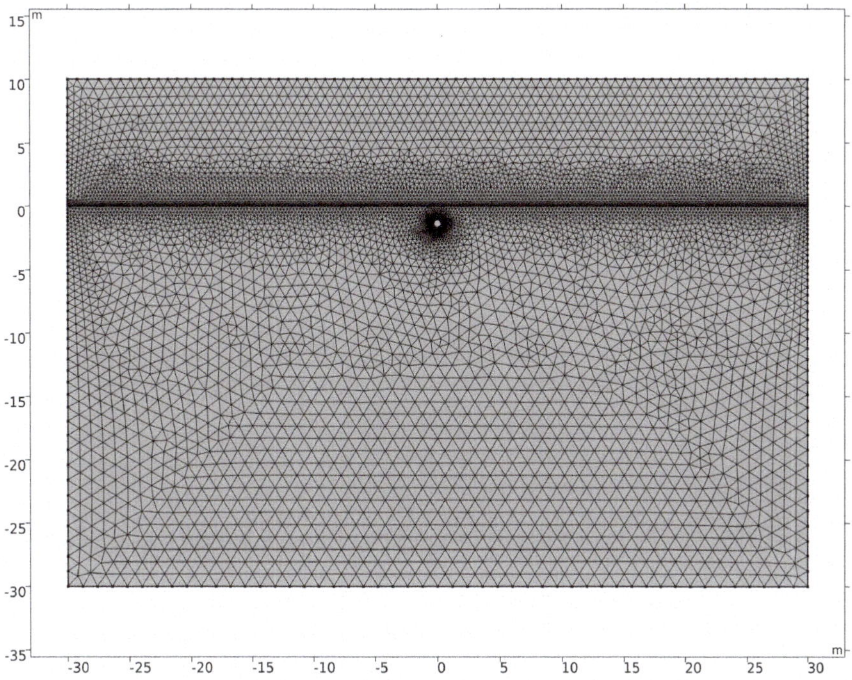


**Fig. 2.31** Cut detail of the numerical model, showing a 2D meshing around a buried pipeline to study the heat transfer between the fluid and the environment

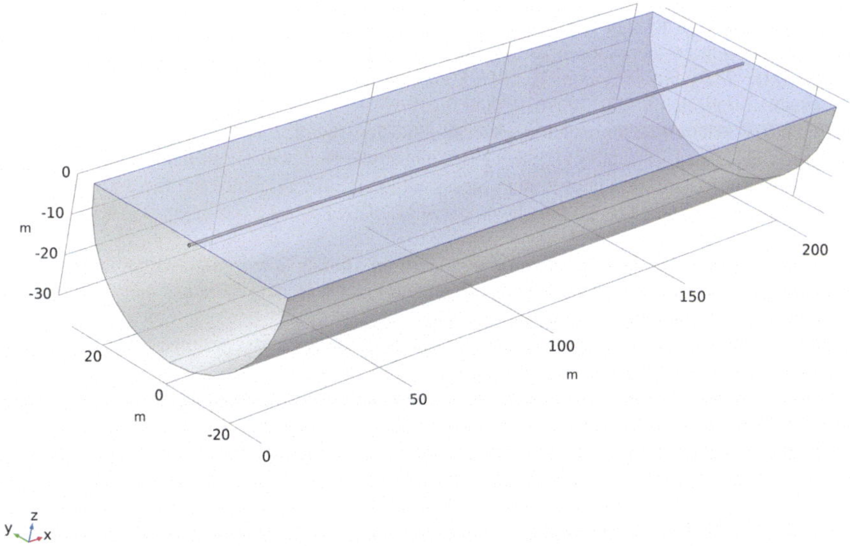


**Fig. 2.32** 3D numerical model of a segment of a pipeline used as standard model to study the heat transmission

- Sizing of the Transport System: Allows knowing the variations of the transport flow according to the seasons of the year. By knowing this and the other transients mentioned above, it allows the sizing of storage tanks to compensate for flow variations or design pumping stations to absorb such variations.

## References

3 - pipe design. In: McAllister EW, editor. Pipeline rules of thumb handbook (Eighth Edition), 8th edn. Boston: Gulf Professional Publishing; 2014. p. 93–123. ISBN 978-0-12-387693-5. https://doi.org/10.1016/B978-0-12-387693-5.00003-6. https://www.sciencedirect.com/science/article/pii/B9780123876935000036.

Bazán FAV, Beck AT. Optimal design of buried pipelines considering failure by mechanical damage. Struct Infrastruct Engin. 2016;12(11):1477–86. https://doi.org/10.1080/15732479.2016.1144618. https://doi.org/10.1080/15732479.2016.1144618

Bellos V, Nalbantis I, Tsakiris G. A new strategy for the estimation of friction parameters in hydrodynamic modelling: implementation to an ephemeral river. In: River flow 2020. CRC Press; 2020. p. 1246–1253.

Bratland O. Pipe flow 1: single-phase flow assurance. Chapter. 2009;2:21–92.

Brkić D. Review of explicit approximations to the colebrook relation for flow friction. J Petrol Sci Eng. 2011;77(1):34–48.

Chaczykowski M. Transient flow in natural gas pipeline - the effect of pipeline thermal model. Appl Math Model. 2010;34(4):1051–67. ISSN 0307-904X. https://doi.org/10.1016/j.apm.2009.07.017. https://www.sciencedirect.com/science/article/pii/S0307904X09002133.

Cheng N-S. Formulas for friction factor in transitional regimes. J Hydraul Eng. 2008;134(9):1357–62.

Chung M, Jung P-S, Rangel RH. Semi-analytical solution for heat transfer from a buried pipe with convection on the exposed surface. Int J Heat Mass Trans. 1999;42(20):3771–86. 1999. ISSN 0017-9310. https://doi.org/10.1016/S0017-9310(99)00046-0. https://www.sciencedirect.com/science/article/pii/S0017931099000460.

Churchill SW. Friction-factor equation spans all fluid-flow regimes. 1977.

Colebrook CF, Blench T, Chatley H, Essex EH, Finniecome JR, Lacey G, Williamson J, Macdonald GG. Correspondence. turbulent flow in pipes, with particular reference to the transition region between the smooth and rough pipe laws. (includes plates). J Inst Civil Engin. 1939;12(8):393–422.

Das aehnlichkeitsgesetz bei reibungsvorgängen in flüssigkeiten. In: Mitteilungen über Forschungsarbeiten auf dem Gebiete des Ingenieurwesens: insbesondere aus den Laboratorien der technischen Hochschulen. Springer. p. 1–41.

Deaton WM. Gas hydrates and their relation to the operation of natural gas pipe lines. US Bureau of mines monograph vol. 8. 1946.

Deaton WM, Frost EM. Temperatures of natural-gas pipe lines and seasonal variations of underground temperatures, vol. 3590. Bureau of Mines: US Department of the Interior; 1941.

Derumaux M. Moody diagram. 2016. https://commons.wikimedia.org/wiki/File:Moody_EN.svg.

Díaz-Damacillo L, Plascencia G. A new six parameter model to estimate the friction factor. AIChE J. 2019;65(4):1144–8.

Dixit US, Shufen R. 2 - finite element method modeling of hydraulic and thermal autofrettage processes. In: Silberschmidt VV, editor. Mechanics of materials in modern manufacturing methods and processing techniques. Elsevier series in mechanics of advanced materials. Elsevier; 2020. p. 31–69. ISBN 978-0-12-818232-1. https://doi.org/10.1016/B978-0-12-818232-1.00002-3. https://www.sciencedirect.com/science/article/pii/B9780128182321000023.

Duan JM, Wang W, Zhang Y, Zheng LJ, Liu HS, Gong J. Energy equation derivation of the oil-gas flow in pipelines. Oil and Gas Science and Technology-Revue d'IFP Energies nouvelles. 2013;68(2):341–53.

Fricke BA, Misra A, Becker BR, Stewart WE. Soil thermal conductivity: effects of saturation and dry density. In: Thermal performance of the exterior envelopes of whole buildings v international conference, Clearwater Beach, Florida, USA; 1992.

Gupta K. Materials forming, machining and post processing. materials forming, machining and tribology. Springer International Publishing; 2020. ISBN 9783030188559. https://books.google.nl/books?id=suCfzQEACAAJ.

Gupta R, Singh SN, Sehadri V. Prediction of uneven wear in a slurry pipeline on the basis of measurements in a pot tester. Wear. 1995;184(2):169–78. ISSN 0043-1648. https://doi.org/10.1016/0043-1648(94)06566-7. https://www.sciencedirect.com/science/article/pii/0043164894065667.

Hopkins P. Pipelines: past, present, and future. In: The 5th Asian Pacific IIW international congress, vol 27. Sydney. 2007.

Li J, Su Y, Yu B, Peng W, Dongliang S. Influences of hydrogen blending on the joule-thomson coefficient of natural gas. ACS Omega. 2021;6(26):16722–35. https://doi.org/10.1021/acsomega.1c00248.

Kadivar M, Tormey D, McGranaghan G. A review on turbulent flow over rough surfaces: fundamentals and theories. Int J Thermofluids. 2021;10:100077. ISSN 2666-2027. https://doi.org/10.1016/j.ijft.2021.100077. https://www.sciencedirect.com/science/article/pii/S266620272100015X.

Katsura T, Sakata Y, Ding L, Nagano K. Development of simulation tool for ground source heat pump systems influenced by ground surface. Energies. 2020;13(17). ISSN 1996-1073. https://doi.org/10.3390/en13174491. https://www.mdpi.com/1996-1073/13/17/4491.

Kerstin A, David M, de Lozar Alberto, Marc A, Dwight B, Björn H. The onset of turbulence in pipe flow. Science. 2011;333(6039):192–6. https://doi.org/10.1126/science.1203223. https://www.science.org/doi/abs/10.1126/science.

Menon ES. Transmission pipeline calculations and simulations manual. Elsevier Science; 2014. ISBN 9781856178310. URL https://books.google.nl/books?id=lPtDBAAAQBAJ.

Mileikovskyi V, Tkachenko T. Precise explicit approximations of the colebrook-white equation for engineering systems. In: International scientific conference ecocomfort and current issues of civil engineering. Springer; 2020. p. 303–310.

Mohitpour M, Golshan H, Murray MA. Pipeline design and construction: a practical approach. ASME Press: Pipelines and Pressure Vessels; 2007. https://books.google.nl/books?id=_d1SAAAAMAAJ.

Moody LF. Friction factors for pipe flow. Trans Am Soc Mech Engin. 1944;66(8):671–8.

Nagler J. Radial body forces influence on fgm and non-fgm cylindrical pressure vessels. J Composit. 2016;2016.

Nendra Wibawa LA, Diharjo K, Raharjo WW, Jihad BH. Stress analysis of thick-walled cylinder for rocket motor case under internal pressure. J Adv Res Fluid Mech Thermal Sci. 2020;70(2):106–15.

Nikuradse J et al. Laws of flow in rough pipes. 1950.

Nikuradse J. Gesetzmäßigkeiten der turbulenten strömung in glatten rohren (nachtrag). Forschung auf dem Gebiet des Ingenieurwesens A. 1933;4(1):44–44.

Palyvos JA. A survey of wind convection coefficient correlations for building envelope energy systems' modeling. Appl Thermal Engin. 2008;28(8):801–8. ISSN 1359-4311. https://doi.org/10.1016/j.applthermaleng.2007.12.005. https://www.sciencedirect.com/science/article/pii/S1359431107004036.

Partovi A, Shamili SS. Analysis of autofrettaged high pressure components. 2012.

Plascencia G, Díaz-Damacillo L, Robles-Agudo M. On the estimation of the friction factor: a review of recent approaches. SN Appl Sci. 2020;2(2):163.

Pratheesh Kumar S, Elangovan S, Mohanraj R, Ramakrishna JR. A review on properties of inconel 625 and inconel 718 fabricated using direct energy deposition. Mater Today: Proc. 2021;46:7892–906.

Rakhsha M, Kees CE, Negrut D. Lagrangian vs. eulerian: an analysis of two solution methods for free-surface flows and fluid solid interaction problems. Fluids. 2021;6(12). ISSN 2311-5521. https://doi.org/10.3390/fluids6120460. https://www.mdpi.com/2311-5521/6/12/460.

Rerak M. Selected soil thermal conductivity models. In: E3S web of conferences, vol 13. EDP Sciences; 2017. p. 02003.

Shashi Menon E, Pramila PE, Menon S. Liquid pipeline hydraulics: second edition. Trafford Publishing; 2013. ISBN 9781466977402. https://books.google.nl/books?id=e6VbRPl6PzEC.

Silvano S. Mathematical model of the lame'problem for simplified elastic theory applied to controlled-clearance pressure balances. arXiv preprint arXiv:1007.0813, 2010.

Swiss Proficiency. Tensile test on steel according to iso 6892-1. https://swissproficiency.com/en/product/tensile-test-steel-iso-6892-1/, 2024. [Accessed 28 Feb 2024].

Van der Waals JD. Over de Continuiteit van den Gas-en Vloeistoftoestand, vol. 1. Sijthoff; 1873.

White FM, Xue H. Fluid mechanics. McGraw-Hill Education; 2021. ISBN 9781260575545. https://books.google.nl/books?id=9R42zAEACAAJ.

Xie H, Li C, Jia W, Zhang C. Three-dimensional cfd modeling on the thermal characteristics of buried oil pipeline involving the heat transfer of wax layer. Energies. 2022;15(16). ISSN 1996-1073. https://doi.org/10.3390/en15166022. https://www.mdpi.com/1996-1073/15/16/6022.

Xiuwei X. Natural gas transmission pipeline temperature drop calculation. Adv Petroleum Explor Develop. 2014;8:24–28. https://api.semanticscholar.org/CorpusID:93154450.

Zeghadnia L, Robert JL, Achour B. Explicit solutions for turbulent flow friction factor: a review, assessment and approaches classification. Ain Shams Engin J. 2019;10(1):243–52. ISSN 2090-4479. https://doi.org/10.1016/j.asej.2018.10.007. https://www.sciencedirect.com/science/article/pii/S2090447919300176.

Zhao Y. Effect of pipe diameter on heat transfer characteristics of waxy crude oil pipeline during shutdown. Case Stud Thermal Engin. 2020;19:100628. ISSN 2214-157X. https://doi.org/10.1016/j.csite.2020.100628. https://www.sciencedirect.com/science/article/pii/S2214157X20301027.

# Chapter 3
# Design and Optimization of Pipeline Systems

> When I meet God, I am going to ask him two questions: why relativity? And why turbulence? I really believe he will have an answer for the first.
>
> *Werner Heisenberg*

## 3.1 Fundamental Purpose of This Book

### *3.1.1 Environmental Protection Premises*

Whenever a project for a pipeline transportation system for a certain flow rate and length must be developed, the main variables to consider are the diameter, the maximum and/or minimum pressure, the distance between the compression or pumping plants and the relationship discharge pressure/suction pressure of the latter; all this is conditioned by the fluid-thermodynamic effects of the transported flow and by the mechanical resistance of the material that makes up the pipeline (Cavallo and Frabotta 2003, 2015).

This book deals with the interrelationships between these variables, the flow to be transported and the flow conditions, i.e., whether incompressible (Ihle 2013; Ihle et al. 2013; Brackett and Zietlow 2011) or compressible (Ruan et al. 2009; Johnson and Ogden 2012; Tabkhi et al. 2009), or heterogeneous in a steady state, concluding with the presentation of relatively simple mathematical formulations, which with the convenient adoption of market coefficients allow for a given project - be it an oil pipeline, an aqueduct, a propanoduct, a multi-product, a gas pipeline, a slurry pipeline, etc. - to determine the optimal values of the aforementioned variables.



P. Druetta et al., *Transport by Pipelines and Its Optimization*, Petroleum Engineering,
https://doi.org/10.1007/978-3-031-66566-0_3

It is essential to highlight that all the products and processes carried out by the industry are basically taken from nature, i.e. materials and energy, which are transformed by man to achieve a specific objective. However, all these transformations are qualified, from a thermodynamic point of view, as irreversible; that is, with some resulting ecological or environmental deterioration.

**Definition: Optimization**

In general terms, "system optimization" as an engineering objective means achieving a project with a minimum cost and with a minimum energy consumption, which consequently results in minimal irreversible transformations and minimal ecological deterioration.

Taking the pipelines for the transportation of liquid and gaseous fuels as an example, it is verified that the hydrocarbons industry in the last decades has been actively incorporating environmental protection standards in the different stages from exploration to commercialization, passing through transportation through pipelines.

Many countries elaborated very strict laws on environmental protection, however, many of them later suffered amendments due to the difficulties derived from their compliance. Indeed, transportation by pipelines introduces modifications to the natural and social environment, and for this reason the environmental impacts and the actions that mitigate the derived adverse effects must be adequately evaluated.

In general, the norms and recommendations of public and private energy organizations emphasize construction practices, assuming that in this period the potential impacts on the environment can be more easily identified and neutralized - or at least mitigated in their effects. It is in this part that engineers should start reflecting, since environmental studies carried out at previous stages of the engineering design do not consider the use of materials and energy in its minimum expression, ruling out, to the authors' understanding, the most important part of environmental degradation.

In effect, by limiting as much as possible the use of steel or other materials for pipes, it limits to a great extent all the operations associated with the extractive and metallurgical production of iron, which harm the environment. In a similar way, the application of the minimum energy for the operation entails fewer developments in exploitation, operation and construction of energy supply systems. Saving energy or resources, renewable or not, strongly conditions achieving less environmental degradation (Wang et al. 2014; Knoope et al. 2013).

In order to relate the engineering aspects with an adequate care of the environment, the diagram shown below (Fig. 3.1) offers a global vision of the proposed "ecological" scheme and its functional relationships. It allows visualizing the link between economic optimization and the environment, by minimizing the use of raw materials and their associated energies, in the design of fluid transport systems.

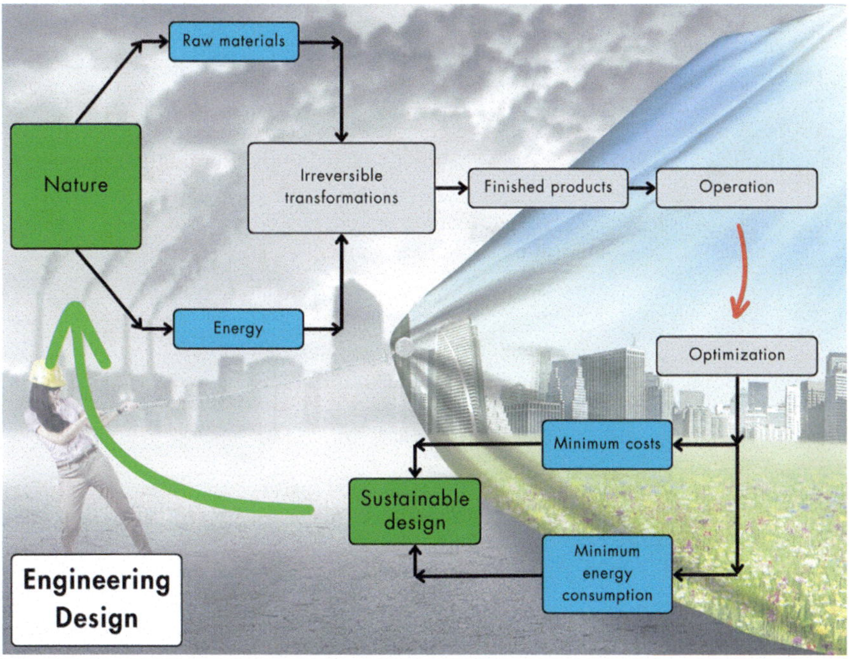


**Fig. 3.1** Flowchart showing the different stages of the engineering design and its connection with one of the engineers' goals, a sustainable design

## 3.2 Optimization of Piping Systems

The optimization of the pipeline project has already been mentioned before and involves a number of techno-economical requirements that any project must comply with for the transmission of fluid substances over long distances, linking with their basic design parameters, namely: flow rate, maximum operating pressure, minimum supply pressure for consumption, and routing through which the pipeline will be laid (Arya et al. 2022). The optimization methodology presented here was developed in the classes on Transport of Fluids through Pipes that the authors taught between 2002 and 2022, for the last year of the Mechanical Engineering degree at the National Technological University in the Buenos Aires Regional Faculty.

With these bases imposed on the project and with the respective mathematical formulation, it is ensured that the project is adequate to transport the required flow, although with this methodology the key objective is still not achieved: the transport the fluid at a *minimum cost*, which -as already said- will be called the "optimal project". It is evident from the many projects carried out in the past, that for a specific flow rate to be transported there will be an optimum diameter, which will transport the fluid under the required conditions with a minimum cost (see Figs. 3.2 and 3.3).

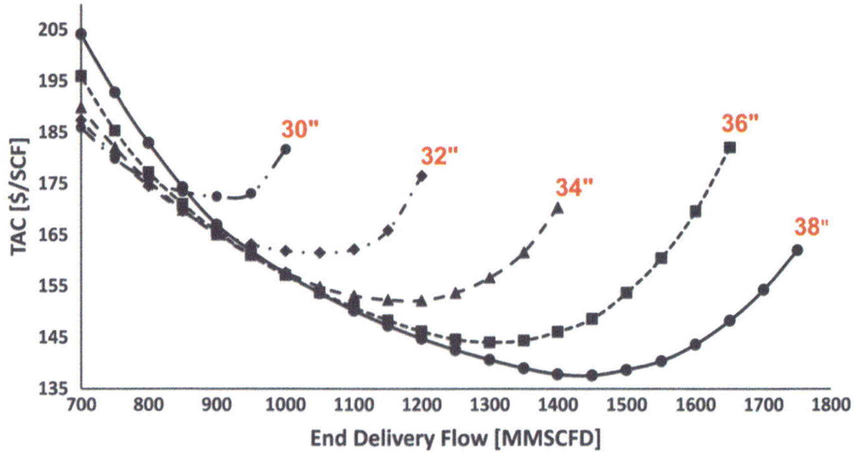


**Fig. 3.2** Effects of the nominal pipe size (NPS) on the total annual costs (TAC) for a wall thickness of 15.88 mm (5/8") as a function of the flow rate in millions of standard cubic feet per day (Ahmadian Behrooz 2019)

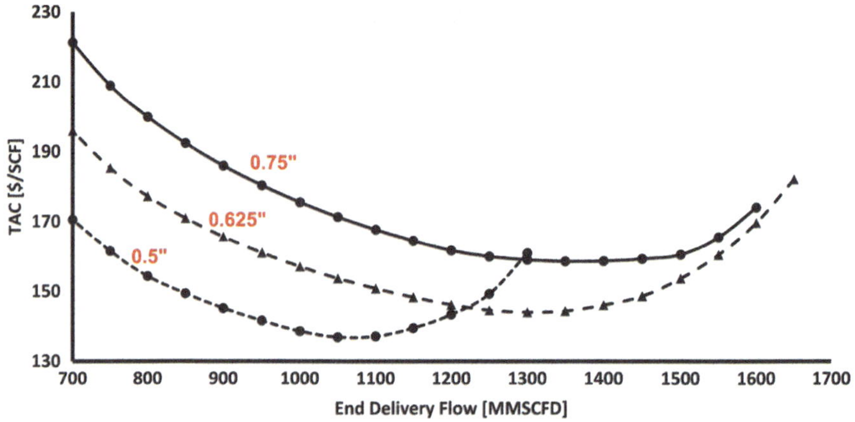


**Fig. 3.3** Effects of pipe thickness on the total annual costs for a pipe of DN 900 (NPS 36") as a function of the flow rate in millions of standard cubic feet per day (Ahmadian Behrooz 2019)

The fundamental objective of this book is to introduce a calculation methodology to obtain this so-called optimum project, which is strictly related to the continuous supply of fluid to consumers, at the lowest cost. The fact of looking for a minimum cost, forces a detailed evaluation of the costs of all the components involved in the system (Tee et al. 2014).

For the aforementioned purposes, the costs that make up a facility for the transportation of fluids can be grouped in the following items:

- Cost of Pipes (CAPEX);
- Cost of Energy Consumed (OPEX);
- Cost of Pump/Compressor Stations (CAPEX);
- Initial Fill Product Cost (OPEX).

With the theoretical and experimental tools of fluid mechanics, the resistance of materials and heat transmission (with a thermodynamic perspective), as described in the previous chapter, cost functions corresponding to all the elements of the installation, e.g., line pipe, compression or pumping plants, accessories, necessary energy, initial charge of product, can be established. Subsequently, with appropriate mathematical methods, for example, search for combined maximums and minimums, the variables are optimized, in such a way that the cost is minimized (Weber and Papageorgiou 2018).

Once the optimal variables have been determined in any analyzed system (transportation of liquids, gases or mixtures), it is advisable to carry out a more precise calculation, considering all the variables and their variations at stake. In this case, given the complexity that an integral calculation would require, it makes it easier to perform an approximate calculation by a finite difference approximation (for predefined short sections or pipe segments), thus obtaining for any section along the pipe, all the parameters that define the state of the fluid. This is however only a pseudo-static analysis and it is only the first part of a more refined verification of the project; the reader will see in the following chapters the verification of the system using a more dynamic approach.

The following graph (Fig. 3.4) shows the format of the cost functions as a function of pipe diameter, which altogether form the total cost function, determining a point of minimum cost called the optimal diameter. The main costs as mentioned are:

- Cost of Pipes ($C_1$);
- Cost of Energy Consumed ($C_2$);
- Cost of Pump/Compressor Stations ($C_3$);
- Initial Fill Product Cost ($C_4$).
- Total Cost = $\sum_i C_i$

Other costs worth of mentioning are, for instance, the costs of the mainline valve stations (generally located every 15 to 30 km, depending on local regulations); metering and pressure regulating stations; costs associated to SCADA (supervisory control and data acquisition) and telecommunications; environmental, right-of-way and general permits; and costs associated to engineering works, inspection and provision of materials.

The so-called optimal or theoretical diameter is associated with the other, also theoretical, optimization parameters (i.e., number of driving plants, maximum operating pressure, pipe thickness, etc.) that simultaneously result in minimum (cost) solutions for the system when solved.

The theoretical optimum of pipe diameter and thickness are generally far from the manufacturing ones, and the theoretical optimum number of stations is generally not a integer number. It will then be necessary to select commercial diameters and

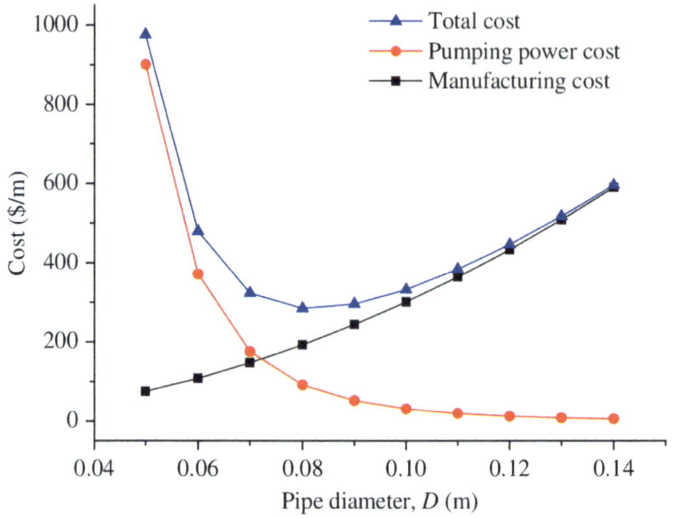


**Fig. 3.4** Variation of the two main types of costs and the total one for a pipeline facility as a function of the pipe size (Asim et al. 2012)

thicknesses (close to the theoretical ones) and a rounded-up number of plants. All this to distinguish between the feasible options and the theoretical optimum, the most appropriate to the system that is intended to be conceptually designed. This last process is called *practical adjustment of the optimal variables*.

It should be noted that there is no "typical" pipeline from a construction cost standpoint; the above mentioned costs depend on the geographic area, the size of the pipe, the number and dimension of the pumping stations and associated installations, and the economic conditions in general (Singh and Mahar 2003; Alkhomairi and Jung 2021).

It is a general rule that the longer the pipeline, the lower the cost per kilometer; a pipeline of a few kilometers will cost considerably more per kilometer than a line of several hundred kilometers, even if they are of the same diameter and laid in a similar environment. Pipeline costs are generally standardized on a "$/inch-meter" basis, to make cost values independent of line size (diameter) and thus obtain comparable values (see Fig. 3.5). Similarly, power plants are generally compared on a "$/W" basis. In a similar way to the case of the ducts, a high-power compressor station installed will have a lower cost per watt than a smaller one.

The United States Federal Energy Regulatory Commission (FERC) statistics indicate some representative costs of transport pipeline construction. These data indicate: materials, 36.6% of the total construction; labor, 38.5%; access permits and damage, 4.4%; and other costs, 20.5% (engineering, supervision, administration, "overhead", interest, contingencies and authorizations). European statistics show that, for large diameter pipes, their material requires 37% of the investment, and installation the remaining 63%. These cases do not take into account the rights of way in the cost of

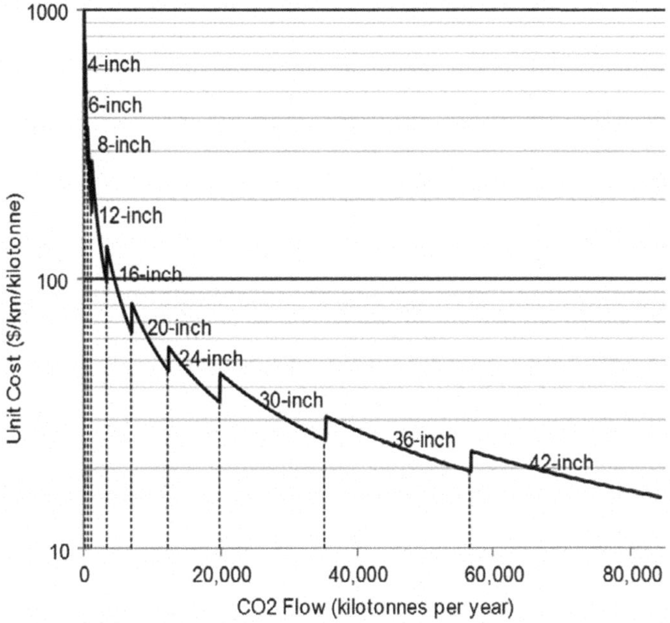


**Fig. 3.5** Plots showing the specific unit cost as a function of the nominal pipe size and the $CO_2$ mass flow rate (Kuby et al. 2011)

the engineering studies, nor supervision of the works that vary between 5 and 12% of the price of the work, depending on the complexity and importance of the investment. In the case of the compressor/pumping stations, the opening of costs shows that the equipment and materials consume 55% of the total investment; labor, 22%; and land, 1%. Miscellaneous costs, same as for ducts, make up the remaining 22%. It should be well understood that in each type of machine there will be variants. Booster stations normally represent between 3 and 15% of the total investment, depending on the lines considered.

Figure 3.6a and b summarize a number of projects carried out in the United States during the first decade of the 21st century (Parker 2004). The first establish a relationship between the cost per km and the pipe diameter. This exhibits a quadratic parabola, almost linear behavior, due to cost functions involved, which will be explained in the following chapters. Which is more interesting in Fig. 3.6b, depicting the total cost of the project, per inch and per km for several project of different pipe sizes. With smaller pipes there is an increase cost due to fixed costs that are unrelated to the size. However, after a certain size, in this case around NPS 10 inches, this costs becomes approximately constant. This may be a good and fast indication to make a tentative cost of any pipeline based on its length and size, before any calculations are done.

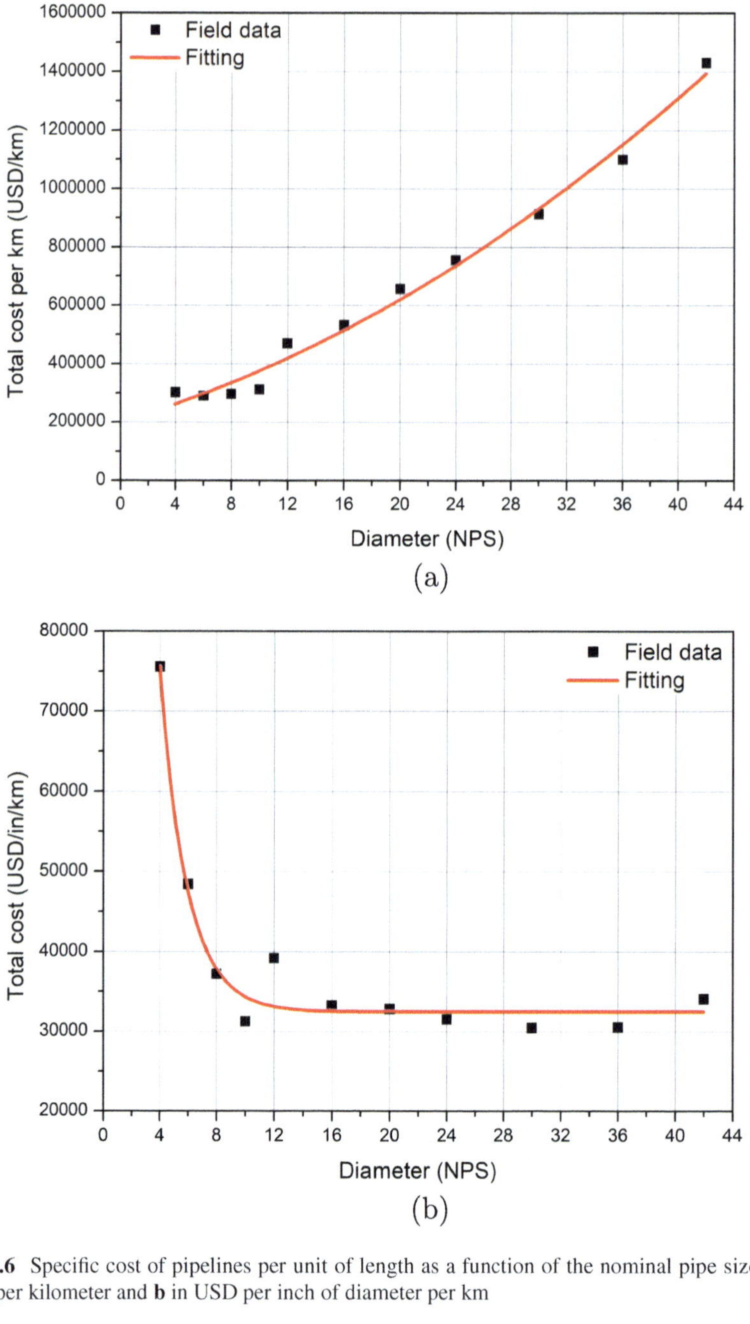


**Fig. 3.6** Specific cost of pipelines per unit of length as a function of the nominal pipe size: **a** in USD per kilometer and **b** in USD per inch of diameter per km

## 3.3 Lagrange Multiplier Method Applied to Several Variables

In optimization problems, the Lagrange multipliers method is a procedure to find the maximum and minimum functions of multiple variables subject to some condition that has been imposed on them (Sharma et al. 2013; Jokić and Zavargó 2001). This method reduces the unrestricted problem with $n$ variables to one with restrictions of $n + \lambda$ variables, where $\lambda$ is equal to the number of restrictions, and whose equations can be solved more easily. These new unknown scalar variables, one for each constraint, are called Lagrange "multipliers".

The proof applies partial derivatives and the chain derivation rule for functions of several variables. It is about extracting an implicit function of the restrictions, and finding the conditions so that the partial derivatives with respect to the independent variables are equal to zero.

The Lagrange multipliers theorem was the first theoretical instrument from the historical point of view, to solve optimization problems with constraints. Its creator, Joseph-Louis Lagrange (Turin, 1736–Paris, 1813), used the multiplier technique for the first time to solve problems of calculus of variations in his famous treatise Méchanique Analitique (1788), in which he endowed Mechanics with a formalism adequate analytic, and later exposed it in his no less famous work on Differential Calculus.[1] The well-known theorem in question deals with the maximization (or minimization) of a function of several variables.

## 3.4 Statement of the Optimization Problem

The methodology for solving an optimal project for a pipeline differs to a certain extent if it is a question of transporting through said pipeline a liquid product, homogeneous or in sequence of different homogeneous products, a unique gaseous product, or heterogeneous fluid mixtures, such as solid + gas or solid + liquid.

In fact, it will be seen that the cost functions of lines for transporting liquids, usually called *Oil Pipelines* or *Multi-product Pipelines*, are practically identical, but differ from the functions and the resolving system applied to the transport of gaseous substances, generally called Gas Pipelines. The same occurs with the functions for the optimization of pipes for the transport of heterogeneous substances, such as the Slurry Pipelines.

In this book it has been generically referred to as Pipelines when dealing with the transport of a single liquid hydrocarbon product, and Multi-product Pipelines in the case of the transport of "batches" of different products transported successively and sequentially, through the same pipe. Therefore, it must be differentiated in the initial definition of the characteristic parameters, whether it is a single product, for example:

---

[1] Théorie des fonctions analytiques contenant les principes du calcul différentiel. Imprimerie de la République, 1797.

oil, or LPG or gasoline; or a succession of liquid products, e.g., gasoline and JP1 and LPG, within the same pipe. The third case refers to the optimization of pipes for the transport of gases, those that transport natural gas are called Gas Pipelines, and the last case to the optimization of pipes for the fluid transport of mineral substances in heterogeneous flow, the so-called Mineral Pipelines. In this last case, the treatment is somewhat different from that of the optimization of homogeneous product flows.

## References

Alkhomairi A, Jung BS. Comprehensive optimization of project cost for long supply pipelines. J Engin Res. 2021;9(3A). https://doi.org/10.36909/jer.v9i3A.8857.

Arya AK, Jain R, Yadav S, Bisht S, Gautam S. Recent trends in gas pipeline optimization. Mater Today: Proc. 2022;57:1455–61. ISSN 2214-7853. https://doi.org/10.1016/j.matpr.2021.11.232. https://www.sciencedirect.com/science/article/pii/S2214785321072400. International chemical engineering conference 2021 (100 glorious years of chemical engineering and technology).

Asim T, Mishra R, Kollar L, Ubbi K. Optimisation of a horizontal capsule transporting pipeline carrying cylindrical capsules. J Phys: Conf Ser. 2012;364:012071. IOP Publishing.

Behrooz HA. A multiobjective robust approach for the design of natural gas transmission pipelines. Chem Product Process Model. 2019;14(4):20190055.

Brackett C, Zietlow D. Optimal design of a pump and piping system. In: 2011 ASEE annual conference and exposition. 2011. p. 22–1126.

Cavallo G, Frabotta R. Optimisation of long distance gas pipeline projects. In: 22nd world gas conference. WGC, 2003. WGC-03 proceedings of the 22nd world gas conference, 1–5 June 2003, Tokyo, Japan. p. WGC–IGU–03.

Cavallo G, Frabotta R. Optimisation of long distance gas pipeline projects. In: 26th world gas conference. WGC, 2015. WGC-15 proceedings of the 26th world gas conference, 1–5 June 2015, Paris, France. p. WGC–IGU–15.

Ihle CF. A cost perspective for long distance ore pipeline water and energy utilization. part i: Optimal base values. Int J Mineral Process. 2013;122:1–12. ISSN 0301-7516. https://doi.org/10.1016/j.minpro.2013.04.002. https://www.sciencedirect.com/science/article/pii/S0301751613000859.

Ihle CF, Montserrat S, Tamburrino A. A cost perspective for long distance ore pipeline water and energy utilization. part ii: effect of input parameter variability. Int J Mineral Process. 2013;122:54–8. ISSN 0301-7516. https://doi.org/10.1016/j.minpro.2013.04.004. https://www.sciencedirect.com/science/article/pii/S0301751613000872.

Johnson N, Ogden J. A spatially-explicit optimization model for long-term hydrogen pipeline planning. Int J Hydrog Energy. 2012;37(6):5421–33. ISSN 0360-3199. https://doi.org/10.1016/j.ijhydene.2011.08.109. https://www.sciencedirect.com/science/article/pii/S0360319911020349. Optimization approaches to hydrogen logistics.

Jokić A, Zavargó Z. Optimization of pipeline network for oil transport. Hungarian J Ind Chem 2001;113–7.

Knoope MMJ, Ramírez A, Faaij APC. Economic optimization of co2 pipeline configurations. Energy Proc. 2013;37:3105–12. ISSN 1876-6102. https://doi.org/10.1016/j.egypro.2013.06.196. https://www.sciencedirect.com/science/article/pii/S1876610213004396. GHGT-11 proceedings of the 11th international conference on greenhouse gas control technologies, 18–22 November 2012, Kyoto, Japan.

Kuby MJ, Middleton RS, Bielicki JM. Analysis of cost savings from networking pipelines in ccs infrastructure systems. Energy Proc. 2011;4:2808–15.

Parker N. Using natural gas transmission pipeline costs to estimate hydrogen pipeline costs. UC Davis: Technical report. Institute of Transportation Studies; 2004.

Ruan Y, Liu Q, Zhou W, Batty B, Gao W, Ren J, Watanabe T. A procedure to design the mainline system in natural gas networks. Appl Math Model. 2009;33(7):3040–51. ISSN 0307-904X. https://doi.org/10.1016/j.apm.2008.10.008. https://www.sciencedirect.com/science/article/pii/S0307904X08002655.

Sharma P, Shelza GSR, Virk RS. Cost optimization of pipeline systems using genetic algorithm. Int J Comput Engin Technol (IJCET). 2013;4(4):485–94.

Singh RP, Mahar PS. Optimal design of multidiameter, multioutlet pipelines. Journal of water resources planning and management. 2003;129(3):226–33.

Tabkhi F, Pibouleau L, Azzaro-Pantel C, Domenech S. Total cost minimization of a high-pressure natural gas network. J Energy Resourc Technol. 2009;131(4):043002, 11. ISSN 0195-0738. https://doi.org/10.1115/1.4000325. https://doi.org/10.1115/1.4000325.

Tee KF, Khan LR, Chen HP, Alani AM. Reliability based life cycle cost optimization for underground pipeline networks. Tunnel Underground Space Technol. 2014;43:32–40. ISSN 0886-7798. https://doi.org/10.1016/j.tust.2014.04.007. https://www.sciencedirect.com/science/article/pii/S088677981400056X.

Wang Z, Fimbres Weihs GA, Cardenas GI, Wiley DE. Optimal pipeline design for ccs projects with anticipated increasing co2 flow rates. Int J Greenhouse Gas Control. 2014;31:165–74. ISSN 1750-5836. https://doi.org/10.1016/j.ijggc.2014.10.010. https://www.sciencedirect.com/science/article/pii/S1750583614003119.

Weber AC, Papageorgiou LG. Design of hydrogen transmission pipeline networks with hydraulics. Chem Engin Res Design. 2018;131:266–78. ISSN 0263-8762. https://doi.org/10.1016/j.cherd.2018.01.022. https://www.sciencedirect.com/science/article/pii/S0263876218300248. Energy systems engineering.

# Chapter 4
# Transport of Liquid Substances by Pipeline Systems

> The ideal engineer is a composite ... He is not a scientist, he is not a mathematician, he is not a sociologist or a writer; but he may use the knowledge and techniques of any or all of these disciplines in solving engineering problems.
>
> *Nathan Dougherty*

## 4.1 Introduction

In ancient times, the problem of transporting liquids, mainly water, was solved mainly by carrying containers (e.g., vessels or barrels) on carts and animal backs. However, the use of open channels and pressure conduits for the transport of vital liquids dates back a long time and there are plenty of historical examples showing the developments of Egyptian and Roman engineers, basically applying the effects of the force of gravity to enable the flow between two points (Mohitpour et al. 2007).

With the industrial revolution and the development of mechanical pumps, the transport of liquids through pipelines became more and more frequent and the first pipelines, in the early 1900s, began to be laid-down to transport water or to transport oil over long distances,[1] incorporating intermediate pumping stations. The first pipelines were built with diameters of less than 0.508 m (20" or DN 500) and maximum operating pressures of less than 35 bar (500 psi), when the construction of steel pipelines for oil transportation began, firstly in the US in 1920 and then in Europe and Argentina in 1950. Currently, pipelines of up to 1.420 m (56" or

[1] For example, the Catskill Aqueduct, part of the water supply to New York City, was built between 1907 and 1916 and then completed in 1924, with a length of 262 km.



P. Druetta et al., *Transport by Pipelines and Its Optimization*, Petroleum Engineering,
https://doi.org/10.1007/978-3-031-66566-0_4

**Fig. 4.1** Alaska oil pipeline running above-ground. Please check the special supports built to considers the different temperatures between summer en winter, photo by Kenneth Gill (2018)

DN 1400) in diameter and with pressures of up to 140 bar have been built, for instance, the Alaska—oil pipeline (see Fig. 4.1).

The transport of liquids through pipelines is also applied to derivative liquid hydrocarbon products, e.g., fuel-oil, gas-oil, naphtha, gasoline, and to products that are gaseous at atmospheric pressure but liquid under transportation pressures, such as ethane, propane, butane, mixtures (LPG), or at very low temperature (LNG). In each case, names derived from the product transported are applied to name the pipeline, such as gas pipeline, naphtha pipeline, ethane pipeline, propane pipeline, etc.

## 4.2 Pure Substances or Homogeneous Mixtures

In order to develop the optimization methodology, in the first instance one is dedicated to the consideration of the flows of pure liquid substances or homogeneous mixtures of miscible liquids one in the other, i.e., forming solutions. In both cases, what is required is to be able to know the parameters of the pure fluid or of the mixture, as unique data—for the fluid in question—at the temperatures and transport pressures that are defined.

In addition, it is initially required to obtain as much data on the fluid and the flow (such as viscosity, density, vapor pressure, rheology behavior, etc.) and its variations with respect to temperature and pressure. If the data were not available, a laboratory test is recommended, given the implication that these parameters have on the development of the optimal design. In fact, since the calculation of pressure losses is very sensitive to the average viscosity and average density adopted for the calculation of the optimum pipe size, both dependent above all on the flow temperature and somewhat less on the pressure, this is considered of the utmost importance. This will be done the first, optimization stage, by estimating average pressure and temperatures, and introducing these parameters in the optimization equations.

## 4.3 Application of the Optimization Method to an Indeterminate Incompressible and Isothermal Flow

Before stepping into the equations and the optimization process, it is important to remember what an incompressible fluid stands for:

**Definition: Incompressible Fluid**

Incompressible fluids are those whose densities remain ideally constant regardless of the pressure to which they are subjected.

However, taking oil as an example, like any fluid called "incompressible", it is actually slightly compressible. On the other hand, the properties of any crude oil type, i.e., the density and the viscosity, change with the pressure and the temperature; as the pressure and temperature change along the pipeline, the characteristics of the oil also change. To optimize the variables, as a first approximation, sufficient for practical purposes, it is advisable to consider constant the characteristics of the oil and the temperature along the pipeline, both adopted on solid hypotheses (Hafner and Luciani 2022).

In general, the problem is stated as: *given an oil, or another pure liquid or a homogeneous mixture, of known density $\rho$ and viscosity $\mu$, both assumed to be constant, as mean values, it is desired to transport it a distance L, following a determined topography, for its delivery at a defined pressure.* In addition, being **m** the mass to be transported in a time **t**, a mass flow rate to be transported is defined as the ratio $\dot{m} = m/t$ measured in kg/m$^3$ (SI system).

To achieve this objective, it is generally necessary to place, in addition to the head station, a certain number $\chi - 1$ of intermediate pumping stations, where for calculation purposes, one considers the total length and the total height, divided by $\chi$ ($\chi$ = header station + number plants), thus defining the length and height variation of each pipe section.

According to fluid mechanics, the principle of conservation of mass, i.e., the continuity equation under steady state conditions, results for the duct that transports an incompressible fluid through a pipe of area $A$,

$$\dot{m} = \rho AV = (\rho AV)_1 = (\rho AV)_1 = \cdots = (\rho AV)_n \tag{4.1}$$

Being $\rho_1 = \rho_2 = \rho_n =$ constant (approximation for liquids). The application of the conservation of energy, i.e., the so-called Bernoulli equation for liquids, yields:

$$\left[\frac{p}{\rho} + \frac{V^2}{2} + gz\right]_1 = \left[\frac{p}{\rho} + \frac{V^2}{2} + gz\right]_2 + \frac{\Delta p_{1-2}}{\rho} \quad (4.2)$$

Being $\frac{\Delta p_{1-2}}{\rho}$ the summation term of friction losses, including pipes and accessories, in each section. These are mainly caused by the irreversible transformation of useful energy, or exergy, into heat, due to the viscous dissipation in the pipeline. Neglecting the minor losses caused by fittings, e.g., flanges, valves, filters, it is assumed that the losses are caused entirely by the pipeline, expressed using the friction factor as,[2]

$$\frac{\Delta p}{\rho} = f\frac{V^2}{2}\frac{L_t}{D_{int}\chi} \quad (4.3)$$

Where $L_t$ is the total length of the pipeline; since the interior area of the duct is practically constant $V_1 = V_2 = V$ and replacing the velocity with the mass flow rate, it renders,

$$\frac{\Delta p}{\rho} = f\frac{V^2}{2}\frac{L_t}{D_{int}\chi} = 8f\frac{\dot{m}^2}{\pi^2\rho^2}\frac{L_t}{D_{int}^5\chi} \quad (4.4)$$

Operating with Bernoulli's equation, taking as reference level the first pumping station, then $z_2 = H/\chi$; and $f$ as an average value, which can be obtained by graphical methods, or by implementing one of the several empirical relationships described in this book, it yields,

$$p_1 = p_2 + \rho g\frac{H}{\chi} + \frac{8f}{\pi^2}\frac{\dot{m}^2}{\rho D_{int}^5}\frac{L}{\chi} \quad (4.5)$$

in which:

- $p_1$: initial pressure in the pipe section, i.e., outlet pressure of the pumping plant.
- $p_2$: final pressure in the pipe section, generally suction pressure in the next pumping plant.
- $D_{int}$: inside diameter of the pipe.
- $L$: length, in some cases equivalent length considering losses in the internal seam of welds and accessories.

For these reasons, the mass flow rate $\dot{m}$ of a certain liquid that can be transported through a pipe of length $L$ is a function of the internal diameter $D_{int}$; of the number of pumping plants $\chi$, and of the pressure $p_1$.

[2] This is because "full bore" valves and fittings are used to allow cleaning and control with internal scrapers. Some authors do not neglect girth weld that occur every 12 meters and include a correction factor.

Thus, there is a single equation with three unknowns, which admits infinitely solutions. The problem is then to select among the infinite solutions the one that minimizes the cost of the entire installation and its operation, that is, the "optimal solution". For this purpose, the cost functions of all the elements involved in the installation must be created.

## 4.4 Application of the Lagrange Method—Cost Functions

As explained at the beginning of the chapter. to facilitate the mathematical work in obtaining the functions, the cost of the entire installation (capital and operational) is divided into two large groups: line steel pipes and initial filling product, directly proportional to the pipe size; and the cost of the pumping plants and the cost of the energy consumed in pumping, which are inversely proportional to the pipe size (Alkhomairi and Jung 2021).

### *4.4.1 Cost of the Pipeline*

To define a function that encompasses the costs of the installed line, the costs related to the weight of the pipe material, i.e., in most of cases carbon steel, have been considered, separated from the costs related to the external surface of the pipe (Cookenboo 1954; Grundfos Ecademy 2014; Jenkins and Murray 2005).

In effect, there are material and installation costs to a greater extent related to the unit weight of the pipe, such as the amount of steel itself, lifting and transport, welding (partially), etc. On the contrary, other costs basically depend on the outer surface of the pipe, for example coatings, opening tracks, trenching, soil movement, etc.

As it was discussed, the cost of the pipe ($c_1$) can be assumed to be made up of two parts, one function of the weight of the pipe and another function of its exterior surface,

$$c_1 = c_{11} + c_{12} \tag{4.6}$$

where $c_{11}$ is a function of the material weight, which for practical purposes one assumes to be a linear function, valid within a given range of diameters.

$$c_{11} = \kappa_{11} \pi D_m t L g \rho_p \tag{4.7}$$

Being:

- $\kappa_{11}$: market constant corresponding to the cost of the material $\frac{EUR}{N}$.
- $\rho_p$: density of the pipe material $\frac{kg}{m^3}$.

- $D_m$: medium diameter of the pipeline, equal to $\frac{D_{ext}+D_{int}}{2}$ $[m]$.
- $t$: pipeline thickness $[m]$.
- $L$: pipeline total length $[m]$.

On the other hand, $c_{12}$ is a function of the outer surface of the pipe and for the purposes of the calculation it is assumed to be a linear function of the outer surface, valid for a certain range of diameters.

$$c_{12} = (\kappa_{12} + \kappa_{13})\,\pi D_{ext} L \tag{4.8}$$

Being:

- $\kappa_{12}$: market constant corresponding to the cost of the thermal and/or anti-corrosive insulation per m$^2$ $\frac{EUR}{m^2}$.
- $\kappa_{13}$: market constant corresponding to the cost of the pipeline installation and lay-down per m$^2$ $\frac{EUR}{m^2}$.

Furthermore, taking into account the expression of the resistance of materials for thin cylindrical steel casings, also known as the "Barlow" equation, normally adopted by the applicable regulations ASME B31.4/B31.8/B31.11 and Res. SE 1460/2006 (Argentina)[3]:

$$\sigma = \frac{p_{int} D_{ext}}{2 \cdot t} \tag{4.9}$$

Being $\sigma$ the admissible stress of the steel, which in general is the specified minimum yield stress—SMYS—affected by a safety factor $F$ called "design factor", with values ranging between 0.8 and 0.5 or less, depending on the type of fluids transported and in some cases the location of the pipeline and the population density of the adjacent area. Summarizing, the total cost of the installed pipe can be expressed as,

$$c_1 = \kappa_{11}\pi D_m t L g \rho_p + (\kappa_{12} + \kappa_{13})\,\pi D_{ext} L \tag{4.10}$$

### Diameter Factor

In the cost expressions several diameters appear, namely: the mean diameter $D_m$ in the simplified weight equation; the outer diameter $D_{ext}$, in the Barlow equation; and the inner diameter $D_{int}$, in the flow equation.

Indeed, to avoid developing the equations in three different diameters, one must try to eliminate one of these three variables. To do this, a new "variable" is created, but with very little variation, i.e., a "quasi-constant"—called $K_D = D_{int}/D_{ext}$, that is, the ratio of the internal diameter to the external diameter of the pipe.

[3] Other codes, such as EN 13480-3 and ASME B31.12 use a different version of this equation, which should be considered in new optimization models.

This factor makes it possible to keep only the outer diameter in the expression, by initially estimating the $K_D$ and then verifying it, hence, a single unknown is used in the procedure.

$$D_m = \frac{D_{int} + D_{ext}}{2} = \frac{D_{ext}\left(\frac{D_{int}}{D_{ext}} + 1\right)}{2} = \frac{D_{ext}\,(1 + K_D)}{2} \tag{4.11}$$

The $K_D$ coefficient generally takes values between 0.92 and 0.98 for rolled steel pipes, manufactured under the API 5L standard. Taking this factor to the cost equation, it renders,

$$c_1 = \kappa_{11}\frac{1 + K_D}{2} D_{ext}^2 \frac{t}{D_{ext}} \pi g \rho_p L + (\kappa_{12} + \kappa_{13})\,\pi D_{ext} L \tag{4.12}$$

Replacing the thickness with the Barlow equation, it yields,

$$c_1 = \kappa_{11}\frac{1 + K_D}{2} D_{ext}^2 \frac{p}{2 \cdot \sigma_{adm}} \pi g \rho_p L + (\kappa_{12} + \kappa_{13})\,\pi D_{ext} L \tag{4.13}$$

Where,

- $\kappa_{11}$: Market constant, corresponding to the cost of the pipe, usually carbon steel type API 5L, per unit of weight, i.e., € per Newton [EUR/N].
- $\kappa_{12}$: Market constant corresponding to the cost of the thermal insulating and/or anti-corrosive coating per $m^2$ [EUR/$m^2$].
- $\kappa_{13}$: Market constant corresponding to the cost of pipe installation per $m^2$ [EUR/$m^2$].
- $\rho_p$: density of the pipeline, e.g., in case of carbon steel for pipelines this value is approximately 7850 kg/$m^3$.
- $D_m$: Average diameter of the pipe in meters [m].
- $D$: Outside diameter of the pipe in meters [m].
- $K_D = D_{int}/D_{ext}$, that is, the ratio of the inside diameter to the outside diameter of the pipe. This factor allows keeping in the expression only the external diameter as unknown, previous an initial estimation of $K_D$ and a subsequent verification, using one unknown during the mathematical optimization.

With respect to $\kappa_{11}$, it has been observed that, though considered in the calculations as a constant, in fact depends on the pipe size, the longitudinal weld, if any (e.g. seamless, SAW or ERW), and the material grade, and independent of the pipe schedule or wall thickness. Based on information gathered in 2023, this constant can be evaluated using the following function for seamless pipes,

$$\kappa_{11} = -33.243 + 32.642 \cdot SMYS^{0.006} - 0.009 \cdot NPS^{2.421} + 0.009 \cdot SMYS^{0.006} \cdot NPS^{2.421} \tag{4.14}$$

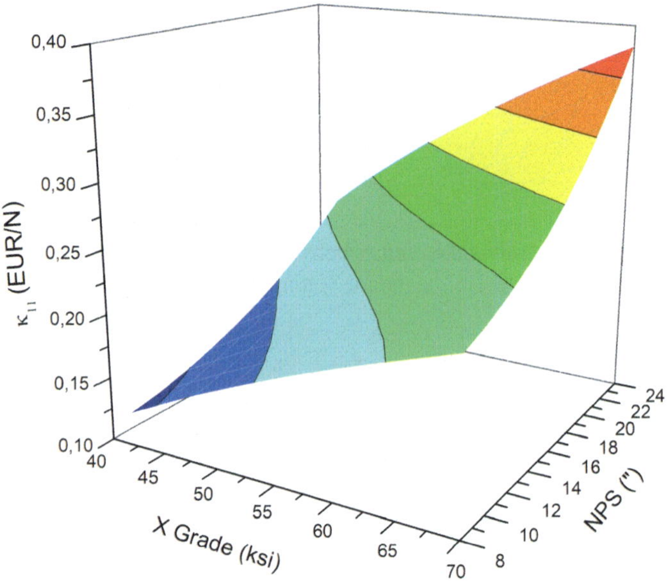


**Fig. 4.2** 3D plot of equation 4.14 as a function of the pipe size and the material grade

with $SMYS$ expressed in ksi (thousands of pounds per square inch) and $NPS$ expressed in inches (see Fig. 4.2). Regarding $\kappa_{12}$, a similar study can be made obtaining a dependency on the pipe size and the schedule/wall thickness. This function, and its graphical representation in Fig. 4.3, to approximate this constant is,

$$\kappa_{12} = 8.202 + 0.314 \cdot SCH^{1.358} + 0.018 \cdot NPS^{1.736} + 3.35 \times 10^{-4} \cdot SCH^{1.358} \cdot NPS^{1.736} \quad (4.15)$$

## 4.4.2 Cost of Energy Consumed by the Pumping Facilities

The energy consumed during the operation of the system depends on the power applied to drive the fluid in the pumping stations. Is it noteworthy mentioning that the installed power must be differentiated from the power applied during the operation of the system. In order to add this cost, which is normally paid to the supplier in monthly or bimonthly periods during the system's operational lifetime, it will be necessary to know its present value, applying an appropriate interest rate or correction factor depending, among others, of the inflation rate in the country where the project will be located. Being the expression of the power for pumping liquids,

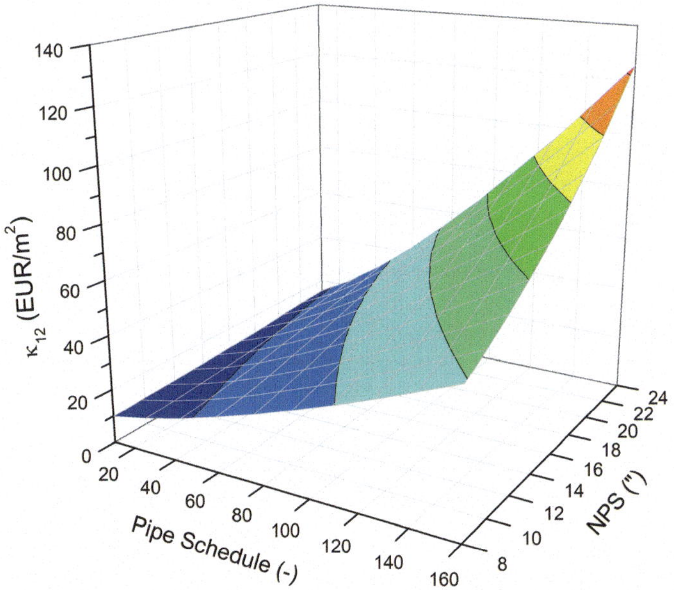


**Fig. 4.3** 3D plot of equation 4.15 as a function of the pipe size and its schedule

$$N_{pump} = \frac{\dot{m}}{\eta_{pump}}\left(\frac{p_d - p_s}{\rho}\right)\chi \tag{4.16}$$

The cost of the energy consumed is then proportional to the power consumed,

$$c_2 = \kappa_{02}\kappa_2 \frac{N_p}{\eta_m}\chi \cdot time \tag{4.17}$$

$$c_2 = \kappa_{02}\kappa_2 \frac{time}{\eta_{pump}\eta_m}\frac{\dot{m}}{\rho}\left(\rho g\frac{H}{\chi} + \frac{8f}{\pi^2}\frac{\dot{m}^2}{\rho\chi}\frac{L}{K_D^5 D^5}\right)\chi$$

$$c_2 = \kappa_{02}\kappa_2 \frac{time}{\eta_{pump}\eta_m}\frac{\dot{m}}{\rho}\left(\rho g H + \frac{8f}{\pi^2}\frac{\dot{m}^2}{\rho}\frac{L}{K_D^5 D^5}\right) \tag{4.18}$$

Being:

- $\eta_{pump}$: pump performance, total hydraulic at the point of operation, in general it is the manufacturer's data, eventually corrected for high viscosity fluids.
- $\eta_m$: performance of the drive motor (e.g., electric, thermal) that relates the energy or fuel delivered to the engine with the energy delivered to the shaft of the driven machine, in this case, the pump.
- $N_{pump}$: theoretical power of the pumps in watts [W] applied to the pump during operation (this does not consider stand-by power).

- $\chi$: Number of pumping plants, including the header station.
- $time$: operating time of the pumping installations (in hours). In general, the operating lifetime that has been determined for the project.
- $\kappa_2$: unit cost of energy to be consumer, in EUR per W-hour [€/W-h].
- $\kappa_{02}$: coefficient of present value of the cost of energy, corresponding to the monthly or bimonthly payment, as appropriate, throughout the entire operational useful life, calculated based on an appropriate interest rate.

Assuming that the energy is paid monthly, if the monthly interest rate is $i$ and the number of periods in months is $n$, the value of $\kappa_{02}$ is obtained from the following financial mathematics formulation,

$$\kappa_{02} = \frac{(1+i)^n - 1}{ni(1+i)^n} \tag{4.19}$$

As the interest rate $i$ is generally fixed annually, the monthly rate to apply in the equation is obtained from,

$$i_{monthly} = (1 + i_{yearly})^{1/12} - 1 \tag{4.20}$$

### 4.4.3 Cost of the Pumping Stations

Obtaining data from the market or region in question, it is possible to verify that the cost of a pumping plant is a function—which is not linear—in relation to the power of the pumps, being understood as "power" for the calculation the total installed pumping power. However, to simplify the calculations, it is assumed that the cost is a linear function, valid within a certain range of (installed) powers, as depicted in Fig. 4.4.

In this case, the increase of the theoretical power to the nominal power, and of the sum of the powers of each machine must be foreseen. In the case of several machines in parallel and/or series arrangement, also considering the existence of "stand-by"—or reserve—power in each facility,

$$c_3 = \left(\kappa_{03} + \kappa_3 N_{pump}\right) \chi \tag{4.21}$$

Being the power of each pumping plant given by,

$$N_{pump} = \frac{\dot{m}}{\eta_{pump} K_{SB}} \frac{\Delta p}{\rho} = \frac{\dot{m}}{\eta_{pump} K_{SB} \rho} \left( \rho g \frac{H}{\chi} + \frac{8 f}{\pi^2} \frac{\dot{m}^2}{\rho \chi} \frac{L}{K_D^5 D^5} \right) \tag{4.22}$$

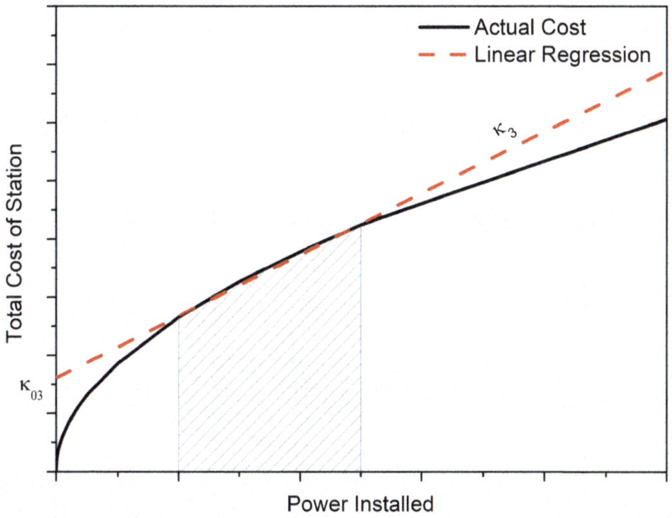


**Fig. 4.4** Plot of the installation cost as a function of the power, showing the linearization for a specific range of powers

Hence, the cost for the pumping plants for the project is equal to,

$$c_3 = \left[\kappa_{03} + \kappa_3 \frac{\dot{m}}{\eta_{pump} K_{SB} \rho} \left( \rho g \frac{H}{\chi} + \frac{8f}{\pi^2} \frac{\dot{m}^2}{\rho \chi} \frac{L}{K_D^5 D^5} \right) \right] \chi \tag{4.23}$$

Where:

- $\kappa_{03}$: figure independent to a certain extent from power, which mainly represents the cost of land and auxiliary facilities indirectly related to installed power, expressed in [€].
- $\kappa_3$: market constant corresponding to the average unit cost of power of the pumping stations (drive units: motors + pumps, assembled and running) for the resulting power range, commonly expressed in [€/HP] or [€/W].
- $K_{SB}$: constant that indicates the established reserve power.[4]
- $N$: installed power of the pumping equipment, including the units in stand-by or redundant, expressed in [W].
- $\chi$: number of pumping plants, including the header.

Since the power to be applied to each plant is not originally known, it must be initially assumed in the calculation with approximate values of $\kappa_{03}$ and $\kappa_3$. The verification of the cost per plant can be done with the following practical expression, obtained

[4] Power in Stand-By (or in reserve) means that, e.g., if 50% is defined, the total power to be applied is divided between two drive machines (50%) each and a third machine of equal power is installed (2+1); or 33% is defined in reserve, the power to be applied is divided between three machines of equal power each (33%) and a fourth additional machine of equal power is installed (3+1).

as a regression of market data, once the power involved in each plant is known. The coefficient $n_{c3}$ depends on several economical and market parameters, with its value going from 0.4 to 0.5.

$$K_{03} = \frac{C_0}{2}(N)^{n_{c3}} \tag{4.24a}$$

$$K_3 = \frac{C_0}{2}\frac{1}{(N)^{n_{c3}}} \tag{4.24b}$$

### 4.4.4 Cost of the Product Used to Fill the Pipeline

It is also necessary to evaluate the cost of the initial filling product, a capital expenditure without which a fluid transport system could not function. In fact, for a molecule of fluid that enters the head of the pipe to be delivered to the opposite end of the pipe, it must be completely filled with the fluid to be transported.

At the end of the useful life before the abandonment of the pipeline the product must be recovered due to regulatory, environmental and safety requirements. In general, current international regulations require the emptying and inertization of the pipeline prior to its abandonment. The final recovery of the fluid that fills the pipeline has a residual value that must be considered in the calculations.

In the case of a liquid, the quantity necessary to completely fill the duct will depend exclusively on the internal volume of the pipe,

$$c_4 = K_4 K_{04} \frac{\pi D_{int}^2}{4} L = K_4 K_{04} K_D^2 \frac{\pi D^2}{4} L \tag{4.25}$$

Being:

- $\kappa_4$: unit cost of the transported product [€/m$^3$]. In the case of oil, in general, the price is set in dollars per barrel.[5]
- $\kappa_{04}$: coefficient of present value of the cost $\kappa_4$. At the end of the operational life and prior to the abandonment of the pipeline, the initial filling product must be extracted, recovering at that moment part of the initial filling cost, which must be evaluated with expressions from financial mathematics, setting an appropriate discount rate $i$ and for the corresponding number of periods (generally annual rate, applied to the number of years—expectation—of the pipeline in operation).

$$K_{04} = 1 - \frac{1}{(1+i)^n} \tag{4.26}$$

[5] Unit of volume used in the oil industry equivalent to 42 US gallons or 35 imperial gallons, in the SI equivalent to 158.987 liters.

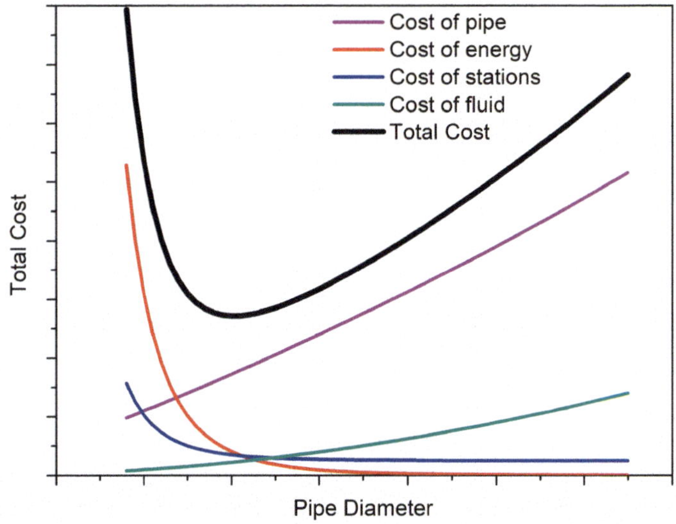


**Fig. 4.5** Total cost of a pipeline facility as a function of the pipe size, showing as well the different costs studied in this book

## *4.4.5 Total Cost of the System*

The total cost $c_T$ of the transportation system, at present or current value, considering the installation of the pipeline and its plants, and the operation and maintenance, is obtained from the sum of the four aforementioned costs, shown as well in Fig. 4.5,

$$c_T = \sum_i c_i = c_1 + c_2 + c_3 + c_4 \tag{4.27}$$

The goal is then to achieve the economic optimization of the pipeline, finding minimum total cost of the project, conditioned on the pressure loss function, expressed in the terms of the Bernoulli equation for Newtonian, incompressible fluids at steady-state (Eq. 4.5).

The pressure $p_1$ is coincident with the outlet pressure of the pumping plants and the pressure $p_2$ is coincident with the suction pressure of the pumps, which is why its minimum value must be limited for reasons of cavitation in the pumps and/or pipes as well as for other operative reasons within the pumping stations themselves.

In energy terms, only the total difference in level between the header and the end of the line and the length (for losses) matter. The intermediate ascents or descents (i.e. altimetry) do not matter and that is why it is assimilated to a straight pipe with a certain slope between the plants. However, if you have to overcome a high hill then you have to do some arrangement of the location of the pumping plants. The pumps must be located in such a way that the discharge pressure is not exceeded and thus does not modify the Total Cost, affecting the pipeline wall thickness. The arrival pressure should be as low as possible, but it is necessary to know what is the minimum pressure that the pump admits. In general, the pumps require a suction pressure $\approx$ 1 to 2 bar and therefore the final pressure is also of the order of 2 to 2.5 bar, in order to make it possible to enter tanks with a maximum level of 20/25 meters when they are full.

To simplify the mathematical developments, in the condition equation, as it was already done in the cost functions $c_1$, $c_2$, $c_3$ and $c_4$, one replaces the internal diameter of the pipe using the diameter relationship explained before. Hence, the function is of the form,

$$p_1 - p_2 = \rho g \frac{H}{\chi} + \frac{8f}{\pi^2} \frac{\dot{m}^2}{\rho K_D^5 D^5} \frac{L}{\chi} \tag{4.28}$$

The technique chosen to find the minimum cost is the Lagrange Multipliers method.[6] This is a mathematical tool to find the extrema of a certain function, when it is subjected to a number of constraints, represented also as functions of different variables. The main idea with all the functions involved consists in creating a new function, the Lagrangian, written as linear combination of the aforementioned functions. Thus, for functions of one variable,

$$\mathcal{L}(x, \lambda) = f(x) + \lambda \cdot \phi(x) \tag{4.29}$$

where $f(x)$ is the function to be maximized/minimized, subjected to a constraint function $\phi(x) = 0$. This means that to find the critical point, all the partial derivatives in the system should be equal to zero,

$$\frac{\partial \mathcal{L}}{\partial x} = 0 = \frac{\partial f(x)}{\partial x} + \lambda \frac{\partial \phi(x)}{\partial x} \tag{4.30a}$$

$$\frac{\partial \mathcal{L}}{\partial \lambda} = 0 \tag{4.30b}$$

The constraint equation is then in this case,

$$\phi(p_1, \chi, D) = (p_1 - p_2) - \rho g \frac{H}{\chi} - \frac{8f}{\pi^2} \frac{\dot{m}^2}{\rho K_D^5 D^5} \frac{L}{\chi} = 0 \tag{4.31}$$

[6] Named after Joseph-Louis Lagrange (1736–1813), he was an Italian (later naturalized French) mathematician, physicist and astronomer. He was a prolific researcher in the fields of number theory, differential calculus and classical mechanics.

Moreover, the total cost function is,

$$c_T = \kappa_{11}\frac{1+K_D}{2}D^2\frac{p}{2\cdot\sigma_{adm}}\pi g\rho_p L + (\kappa_{12}+\kappa_{13})\pi DL + \ldots$$
$$+\kappa_{02}\kappa_2\frac{time}{\eta_{pump}\eta_m}\frac{\dot{m}}{\rho}\left(\rho gH + \frac{8f\,\dot{m}^2}{\pi^2}\frac{L}{\rho}\frac{L}{K_D^5 D^5}\right) + \ldots$$
$$+\left[\kappa_{03} + \kappa_3\frac{\dot{m}}{\eta_{pump}K_{SB}\rho}\left(\rho g\frac{H}{\chi} + \frac{8f\,\dot{m}^2}{\pi^2\,\rho\chi}\frac{L}{K_D^5 D^5}\right)\right]\chi + \ldots$$
$$+ K_4 K_{04} K_D^2\frac{\pi D^2}{4}L \tag{4.32}$$

## Optimization of the Variables—Case with Liquid Pipelines

It was already expressed in the previous point that in the case of liquid transport systems through pipelines, the unknowns to optimize are: the pipe diameter $D$; the maximum operating pressure $p_1$; and the number of pumping stations $\chi$, including the header (Brackett and Zietlow 2011; Jokić and Zavargó 2001; Sharma et al. 2013).

It should be remembered that the outer diameter has been chosen for optimization purposes, so other diameters are not included in the optimization equations. With the outer diameter and the discharge pressure, it is feasible to calculate the minimum theoretical wall thickness and the theoretical inner diameter, similarly the inner diameter can be inferred with the $K_D$ and the outer diameter.

In order to find the minimum of the summation $c_T$ function of the costs, conditioned to the function $\phi(p_1, \chi, D)$, the Lagrange multipliers method is applied, for which the following system of equations is formed,

$$\frac{\partial c_T(p_1,\chi,D)}{\partial D} + \lambda\frac{\partial\phi(p_1,\chi,D)}{\partial D} = 0 \tag{4.33a}$$

$$\frac{\partial c_T(p_1,\chi,D)}{\partial p_1} + \lambda\frac{\partial\phi(p_1,\chi,D)}{\partial p_1} = 0 \tag{4.33b}$$

$$\frac{\partial c_T(p_1,\chi,D)}{\partial \chi} + \lambda\frac{\partial\phi(p_1,\chi,D)}{\partial \chi} = 0 \tag{4.33c}$$

$$\phi(p_1,\chi,D) = (p_1 - p_2) - \left(\rho g\frac{H}{\chi} + \frac{8f}{\pi^2}\frac{\dot{m}^2}{\rho K_D^5 D^5}\frac{L}{\chi}\right) = 0 \tag{4.33d}$$

With the proposed method these equations form a system of four equations with four unknowns, allowing to find a unique solution. Note that the Lagrange multiplier $\lambda$ is just an ancillary variable and only the remaining three are of interest for optimization purposes. Executing the partial derivatives according to each unknown and rearranging the equations, the following system is obtained,

$$\frac{1+K_D}{2}\kappa_{11}D^2\rho_p g L\frac{\pi}{2\sigma_{adm}}+\lambda=0 \tag{4.34a}$$

$$\kappa_{03}+\frac{\lambda}{\chi^2}\left(\rho g H+\frac{8f}{\pi^2}\frac{\dot{m}^2}{\rho}\frac{L}{K_D^5D^5}\right)=0 \tag{4.34b}$$

$$p_1=p_2+\rho g\frac{H}{\chi}+\frac{8f}{\pi^2}\frac{\dot{m}^2}{\rho K_D^5D^5}\frac{L}{\chi} \tag{4.34c}$$

$$\left(\frac{1+K_D}{2}\kappa_{11}\frac{\pi}{\sigma_{adm}}p_1\rho_p g+K_4K_{04}K_D^2\frac{\pi}{2}\right)D+(\kappa_{12}+\kappa_{13})\pi-\ldots$$
$$\ldots-\left(\frac{\kappa_{02}\kappa_2 time}{\eta_{pump}\eta_m}+\frac{\kappa_3}{K_{SH}\eta_{pump}}\right)\frac{40f\dot{m}^3}{\pi^2\rho^2K_D^5D^6}+\lambda\frac{40f\dot{m}^2}{\pi^2\rho K_D^5\chi D^6}=0 \tag{4.34d}$$

This renders a highly non-linear system of equations, which can be solved by any traditional technique, e.g. Newton or Quasi-Newton approaches. In this chapter, the authors have decided to solve it, for the seek of clarity, by successive approximations. During the cycle, the $K_D$, the friction factor $f$, calculated from the corresponding Reynolds number, the cost constants $\kappa_{11}, \kappa_{12}, \kappa_3, \kappa_{03}$ that depend on the resulting power and the pipe size must also be verified.

The results obtained obey a mathematical optimization, and the approximation to commercial or regulatory values poses, *a posteriori*, a new optimization in terms of commercially used or standardized variables (e.g., pipe external diameters and thicknesses), discrete variables (e.g. pumping stations), or dependent on current national/international regulations (e.g., maximum operating pressures), normally achieved through a series of detailed calculations of the pipeline that is being projected. The diameter $D$ of the pipe will be selected within the standardized series: the diameter closest to the calculated optimum is chosen, whether it is larger or smaller, which is discussed in the next section. See the following pseudo-code as a general guideline of the complete optimization process.

### 4.4.6 Practical Adjustment of the Optimization Variables

Since the variables $D$ and $p_1$, and with them the thickness, are obtained from the theoretical formulation, the next step consists in selecting the commercial dimensions, which can be called the practical adjustment of the results. The result of the pumping stations must be operated in the same way, since the number of plants must be an integer. Therefore, to solve a practical alternative for a pipeline for liquids by

**Algorithm 4.1:** Solution of optimization for incompressible pipelines

**Data**: $\dot{m}, L, p_2, H, \kappa_{ii}, \sigma_{adm}, \rho, \rho_p, \eta_i, time$
**Result**: Optimum diameter, discharge pressure, number of pumping stations
Estimate friction factor, $K_D$ and Diameter;
Calculate $\kappa_{11}, \kappa_{12}$;
**while** *Eq. 4.34d* $\neq 0$ **do**
  Get $\lambda$ from Eq. 4.34a, $\lambda = -\frac{1+K_D}{2}\kappa_{11}D^2\rho_p g L \frac{\pi}{2\sigma_{adm}}$;
  Get $\chi$ from Eq. 4.34b, $\chi = \sqrt{\frac{-\lambda}{\kappa_{03}}\left(\rho g H + \frac{8f}{\pi^2}\frac{\dot{m}^2}{\rho}\frac{L}{K_D^5 D^5}\right)}$;
  Get $p_1$ from Eq. 4.34c, $p_1 = p_2 + \rho g \frac{H}{\chi} + \frac{8f}{\pi^2}\frac{\dot{m}^2}{\rho K_D^5 D^5}\frac{L}{\chi}$;
  Calculate $N_{pump}, \kappa_{03}, \kappa_3$;
  Verify Eq. 4.34d = 0;
  **if** *Eq. 4.34d = 0* **then**
    Proceed to next verification;
  **else**
    Estimate $D$ with Eq. 4.34d with $\lambda$, $p_1 \wedge \chi$;
    Recalculate $f, \kappa_{11}, \kappa_{12}, K_D$
  Compare $N_{pump}, \kappa_{11}, \kappa_{12}, \kappa_{03}, \kappa_3, t, K_D$ and $f$;
  **if** *all these do not change* **then**
    Proceed to next verification;
  **else**
    Take the calculate values and start again the optimization;
Calculate $D/t$;
Get critical $D/t$ value from graph based on the D;
**if** $D/t >$ *critical* **then**
  Minimum Thickness Calculation - See point Sect. 4.4.6;
  Adopt $D/t$ = critical;
  Calculate $K_D$;
  Get $p_1$ from Eq. 4.36, $p_1 = (1 - K_D)\sigma_{adm}$;
  Get $D$ from Eq. 4.37, $\left(\frac{1-K_D^2}{2}\kappa_{11}\rho_p g + \pi\kappa_{04}\kappa_4 K_D^2\right)D^7 + (\kappa_{12} + \kappa_{13})\pi D^6 - \ldots$
  $-\left\{\left(\frac{\kappa_{02}\kappa_2 time}{\eta_{pump}\eta_m} + \frac{\kappa_3}{\eta_{pump}}\right)\frac{40 f \dot{m}^3}{\pi^2\rho^2 K_D^5} + \frac{\kappa_{03}}{[(1-K_D)\sigma_{adm} - p_2]}\frac{40 f \dot{m}^2}{\pi^2\rho^2 K_D^5}\right\} = 0$;
  Get $\chi$ from Eq. 4.38, $\chi = \frac{1}{(p_1 - p_2)}\left(\rho g H + \frac{8f}{\pi^2}\frac{\dot{m}^2}{\rho K_D^5}\frac{L}{D^5}\right)$;
  Calculate the friction factor $f$; **if** *$f$ did not change* **then**
    Proceed to next verification;
  **else**
    Recalculate $f$ and start over the Minimum Thickness Calculation;
**else**
  Export all the results;
  Adopt commercial, integer and technical values for all the parameters;
  Set the 4 different alternatives based on $D$ and $\chi$ and choose the optimum alternative;

the optimization method developed here, the following logical adjustment sequence of the variables must be followed:

1. Correctly solve, with adjustment loops, the complete equations for obtaining $p_1$, $\chi$, $D$, achieving the verification of the $K_D$ relationship adopted, the friction factor and the cost constants $\kappa_{03}$, $\kappa_3$.
2. Select with respect to the theoretical optimum $D$, the largest and smallest commercial diameter. For example, if the $D$ is 13.3", select 12.75" and 14".
3. Select, with respect to the "non-integer" $\chi$, the largest and smallest integer (for example, if $\chi = 3.6$ pumping plants, then select $\chi = 3$ and $\chi = 4$).
4. Define several projects selecting the one with the largest diameter with the least number of pumping plants, and the largest diameter with the largest number of plants, and vice versa—the smallest diameter with the two numbers of pumping plants, that is, developing at least four alternatives to evaluate technical performance and total cost.
5. Calculate the pressure $p_2$ for the commercial alternatives defined in the previous point, applying the liquid flow equation at a steady state, for the defined sections ($L/\chi$ integer) for each case of commercial $D$, and therefore of thickness, also those selected from the "commercial" tables. Then calculate the pumping power, using the fixed $p_1$ and the calculated $p_2$.
6. For all cases, with each diameter and commercial thickness, the pressures $p_1$, $p_2$ and the rounded-up quantity $\chi$, calculate with the summation expression of the four costs, the total cost of each alternative and select as optimal the one that results in lower cost or, in the case of comparable solutions in terms of cost, the one that meets some additional requirement that makes it more suitable for the effects sought. This may include for instance, the prospective changes in the flow rate due to expansions, that may lead to choose the bigger pipes sizes rather than the smaller ones.
7. Once the alternative to be developed has been selected—the one with the lowest total cost—proceed with a quasi-static analysis of the system, dividing the pipe in small sections, considering then small variations in the variables that were not considered during the optimization process. This also allows relocating the plants in order to optimize their position.

It should be noted that any of these alternatives defined in this list will be more optimal than the one that theoretically solves the system of equations, even if it is also the one that results in the lowest cost among all of them. In addition, if a corrosion allowance thickness is applied, the procedure must be adapted as indicated in the following sections.

### Pipe Thickness Less Than the Minimum Manufactured

Pipe manufacturers normally limit the manufacture of tubes to certain minimum thicknesses for practical reasons (rigidity, ovalization, handling, manufacturing method, etc.). Likewise, the construction of pipelines with pipes with very small

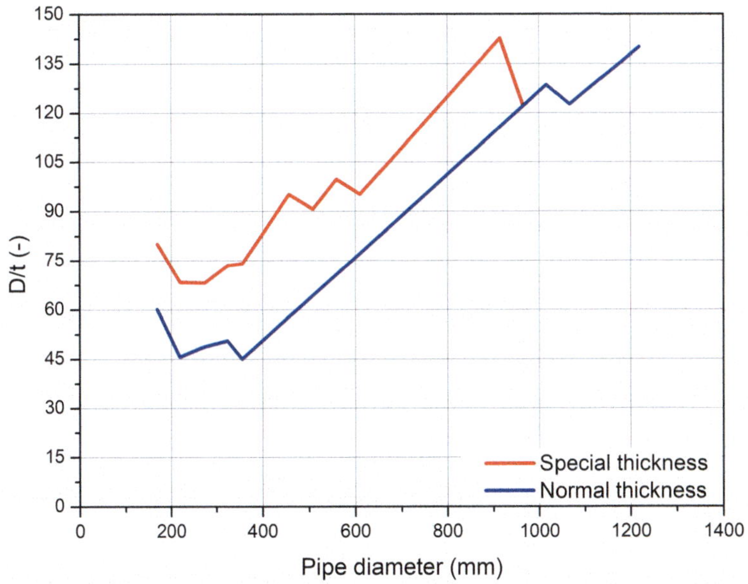


**Fig. 4.6** Plot showing the maximum relationships diameter/thickness, as a function of the pipe size

thicknesses in relation to their diameter entails construction difficulties, e.g., dents due to support in the bottom of an irregular trench and difficulties during welding and security against third parties, such as the possibility that impacts from adjacent works cause ruptures with fluid leaks.

For this reason, it is of a technical practice to adopt an adequate manufacturing relationship between thickness and diameter, given by,

$$\frac{t}{D} = \frac{1 - K_D}{2} = \frac{p_1}{2\sigma_{adm}} \tag{4.35}$$

The inverse relationship $D/t$ is commonly established with values in the order of 100 to 120 (see Fig. 4.6), although API pipe manufacturers' tables generally do not go beyond $D/t = 80$ or 90.[7]

In one way or another, the maximum operating pressure is implicitly defined when setting the $K_D$, or the $D/t$ ratio, and the grade of the resulting material, yielding,

$$p_1 = (1 - K_D)\sigma_{adm} \tag{4.36}$$

[7] ASME B31.4, which is applicable to lines for the transportation of liquid hydrocarbons, states in its article 403.2.5 "Use of High D/t Ratio" that the designer must take additional precautions if the D/t ratio is greater than 100.

and when the operating pressure is prefixed (by fixing the thickness/diameter ratio) the cost functions are formulated again, with one less variable, since $p_1$ is prefixed, and the optimal diameter is determined, resulting in the following expression,

$$\left(\frac{1-K_D^2}{2}\kappa_{11}\rho_p g+\pi\kappa_{04}\kappa_4 K_D^2\right)D^7+(\kappa_{12}+\kappa_{13})\pi D^6-\dots$$
$$-\left\{\left(\frac{\kappa_{02}\kappa_2 time}{\eta_{pump}\eta_m}+\frac{\kappa_3}{\eta_{pump}}\right)\frac{40f\dot{m}^3}{\pi^2\rho^2K_D^5}+\frac{\kappa_{03}}{[(1-K_D)\sigma_{adm}-p_2]}\frac{40f\dot{m}^2}{\pi^2\rho^2K_D^5}\right\}=0 \tag{4.37}$$

From this equation, one finds the diameter $D$. Then, with Eq. 4.36, the maximum pressure allowed in the line can be calculated, and using the Bernoulli equation, the number of pumping stations is finally obtained,

$$\chi=\frac{1}{(p_1-p_2)}\left(\rho g H+\frac{8f}{\pi^2}\frac{\dot{m}^2}{\rho K_D^5}\frac{L}{D^5}\right) \tag{4.38}$$

It is then set to an integer and then the thickness is set to a normalized value; finally, the maximum pressure $p_1$ is calculated again more precisely, and the thickness is verified.

### Pipe Requiring a Corrosion Allowance

In the event that the characteristics of the fluid to be transported and the type of pipe material are likely to cause pipe corrosion and therefore advise the adoption of an extra thickness due to corrosion,[8] some of the expressions above must be modified. In fact, the thickness of the pipe will be in these case,

$$t_{pipe}=t_{press}+t_{corr} \tag{4.39}$$

It should be noted that to determine the $D/t$ ratio, the excess thickness due to corrosion must be considered; this is for obvious practical reasons, manufacturing and installation handling will benefit from this excess thickness, which makes the pipes more rigid.

As it was presented, the cost of the pipe can be assumed to be made up of two parts, one function of the weight of the pipe and another function of the exterior surface. If one wants to consider the extra costs due to the corrosion allowance, the extra material used can be written according to the following expression,

---

[8] Generally this value goes from 1.59 to 3.125 mm in countries using the Imperial System, whilst in Europe the commonly found values range from 1 to 3 mm.

$$c_{corr} = \kappa_{11} t_{corr} \pi D_m \rho_p g L = \kappa_{11} t_{corr} \pi \frac{1 + K_D}{2} D \rho_p g L$$
$$= \kappa_{11} t_{corr} \frac{1 + K_D}{2} \rho_p g \pi D L = \kappa_{14} \pi D L \quad (4.40)$$

One can then introduce this cost in the second part of the costs related to the pipe, the part of the function depending linearly on the pipe diameter. Thus,

$$c_1 = \kappa_{11} \frac{1 + K_D}{2} D^2 \frac{p}{2 \cdot \sigma_{adm}} \pi g \rho_p L + (\kappa_{12} + \kappa_{13} + \boldsymbol{\kappa_{14}}) \pi D L \quad (4.41)$$

The rest of the cost functions do no suffer significant alterations due to this addition. With respect to the optimization, the procedure is quite similar to the presented previously, since the only equation to be modified is Eq. 4.34d,

$$\left( \frac{1 + K_D}{2} \kappa_{11} \frac{\pi}{\sigma_{adm}} p_1 \rho_p g + K_4 K_{04} K_D^2 \frac{\pi}{2} \right) D + (\kappa_{12} + \kappa_{13} + \boldsymbol{\kappa_{14}}) \pi - \ldots$$
$$\ldots - \left( \frac{\kappa_{02} \kappa_2 time}{\eta_{pump} \eta_m} + \frac{\kappa_3}{\eta_{pump}} \right) \frac{40 f \dot{m}^3}{\pi^2 \rho^2 K_D^5 D^6} + \lambda \frac{40 f \dot{m}^2}{\pi^2 \rho K_D^5 \chi D^6} = 0 \quad (4.42)$$

In the event that the result of the optimization determines a pipe with a thickness less than the manufacturing minimum, the minimum thickness equation shall be applied after this optimization. When this done, the extra thickness due to corrosion must not be considered when setting the maximum pressure $p_1$.

### 4.4.7 Preliminary Conclusions Regarding the Calculation of Optimum Values

As seen, to focus on the problem, it is first necessary to carry out a limited investigation in the market in order to obtain all the costs of the elements that intervene in the installation, e.g., materials, labor and equipment, and then develop appropriate mathematical regressions to obtain the cost functions required by the method.

Next, the equation that technically conditions the project must be considered, such as the flow equation of viscous liquids, in which all the variables to be determined are involved. An important issue is to adequately set the viscosity and density of the product to be transported considering the average temperature of the ground in which it will be buried, at depth and in general in winter, and estimated average pressure.

Finally, with the total cost function, the minimum cost conditioned to the flow equation is sought, using the Lagrange Multipliers Method. Then the optimum values must be adjusted to practical values, that is, diameter and maximum operating pressure.

## 4.5 Influence of Great Height Differences in Liquid Transport Pipelines

The great elevations and depressions of a certain route[9] cause the pipes to be designed with extra thickness due to the hydrostatic column when they are out of service; while when it is in operation, the over-pressure often disappears due to the loss of charge, and consequently, a large part of the material in the pipe remains "unusable".

To improve the solution, one possible way is to sectorize the pipe segments and correspondingly the heights, either up or down. For the optimization calculation carried out so far, only the height differences between the beginning of the duct and the end are involved, regardless of what happens along the way. This calculation method has two limitations:

- The power of the pumping stations is greater than or equal to zero ($N_{pump} \geq 0$).
- The pressure in certain points of the pipeline is higher than the vapor pressure of the transported fluid ($p \geq p_v$).

Failure to comply with these limitations occurs when the rises and/or falls in the pipe path are significant.

### 4.5.1 Pipes Uphill

In general, pipes with a single diameter are designed to simplify the passage of plugs ("scraper") for cleaning or of an intelligent type. In order to use the optimal calculation developed, an average height is taken between the ends of the pipe, and the optimal variables are resolved.

Once resolved, it is sectorized into sections of pipe length between pumping stations, consequently sections of different heights, and the respective thicknesses are calculated, taking care that both the thicknesses and the lengths correspond to standard dimensions and economic manufacturing batches.

### 4.5.2 Pipes Downhill

In this case the primary calculation may result in one or more pumping stations being needed, or none at all. In the case of not needing pumping stations, the optimal calculation is not carried out, only the calculation of the pipe with the minimum possible diameter is made.

[9] An example of this are some pipelines operated in Peru, with total slopes of almost 4,000 meters between the headwaters; for example, the Camisea field is in the heights of the Andes, while the supplied process plants are located almost at sea level, on the Pacific Ocean coast.

If pumping stations are necessary, the optimal calculation must be made. Similar to what was said for the rising pipe, for the calculation an average height difference is taken between the beginning and the end, in order to determine the optimal variables.

Once this calculation is made, the pipe length sections between pumping stations and consequently the height are subdivided, calculating the respective thicknesses, considering manufacturing standards and economic lots. If this does not happen, longer sections must be taken until reaching the economic lot.

## 4.6 Calculation by Segments in Pipelines for Liquid Transportation

It has been verified that inside a pipe a fluid, whether it is crude oil or derivatives, other more or less viscous liquids, or even water, the pressure changes as it circulates and consequently—section by section—the density and viscosity change to a greater or lesser extent. This change is also affected by the heat transfer between fluid and terrain/air, modifying also the temperature. Being $\rho = f(p, T)$ and $\mu = f(p, T)$, the Reynolds number will also be affected and hence the friction factor. In addition, if the mass flow rate remains constant, then if the density changes, the velocity varies as well. Additionally, the diameter of the pipe is modified by the thermal expansion and the elasticity of the steel affected by the internal pressure (McAllister 2014; Tee et al. 2014).

In fact, when changing the characteristics of the fluid, the Reynolds number changes and, as is known, the friction factor is a function of the Reynolds number and the relative roughness. In addition, the dimensions of the pipe change—the diameter and therefore the cross-sectional area—due to the simultaneous effect of temperature and pressure because it is the diameter. Since the magnitude of each variable cannot be known from the beginning, all of them must be initially considered and only later those with little influence can be neglected (Menon 2005; Menon and Menon 2010; Shashi Menon et al. 2013).

In order to carry out the verification by segments, it is necessary to divide the pipeline in a certain number of elements, calculating the parameters in the section "n", continuing then to the section "n+1". In these sections, it is necessary to choose which variables are kept constant and which variables are can vary, since varying all of them at the same time, i.e. a differential calculation, it is not possible to solve by numerical means. In the case of liquids, one assumes the speed as constant within every section.

In the case of crude oil, if the temperature of the oil is very variable, in general so is its viscosity with the temperature, and the segments must be taken of adequate length, making them shorter. Even if the oil were heated, the fluid is losing temperature, although the oil can be heated with intermediate heaters.

In section "n" all the variables are known and applying Bernoulli's equation considering the friction losses due to viscosity,

$$\frac{p_n}{\rho_n} + \frac{v_n^2}{2} + gz_n = \frac{p_{n+1}}{\rho_{n+1}} + \frac{v_{n+1}^2}{2} + gz_{n+1} + f_n \frac{v_n^2}{2} \frac{\Delta x}{D_{int,n}} \tag{4.43}$$

Considering the pipe with a constant section ($dA \approx 0$), without extractions or injections in the section considered ($d\dot{m} = 0$), and for a relatively small section of pipe, due to the difficulties of developing an integral calculation, a parameter will be maintained that will be varied by steps for each section, choosing in this case the flow velocity. It is here in the friction loss term that the problem begins, about whether to base it on the $v_n$ or $v_{n+1}$ and also what diameter should be considered.

Assuming constant speed and diameter (i.e., $V_n \approx V_{n+1}$ and $D_n \approx D_{n+1}$) for the section in question, and that $\rho_n \approx \rho_{n+1}$, the expression to calculate the pressure at the end of the segment becomes:

$$\frac{p_{n+1}}{\rho_{n+1}} = \frac{p_n}{\rho_n} + g(z_n - z_{n+1}) - \frac{8 f_n}{\pi^2} \frac{\dot{m}^2}{\rho_n^2} \frac{\Delta x}{D_{int,n}^5} \tag{4.44}$$

$$p_{n+1} = p_n + g\rho_n(z_n - z_{n+1}) - \frac{8 f_n}{\pi^2} \frac{\dot{m}^2}{\rho_n} \frac{\Delta x}{D_{int,n}^5} \tag{4.45}$$

All features that have been assumed to be constant for the optimal calculation must now be adjusted. In fact, the temperature varies by heat transfer or by the existence of an internal heat source. Inside the segment there can be an energy transformation but, as in any engineering process, there are irreversible transformations inside the pipe, which comes from the transformation of kinetic into internal energy, i.e., the viscous dissipation. From the point of view of temperature it should be analyzed how it is considered. In the energy conservation equation this is commonly found as,

$$\dot{E}_{vd} = \underline{\underline{T}} : \underline{\underline{\nabla v}} = \begin{pmatrix} T_{xx} & T_{xy} & T_{xz} \\ T_{yx} & T_{yy} & T_{yz} \\ T_{zx} & T_{zy} & T_{zz} \end{pmatrix} : \begin{pmatrix} \frac{\partial u}{\partial x} & \frac{\partial u}{\partial y} & \frac{\partial u}{\partial z} \\ \frac{\partial v}{\partial x} & \frac{\partial v}{\partial y} & \frac{\partial v}{\partial z} \\ \frac{\partial w}{\partial x} & \frac{\partial w}{\partial y} & \frac{\partial w}{\partial z} \end{pmatrix} \tag{4.46}$$

Where $\dot{E}_{vd}$ is the viscous energy contribution to the energy conservation equation per unit of volume and per unit of time, $\underline{\underline{T}}$ is the general viscous stress tensor. For a Newtonian fluid, the viscous stress tensor is written as,

$$\underline{\underline{T}} = \zeta(\nabla \cdot v)\boldsymbol{I} + \mu\left[\left(\nabla v + \nabla v^T\right) - \frac{2}{3}(\nabla \cdot v)\boldsymbol{I}\right] \tag{4.47}$$

where $\zeta$ is the bulk or dilatational viscosity, $\mu$ is the dynamic viscosity and $\boldsymbol{I}$ is the identity matrix. Thus, for an incompressible fluid, i.e., $(\nabla \cdot v) = 0$, the viscous dissipation is equal to,

$$\dot{E}_{vd} = \underline{\underline{T}} : \underline{\underline{\nabla v}} = \zeta(\nabla \cdot v)^2 + \mu\left\{2\left[\left(\frac{\partial u}{\partial x}\right)^2 + \left(\frac{\partial v}{\partial y}\right)^2 + \left(\frac{\partial w}{\partial z}\right)^2\right] + \left(\frac{\partial u}{\partial y} + \frac{\partial v}{\partial x}\right)^2 + \ldots \right.$$
$$\left. + \left(\frac{\partial u}{\partial z} + \frac{\partial w}{\partial x}\right)^2 + \left(\frac{\partial w}{\partial y} + \frac{\partial v}{\partial z}\right)^2 - \frac{2}{3}(\nabla \cdot v)^2\right\} \quad (4.48)$$

$$\dot{E}_{vd} = \mu\left\{2\left[\left(\frac{\partial u}{\partial x}\right)^2 + \left(\frac{\partial v}{\partial y}\right)^2 + \left(\frac{\partial w}{\partial z}\right)^2\right] + \left(\frac{\partial u}{\partial y} + \frac{\partial v}{\partial x}\right)^2 + \left(\frac{\partial u}{\partial z} + \frac{\partial w}{\partial x}\right)^2 + \ldots \right.$$
$$\left. + \left(\frac{\partial w}{\partial y} + \frac{\partial v}{\partial z}\right)^2\right\} \quad (4.49)$$

One can also perform a simplified analysis of the terms involved in the viscous dissipation and the pressure loss in the Bernoulli equation to reach some equivalence. Starting with the viscous dissipation,

$$\dot{E}_{vd} \sim \mu\left(\frac{u}{D}\right)^2 \left[\frac{W}{m^3}\right] \rightarrow \dot{E}_{vd} \sim \mu\left(\frac{u}{D}\right)^2 A_p L\,[W] \quad (4.50)$$

Proceeding now with the pressure loss term from the Bernoulli equation, the relationship between both equations can be derived,

$$\frac{\Delta p}{\rho} \sim f u^2 \frac{L}{D},\ f \sim \frac{1}{Re} = \frac{\mu}{\rho u D} \rightarrow \frac{\Delta p}{\rho} \sim \frac{\mu}{\rho u D} u^2 \frac{L}{D} = \mu\left(\frac{u}{D}\right)^2 \frac{L}{\rho u}\left[\frac{m^2}{s^2}\right] \rightarrow$$
$$\frac{\Delta p}{\rho} \sim \mu\left(\frac{u}{D}\right)^2 A_p L\,[W] \quad (4.51)$$

Leaving the theoretical considerations, which can be easily solved for laminar flows, the authors adopted a more practical approach. In what follows, the following sign convention must be taken into account: heat Q is considered positive (+) when the medium delivers heat to the system and negative (–) vice versa. In order to consider this, the authors have incorporated a certain fraction of this heat to the fluid itself. Thus, the total heat exchange in the system will be made of: the heat exchanged with the environment ($Q_1$), the heat part caused by the heat dissipation ($Q_2$), and the heat caused by the pressure change. The first term comes from the heat transfer analysis done in Chap. 2. From this analysis,

$$Q_1 = U_n \pi D_n \Delta x\,(T_{ext} - T_n) \quad (4.52)$$

Assuming that the outside temperature $T_{ext}$ does not change (or will be one extra variable in the analysis); and the losses due to irreversibilities are evaluated using the analysis previously presented, obtaining for the second component,

$$Q_2 = K_\mu \frac{8 f_n}{\pi^2} \frac{\dot{m}^3}{\rho_n^2} \frac{\Delta x}{D_{int,n}^5} \quad (4.53)$$

where $K_\mu$ is a coefficient, which can be established between 0 and 1. For instance, a value of 0.6 represents that 60% of the heat lost in the form of viscous dissipation would be reincorporated into the flow and consequently 40% would be transferred to the environment.

In order to address the last component, one can take the 1st Law of Thermodynamics for open systems and considering the enthalpy $H$ and $\dot{W}$ being the power exchanged (both must be expressed per unit of mass), and therefore $\dot{Q}$ and $\dot{W}$ must be divided by the mass flow rate. Starting from the conservation of energy using the Reynolds Transport Theorem,

$$\left.\frac{dE}{dt}\right|_s = \frac{dQ}{dt} - \frac{dW}{dt} = \frac{d}{dt}\iiint_{CV}\left(\hat{u} + \frac{v^2}{2} + gz\right)\rho d\forall + \ldots$$
$$+ \iint_{CS}\left(\hat{u} + \frac{p}{\rho} + \frac{v^2}{2} + gz\right)\rho\,(v\cdot n)\,dA \quad (4.54)$$

Considering that the pipeline is operating at steady-state conditions and that the difference in the kinetic energy at the inlet and outlet can be neglected, it yields,

$$\frac{dQ}{dt} - \frac{dW}{dt} = \dot{m}\left[(g(z_{n+1} - z_n) + (h_{n+1} - h_n)\right] \quad (4.55)$$

The enthalpy is a state function depending on the pressure and temperature, i.e., $H = f(p, T)$, and it can be written in a differential form as,

$$dH = \left(\frac{\partial H}{\partial T}\right)_p dT + \left(\frac{\partial H}{\partial p}\right)_T dp \quad (4.56)$$

The first partial derivative term, at constant pressure, becomes the specific heat at constant pressure ($c_p$), while the second partial derivative, at constant temperature, can be written for almost incompressible liquids using the volumetric dilatation coefficient,

$$dH = c_p dT + (1 - \beta T)\frac{dp}{\rho}, \quad \beta = \frac{1}{V}\frac{\partial \forall}{\partial T}\left[\frac{1}{K}\right] \quad (4.57)$$

Taking as reference the absolute pressure at zero and the temperature at zero kelvin, the enthalpy is also zero for that reference and considering the constant density and temperature for the segment,

$$h_{n+1} - h_n = c_p\,(T_{n+1} - T_n) + \left[(1 - \beta T_{n+1})\frac{p_{n+1}}{\rho_n} - (1 - \beta T_n)\frac{p_n}{\rho_n}\right] \quad (4.58)$$

If the temperature changes not significantly in the segment under study, then $T_{n+1} \approx T_n$ and relating this to the conservation of energy derived before,

$$h_{n+1} - h_n = c_p\,(T_{n+1} - T_n) + \left[(1 - \beta T_n)\,\frac{(p_{n+1} - p_n)}{\rho_n}\right] = \frac{\dot{Q} - \dot{W}}{\dot{m}} - g\,(z_{n+1} - z_n) \quad (4.59)$$

Leaving the unknown temperature on one side, expressed as function of the known parameters, it renders,

$$T_{n+1} = T_n - (1 - \beta T_n)\,\frac{(p_{n+1} - p_n)}{c_p \rho_n} - \frac{g}{c_p}\,(z_{n+1} - z_n) + \frac{\dot{Q} - \dot{W}}{c_p \dot{m}} \quad (4.60)$$

To introduce the pressure variation in the calculation and taking Bernoulli into account, omitting the velocity,

$$-\frac{p_{n+1} - p_n}{\rho_n} = g(z_{n+1} - z_n) + \frac{8 f_n}{\pi^2}\frac{\dot{m}^2}{\rho_n^2}\frac{\Delta x}{D_{int,n}^5} \quad (4.61)$$

Since there is not work done in the pipeline, one can discard this term in the energy conservation equation, and the heat term is composed by the two heat contributions mentioned before,

$$T_{n+1} = T_n + \frac{(1 - \beta T_n)}{c_p}\left[g(z_{n+1} - z_n) + \frac{8 f_n}{\pi^2}\frac{\dot{m}^2}{\rho_n^2}\frac{\Delta x}{D_{int,n}^5}\right] - \frac{g}{c_p}\,(z_{n+1} - z_n) + \frac{\dot{Q}_1 + \dot{Q}_2}{c_p \dot{m}} \quad (4.62)$$

The authors have evaluated a total transmission coefficient $U$ for the exchanged heat $Q_1$ from the environment, generally from the ground, with temperature $T_{ext}$ to the fluid system with temperature $T_n$. Replacing now $Q_1$ and $Q_2$ in the previous equation,

$$T_{n+1} = T_n + \frac{(1 - \beta T_n)}{c_p}\left[g(z_{n+1} - z_n) + \frac{8 f_n}{\pi^2}\frac{\dot{m}^2}{\rho_n^2}\frac{\Delta x}{D_{int,n}^5}\right] - \frac{g}{c_p}\left(z_{n+1} - z_n\right) + \ldots$$
$$+ \frac{U_n \pi D_n \Delta x\,(T_{ext} - T_n)}{c_p \dot{m}} + \frac{K_\mu \frac{8 f_n}{\pi^2}\frac{\dot{m}^3}{\rho_n^2}\frac{\Delta x}{D_{int,n}^5}}{c_p \dot{m}} \quad (4.63)$$

$$T_{n+1} = T_n + \frac{(1 - \beta T_n)}{c_p} g(z_{n+1} - z_n) + \frac{(1 - \beta T_n)}{c_p}\frac{8 f_n}{\pi^2}\frac{\dot{m}^2}{\rho_n^2}\frac{\Delta x}{D_{int,n}^5} - \ldots$$
$$- \frac{g}{c_p}\,(z_{n+1} - z_n) + \frac{U_n \pi D_n \Delta x\,(T_{ext} - T_n)}{c_p \dot{m}} + \frac{K_\mu}{c_p}\frac{8 f_n}{\pi^2}\frac{\dot{m}^2}{\rho_n^2}\frac{\Delta x}{D_{int,n}^5} \quad (4.64)$$

Although the specific heat at constant pressure varies greatly with temperature, since the temperature of the fluid varies little from one segment to another, this variable can be considered as constant, within a certain range.

$$T_{n+1} = T_n - \frac{\beta T_n}{c_p} g(z_{n+1} - z_n) + \frac{\left(1 + K_\mu - \beta T_n\right)}{c_p} \frac{8 f_n}{\pi^2} \frac{\dot{m}^2}{\rho_n^2} \frac{\Delta x}{D_{int,n}^5} + \ldots$$
$$+ \frac{U_n \pi D_n \Delta x \left(T_{ext} - T_n\right)}{c_p \dot{m}} \tag{4.65}$$

$$T_{n+1} = T_n - \frac{\beta T_n}{c_p} g(z_{n+1} - z_n) + \frac{\Delta x}{c_p \dot{m}} \left[ \left(1 + K_\mu - \beta T_n\right) \frac{8 f_n}{\pi^2} \frac{\dot{m}^3}{\rho_n^2} \frac{1}{K_D^5 D_n^5} + \ldots \right.$$
$$\left. + U_n \pi D_n \left(T_{ext} - T_n\right) \right] \tag{4.66}$$

Once the temperature and the pressure have been calculated for the section $n + 1$, now one can calculate the fluid properties, i.e., the density and viscosity, for the aforementioned conditions. Moreover, also the new diameter $D_{n+1}$ can be calculated, using the equations from Chap. 2. With all these parameters known, the section calculation can be finalized by obtaining the new velocity for the section $n + 1$, using the conservation of mass,

$$v_{n+1} = \frac{4\dot{m}}{\rho_{n+1} \pi D_{int,n+1}^2} \tag{4.67}$$

**Important: To keep in mind**

At the end of the calculation by segments for a section between pumping plants, it must be verified that the arrival pressure at the subsequent pumping plant is similar to that found in the optimization calculation, or what is the same, that the effective power of pumping plants is similar to that used in the optimization procedure.

If the powers of the pumping plants were not similar, the problem is surely related to the viscosity and the average density adopted for the calculation of the flow. If large differences are found, perhaps the optimization calculation should be redone with a different mean viscosity and density and a new analysis of alternatives.

### 4.6.1 Summary of Equations for the Calculation by Segments in Pipelines for Liquids

In summary, it has been seen that the pipe deforms elastically and can expand with temperature, changing its diameter; there is also a heat transmission from the medium to the fluid, or vice versa. Friction generates additional heat that modifies the temperature of the fluid (only a part of the heat due to friction), and the enthalpy variation due to fluid expansion is considered. The length $\Delta x$ should be set in such a way that it is a segment "sensitive" to changes in density and viscosity due to changes in temperature.

---

**Algorithm 4.2:** Calculation by segments

---

**Data**: $\dot{m}, L, p_1, p_2, H, \Delta x, \rho(p,T), \mu(p,T), c_p, D_{p,T}, \chi$

**Result**: Pump stations located, temperature profile known, power for each station

Calculate number of segments $\rightarrow n_s = \frac{L}{\Delta x}$;

Calculate entry parameters to the pipeline;

**for** $i = 1$ **to** $n_s$ **do**

- Get new pressure with Eq. 4.45, $p_{i+1} = p_i + g\rho_i(z_i - z_{i+1}) - \frac{8f_i}{\pi^2}\frac{\dot{m}^2}{\rho_i}\frac{\Delta x}{D^5_{int,i}}$;
- Calculate overall heat transfer coefficient (See Chap. 2), $U_i$;
- Get new temperature at the end of the segment with Eq. 4.66,

  $$T_{n+1} = T_n - \frac{\beta T_n}{c_p} g(z_{n+1} - z_n) + \frac{\Delta x}{c_p \dot{m}}\left[\left(1 + K_\mu - \beta T_n\right)\frac{8f_n}{\pi^2}\frac{\dot{m}^3}{\rho_n^2}\frac{1}{K_D^5 D_n^5} + \ldots + U_n \pi D_n \left(T_{ext} - T_n\right)\right];$$

- Calculate the new diameter with Eq. 2.75, $D_{i+1}$;
- Calculate new density and viscosity with Eqs. 1.9 and 1.10, $\rho_{i+1}$ and $\mu_{i+1}$, respectively;
- Calculate new velocity with Eq. 4.67, $v_{i+1} = \frac{4\dot{m}}{\rho_{i+1}\pi D^2_{int,i+1}}$;
- Calculate new Reynolds and friction factor;
- **if** $p_{i+1} \leq p_2$ **then**
  - Locate a pumping station in this coordinate;
  - $p_{i+1} = p_1$;
  - $T_{i+1} =$ Eq. $4.66 + \Delta T_{pump}$;
  - Calculate real power for the pump;
- **else**
  - No need to locate a pump after this segment;

---

### 4.6.2 Control Points

A situation to consider is the conjunction between the pressure and the altimetry of the pipeline, especially in the case of transporting volatile fluids. In effect, at specific points in a pipeline line, away from the pumps and at high points of the route, the

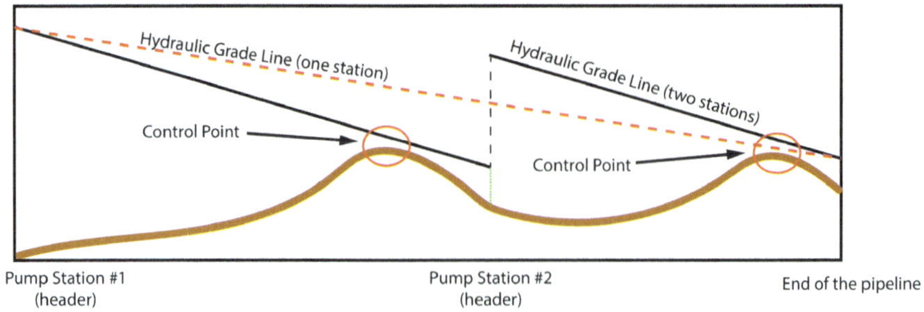


**Fig. 4.7** Simplified cut scheme of a pipeline and the ground level, showing points of attention in the design, which should be verified during the calculation by segments

remaining internal pressure in the line should be verified, since it could happen that said pressure drops below the vapor pressure of the liquid, at the temperature that flows at that point, and a vapor bubble is produced that restricts the passage of the liquid. These high points on the line are known as *control points* and must be verified before fully defining the operating pressure (see Fig. 4.7).

## 4.7 Conclusions Relating to the Optimal Calculation

As seen and discussed in the previous sections, to focus on the problem, it is first necessary to carry out a limited market investigation in order to obtain all the costs of the elements that intervene in the installation (i.e. materials, labor and equipment) and then develop adequate mathematical regressions to obtain the cost functions required by the method.

Next, the equation that technically conditions the project must be considered, in the present case the steady-state Bernoulli flow equation for incompressible fluids in closed lines is used, in which all the variables to be determined are involved. Subsequently, with the total cost function, the minimum cost conditioned to the flow equation is sought, using the Lagrange Multipliers Method. Then the optimum values must be adjusted to practical values, i.e., diameter and (integer) number of pumping plants; finally, execute the detailed verification of the flow by segments.

Indeed, once the number of pumping plants has been adjusted to an integer, the sections between them (all of equal length) and the outlet and arrival pressure in them are determined. The verification by small segments incorporates the effects of the gradual variation of pressure and temperature within the pipeline section as the flow advances, and also the height difference between the ends of the segment. This detailed calculation by segments allows adjusting the optimal project in regards to the location of the pumping plants and the powers involved.

The proposed method aims to meet the requirements of effective use of resources with rationality and effective calculation possibilities, since most of the data used is usually available in the companies of the sector or industry in question, although some of them eventually require, as already stated, a limited market investigation.

Given the major importance of liquid transportation systems through pipelines within the oil industry, and also other liquids in general, particularly water, the efforts made to achieve the greatest economic rationality at the time of projecting will never be enough. As a closing remarks, this method is not the first treatment given to the subject. The optimization methods applied to the transport of fluids and in the hydrocarbons industry in particular, are already long-standing. However, this new approach is that of an updated review in its mathematical language and fundamentally in the number of variables optimized simultaneously.

## 4.8 Verification Procedure for Centrifugal Pumps Applied to Oil Pipelines

### *4.8.1 Introduction*

This calculation process should be included as part of the verification program for an oil pipeline or any pipeline that transports a liquid other than water. It is known that pumps for the impulsion of various fluids are tested in the factory using water as the test fluid, under specific testing conditions (Fig. 4.8). Their test data, operation curves of the pump, must then be corrected if the product to be driven has a higher viscosity than water (Moran 2016).

Once the optimal parameters $D$, $p_1$ and $\chi$ for a given flow rate $Q$, and a fluid of density $\rho$, viscosity $\mu$, and temperature $T$, the height and flow required for the pumps to be installed in the respective pumping plants are determined. The specifications of the equipment offered by the industry must also be verified to determine the number of equipment to operate per pumping plant and its configuration, i.e., series and/or parallel.

To define which type of pump is best adapted to a certain type of fluid, the specific speed ($n_S$) is not very helpful because said number is obtained by applying the similarity to turbo-machines, which does not take into account the effect of viscosity (Fig. 4.9). Being the $n_S$ in its dimensionless expression (for SI units),

$$n_S = \omega \frac{Q^{\frac{1}{2}}}{(gH)^{\frac{3}{4}}} \tag{4.68}$$

When the viscosity increases with respect to certain limits, centrifugal pumps become expensive. Although for positive displacement pumps (D+) there are no similarity criteria, it is possible to give some guidelines. For high viscosity ($\mu > 600$ mPa·s), D+ are applied and for low viscosities ($\mu < 150$ mPa·s) centrifugal or diagonal type are used. For the intermediate zone, both types work and then you have to look at performance and price (Sery et al. 2006).

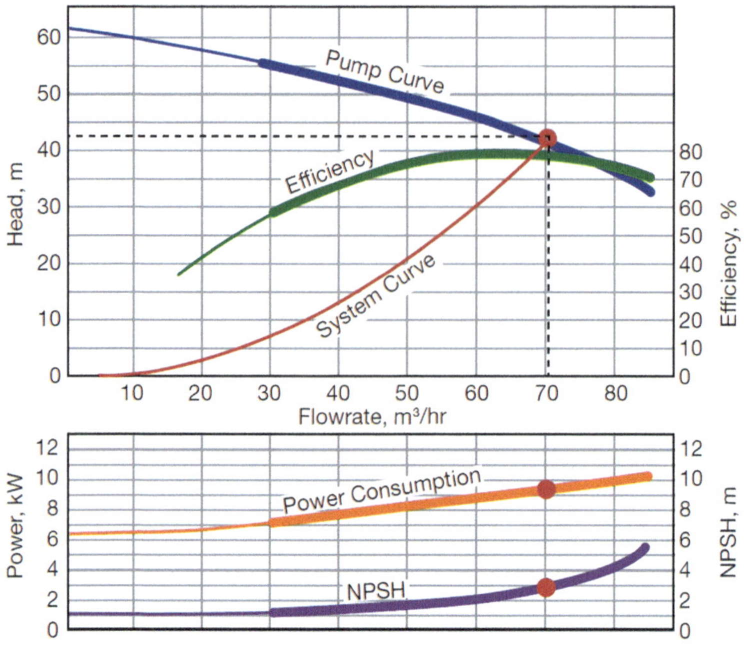


**Fig. 4.8** Characteristic curves for a centrifugal pump: the pressure or head (dark blue), efficiency (green) and typical installation or system curve (in red) (top); and the power consumed (orange) and the net positive suction head (purple) (bottom), all as a function of the flow rate (Image courtesy of Grundfos International) (Grundfos Ecademy 2014)

For oil/multi-product pipelines, both types can work well, but since oil in general is not a clean fluid, it must be filtered very well to use D+. Then, for normal (dirty) oil it is necessary to leave larger gaps and then the internal leaks lower the volumetric efficiency. For centrifugal pumps, the performance is a function of the hydraulic, volumetric and mechanical performance, and almost all of them are a function of the $n_S$ (see graphs at the end of this note); while in D+ it is only a function of volumetric and mechanical performance (hydraulic is equal to 1).

In the event that the pumps are going to be operated with variable rotation speed such as thermal engines, electric motors with electronic frequency inverter, or variable couplings of a mechanical or hydraulic type, the following calculation sequence is carried out:

1. A pump is pre-selected for certain $Q$ (flow), $H$ (loading height), $\eta$ (performance), and $\omega$ (speed of rotation) for water, and the curves are corrected for the fluid of different (higher) viscosity, generally not coincident with the calculation $Q - H$ operating point.
2. By applying similarity laws, the curves for the viscous fluid are corrected and thus obtain the operating point, thus calculating the necessary angular velocity $\omega$.

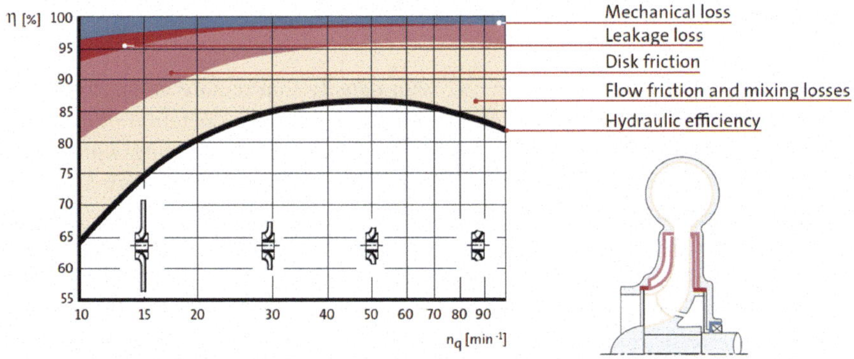


**Fig. 4.9** Plot showing the different types of losses in centrifugal pumps as a function of their specific numbers (Ludwig et al. 2003)

3. With the corrected curves, one obtains new $Q$, $H$, $\eta$ for the viscous fluid, and the power $N$ and the temperature increase when crossing the pump $\Delta T$ are calculated (see Sect. 4.8.4).
4. With the $\Delta T$ obtained, the characteristics of the fluid for the temperature and pressure at the discharge of the pumping plant are calculated and the application of the methodology by segments continues until reaching the other pumping plant, reaching the fluid with $p_2$, $T_2$, $\mu_2$ and $\rho_2$.
5. The pump curves are corrected for viscosity with $\mu_2$, obtaining $\eta$ and $N$. To operate the pump at the required operating point, $\omega$ is calculated.
6. With the new $\eta$ and $N$, the new $\Delta T$ to cross the pump is calculated and the sequence is continued by segments until completing the $\chi$ pipeline pumping plants.

## *4.8.2 Consideration of the Driven Fluids*

The liquid propelled by a centrifugal pump affects: (a) the column and capacity at which the equipment can operate; (b) the power required; and (c) the materials that ensure an adequate useful life. In the problems of pumping homogeneous fluids, three types of relevant substances are distinguished (beyond water): (a) viscous liquids, (b) volatile liquids, and (c) acid or corrosive chemical products (Hydraulic Institute 2010).

**Viscous Liquids**

Viscosity is one of the properties along with density that require consideration in the selection of pumping systems. Viscous liquids affect the operation of centrifugal pumps in three ways, namely: smaller head than when operating with water; lower capacity (flow rate); and higher power required (decreased performance due to increased friction).

The Hydraulic Institute (HI of New Jersey, USA) has published a graph that shows the correction factors based on three coefficients: $c_H$, $c_Q$, and $c_E$, all less than 1.0, that apply to the head, flow rate and performance of the pump, respectively. All these factors are derived from the curves for operation with water (Hydraulic Institute 2010).

**Volatile Liquids**

Fluids that easily vaporize at normal temperatures and atmospheric pressures are called "volatile" (e.g., gasoline, kerosene). The main problem when pumping volatile liquids is cavitation in the pump suction. To verify the operation, it is necessary to know both the detail of the suction installation and the NPSH (Net Positive Suction Head) test curve of the pump, as well as the vapor pressure of the fluid at the operating temperature.

**Chemical Products**

Acidic and/or corrosive fluids are difficult to pump because they attack or corrode various parts of the pump, reducing its life unless proper precautions are taken in the construction of the equipment. There are many factors that influence the selection of the construction materials of the pump: fluid temperature, head per stage, discharge pressure, resistance to corrosion, suspended solids, electrochemical action, type of seals, etc.

### *4.8.3 Effect of Viscosity on the Impeller of a Centrifugal Pump*

Centrifugal pumps are normally selected based on the operating curves reported by the manufacturers for the model in question. These curves are the representation of the impeller performance indicative of the change in the discharge head (pressure) versus the capacity (flow rate) of the pump for a given rotation speed.

These performance curves are derived from laboratory tests using clean water as the test medium. Obviously this does not represent a problem when it comes to

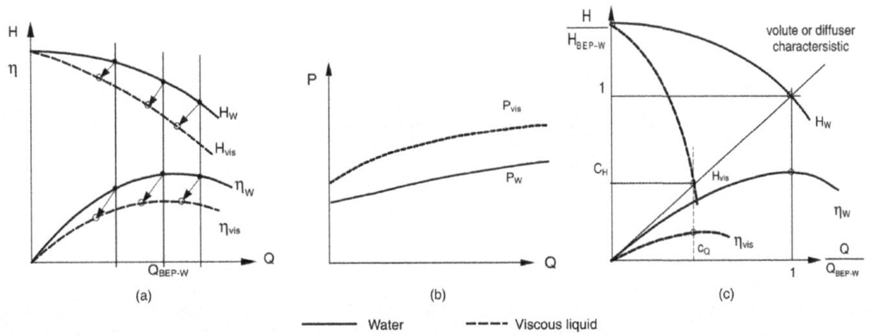


**Fig. 4.10** Modification of pump characteristic curves when pumping viscous liquids (Courtesy of the Hydraulic Institute and Pump Systems Matter, www.pumps.org)

applications for water or other fluids with properties close to those of water. However, when the properties of the fluid to be pumped deviate significantly from those of the water, it is necessary to make corrections to the aforementioned curves, basically in the case of pumping viscous liquids (Fig. 4.10).

In industry, two different techniques are used to consider the influence of the viscosity in the pump's operational behavior: the Ingersoll-Rand method and the approach developed by the Hydraulic Institute Standards (HIS). This last one has developed a analytic/graphical approach to correct the aforementioned values for "conventional" centrifugal pumps that handle viscous liquids based on three factors: head ($c_H$), flow rate ($c_Q$), and performance ($c_E$). Since all factors are less than 1, increasing viscosity negatively affects head, flow, and performance.[10]

Calling with subscripts $\mu$ and $w$, the operating characteristics of the pump with viscous liquid and water, respectively, from the graph (from the HIS), having: $Q_\mu = c_Q Q_w$, and $H_\mu = c_H H_w$. For handling large flows ($Q > 50m^3/h$ and up to $2500m^3/h$), the $c_H$ factor depends on the flow corresponding to the point of maximum performance of the curve for water, also known as BOP for the English "Best Operating Point", that is, the optimum flow rate of the pump for the speed of rotation of the test (Figs. 4.11 and 4.12).

Relating the flow of viscous fluid with the optimal flow of the test with water, a fraction or multiple of that optimal flow will be obtained: 0.6/0.8/1.0/1.2. Hence, there are four correction curves. If you have a different flow and height, you have a new curve, but the origin for ($Q = 0$) does not change. In fact, the preset relationships between the new flow rate and the optimum flow rate allow drawing the new curves without testing the pump.

Something similar happens with performance, $\eta_\mu = c_E \eta_w$. The correction must be considered in stages (i.e. for each rotor). For example, if the pump has 4 stages (series) and assuming $H = 600$ m, the correction must be made for a height of

[10] The methodology presented in this book is only to serve as an example of the HIS and it corresponds to the 2004 edition. The reader is encourage to access later and newer edition of this standard for the updated procedures.

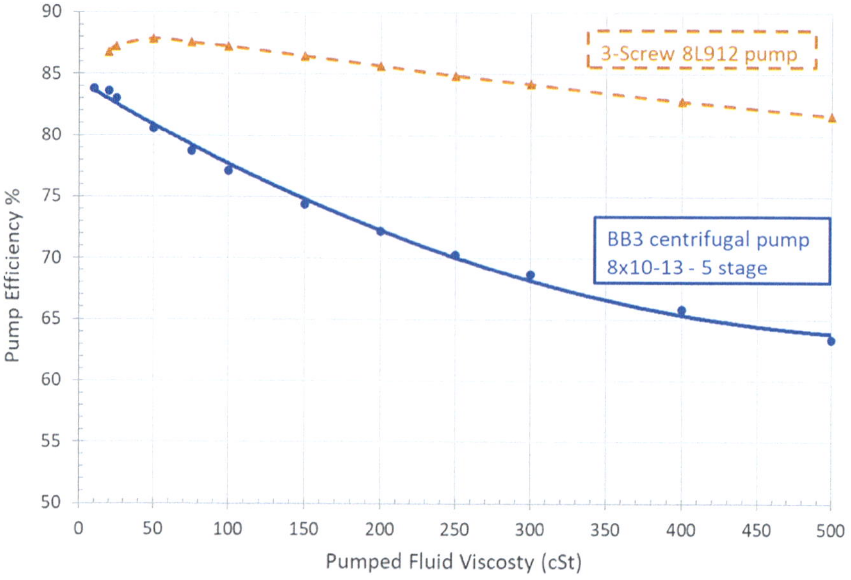


**Fig. 4.11** Plot of the pump's efficiency as a function of the fluid's viscosity (Images courtesy of CIRCOR International)

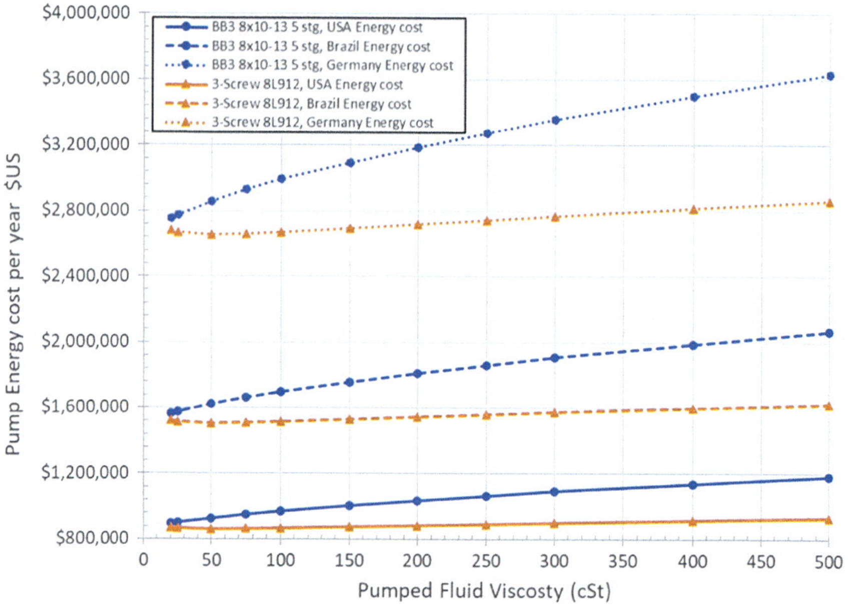


**Fig. 4.12** Operational costs as a function of the fluid's viscosity in US Dollars per year (Images courtesy of CIRCOR International)

150 m. HIS has developed as well an analytical method to calculate the correction using a series of equations, described hereunder. The first step consists in calculate a factor $B$, which considers the properties of the fluid to be pumped based on the best efficiency point (BEP) of the test with water,

$$B = 16.5\frac{\eta_{\mu}^{0.50}\ H_{BEP,w}^{0.0625}}{Q_{BEP,w}^{0.0375}\ \omega^{0.25}} \tag{4.69}$$

where $\eta_{\mu}$ is the kinematic viscosity of the viscous fluid in centiStokes, $H_{BEP,w}$ and $Q_{BEP,w}$ are the head in meters and flow rate in m$^3$/h, respectively, of the best efficiency point with water and $\omega$ the rotation speed in rpm. If $B > 40$ the technique cannot be applied and more detailed studies (e.g. CFD) should be applied; if $B < 1$ then $c_H = c_Q = 1$. For values in between, the following steps should be carried out.

Then, the first correction if for the flow rate, which in this technique is also equal to the factor used to factor to correct the head in the BEP. The correction coefficient $c_Q$ is valid for all the flow rates.

$$c_Q = 2.71^{-0.165\cdot(\log B)^{3.15}} \quad \rightarrow \quad Q_{\mu} = c_Q Q_w \tag{4.70}$$

$$c_{BEP,H} = c_Q \quad \rightarrow \quad H_{BEP,\mu} = c_{BEP,H} H_{BEP,w} \tag{4.71}$$

After the correction to the flow rate, the next step involved the correction to the head for all the flow rates ($c_H$), based on the water flow rates given in the test.

$$c_H = 1 - \left[\left(1 - c_{BEP,H}\right) \times \left(\frac{Q_w}{Q_{BEP,w}}\right)^{0.75}\right] \quad \rightarrow \quad H_{\mu} = c_H H_w \tag{4.72}$$

The final step consists in the correction of the efficiency taking into account the viscosity of the fluid to be pumped. In this case, the correction factor equation depends on the value of the parameter $B$; finally, with all these known values, one can proceed to calculate the power required to pump the viscous fluid.

$$c_E = \left\{\begin{array}{l} 1 < B < 40 \quad \rightarrow \quad c_E = B^{-\left(0.0547\times B^{0.69}\right)} \\ B \leq 1 \quad \rightarrow \quad c_E = \dfrac{1-\left[\left(1-\eta_{BEP,w}\right)\times\left(\frac{\eta_{\mu}}{\eta_w}\right)^{0.07}\right]}{\eta_{BEP,w}} \end{array}\right\} \quad \rightarrow \quad \eta_{mu} = c_E \eta_w \tag{4.73}$$

The explained methodology assumes starting from the laboratory test values. On the contrary, if the correction is started knowing the head and flow rate of the viscous fluid to be pumped, the approach is quite similar, with the following steps to determine the correction factor and the values of operation parameters for water, in order to know what type of pump should be asked to the manufacturers. Starting with the parameter B,

$$B = 2.80\frac{\eta_{\mu}^{0.50}}{Q_{\mu}^{0.25}\,H_{\mu}^{0.125}} \tag{4.74}$$

One more time, if $B > 40$ this procedure should not be followed; if $B < 1$ then $c_H = c_Q = 1$. For values between 1 and 40, the following steps should be done.

Starting as before with the correction factors for the flow rate ($c_Q$) and it is also included the head as well ($c_H$). According to HIS, these factors are approximately the same at a certain flow rate.

$$c_Q = c_H = 2.71^{-0.165\cdot(\log B)^{3.15}} \quad \rightarrow \quad Q_w = \frac{Q_{\mu}}{c_Q} \wedge H_w = \frac{H_{\mu}}{c_H} \tag{4.75}$$

With these two parameters, engineers are now in conditions to select a pump (for water) with those operating conditions. Then, with the performance of this test, i.e., for water, one can calculate the efficiency correction factor for the viscous fluid, which is done using the equations described before.

The last step in the correction or verification of pump curves when used with fluids different than water is the analysis of the net positive suction head required (NPSHr) by the pump. This depends on the properties of the fluid and of course, of the pressure drop from the inlet to the impeller, which ultimately depends on the viscosity. The NPSHr curve is then also affected by a correction coefficient,

$$c_{NPSHr} = 1 + 274 \times 10^{3}\, F_{inlet}\left(\frac{1}{c_H} - 1\right)\left[\frac{NPSHr_{BEP,w}}{Q_{BEP,w}^{0.667}\,\omega^{1.33}}\right] \tag{4.76}$$

$$NPSHr_{\mu} = c_{NPSHr} NPSHr_{w} \tag{4.77}$$

the $F_{inlet}$ is a factor, which depends on the location or type of suction in the centrifugal pump. For pumps with an end suction the factor is equal to 0.1, whilst for side-suction pumps, the value is equal to 0.5. This is related to the way the fluid follows inside the pump.

**Example: Pump for crude oil**

Given the characteristics required by an oil pumping system and the performance curves of a pump tested with water known, the effect of viscosity and the operating characteristics of said pump applied to the pumping of crude oil must be determined.

- Fluid: medium crude oil (°API 25)
- Density = 901 kg/m$^3$
- Dynamic viscosity = 180 cP
- Kinematic viscosity = 199.8 cSt
- cSt = 199.8 m$^2$/s

For the pump selected and tested, the following parameters are known:

- Fluid tested: water
- Speed of rotation = 3000 rpm
- Flow rate = 227.1 $m^3/h$
- Differential height = 40 m
- Number of stages = 2 $\rightarrow \Delta H = 20$ m
- Maximum performance = 0.75 (BEP)

The power required for water is then,

$$N_w = \frac{\rho_w g Q_w H_w}{\eta_w} = 33\,\text{kW} \tag{4.78}$$

The performance of said pump to drive a viscous fluid, the oil of 180 cP is calculated below. Starting with the parameter $B$,

$$B = 16.5 \frac{\eta_\mu^{0.50} H_{BEP,w}^{0.0625}}{Q_{BEP,w}^{0.0375}\ \omega^{0.25}} = 5.188 \tag{4.79}$$

Since this value is between 1 and 40, one can use the steps described in this book. The first correction coefficient is the flow rate,

$$c_Q = 2.71^{-0.165\cdot(\log B)^{3.15}} = 0.944 \tag{4.80}$$

$$Q_\mu = 0.944 \times 227.1\,\text{m}^3/\text{h} = 214.48\,\text{m}^3/\text{h} \tag{4.81}$$

$$c_{BEP,H} = 0.944 \quad \rightarrow \quad H_{BEP,\mu} = 0.944 \times 40\,\text{m} = 37.76\,\text{m} \tag{4.82}$$

Then one proceeds to calculate the head correction coefficient for flow rates at 60, 80, 100 and 120% of the BEP flow rate with water,

$$c_H = \begin{cases} 60\% & \rightarrow \quad 1 - \left[\left(1 - c_{BEP,H}\right) \times \left(\frac{Q_w}{Q_{BEP,w}}\right)^{0.75}\right] = 0.962 \\ 80\% & \rightarrow \quad 1 - \left[\left(1 - c_{BEP,H}\right) \times \left(\frac{Q_w}{Q_{BEP,w}}\right)^{0.75}\right] = 0.953 \\ 100\% & \rightarrow \quad 1 - \left[\left(1 - c_{BEP,H}\right) \times \left(\frac{Q_w}{Q_{BEP,w}}\right)^{0.75}\right] = 0.944 \\ 120\% & \rightarrow \quad 1 - \left[\left(1 - c_{BEP,H}\right) \times \left(\frac{Q_w}{Q_{BEP,w}}\right)^{0.75}\right] = 0.936 \end{cases} \tag{4.83}$$

Finally, calculating the efficiency correction factor,

$$c_E = B^{-(0.0547\times B^{0.69})} = 0.755 \tag{4.84}$$

The power required to pump the viscous fluid at 100% the BEP for water is then,

$$N_w = \frac{\rho_\mu g Q_\mu H_\mu}{\eta_\mu} = 35.13\,\mathrm{kW} \tag{4.85}$$

### 4.8.4 *Heating of the Fluid in a Pump*

The energy delivered to the pump shaft is much greater than the energy actually received by the fluid and therefore, for practical purposes, this loss is included in the performance of the pump.

However, the energy cannot simply disappear, and therefore, it must be incorporated into the flow and into the equipment itself, in the form of heat (i.e., temperature). If, for example, $\eta_p = 0.7$, then the remaining 0.3 is an energy that goes partly to heating the fluid and the equipment.

For a certain design condition, the speed of the flow and the rotation are in their best condition with respect to the rotor blades. If there are shocks because the flow increases or decreases with respect to the design one, the losses increase. The ideal configuration could be maintained by simultaneously changing the speed of rotation and the flow rate.

Regarding performance, most of the losses occur due to internal "shocks" of the fluid and the vanes, so emphasis must be placed on maintaining the incidence of relative velocity to the vane, with variation in rotation speed, to avoid as much as possible shocks that lead to marked drops in performance. The theoretical power from a pump can be obtained from a control volume analysis around the pump, applying the Reynolds Transport Theorem,

$$\left.\frac{dE}{dt}\right|_s = \frac{dQ}{dt} - \frac{dW}{dt} = \frac{d}{dt}\iiint_{CV}\left(\hat{u} + \frac{v^2}{2} + gz\right)\rho d\forall + \ldots$$
$$+ \iint_{CS}\left(\hat{u} + \frac{p}{\rho} + \frac{v^2}{2} + gz\right)\rho\,(v\cdot n)\,dA \tag{4.86}$$

Considering that the pump is operating at steady-state conditions and that the difference in kinetic energy and potential energies can be neglected, it yields,

$$\frac{dQ}{dt} - \frac{dW}{dt} = \dot{m}\left[\frac{(p_d - p_s)}{\rho} + \left(\hat{u}_d - \hat{u}_s\right)\right] \tag{4.87}$$

For an ideal pumping system, there is no exchange of heat, hence no variation in the internal energy, and then the equation becomes,

$$-\frac{dW}{dt} = \dot{m}\frac{(p_d - p_s)}{\rho} \tag{4.88}$$

The work per unit of time, or power, results negative according to the sign convention since it is a power given to the system, not extracted from the system. However, this is a theoretical value; it is known that there are minor losses of different origins, e.g., mechanical friction in the bearings, pressure loss due to viscosity, etc. All this causes that the real power needed in the pump to be higher than the theoretical. This is accounted for by considering a performance or hydraulic efficiency,

$$N_{real} = \frac{N_{theory}}{\eta_p} \tag{4.89}$$

The losses are transformed into thermal energy, causing a jump in the temperature. The temperature variation due to the pressure change should also be considered, just as in the study per segments for the main pipeline. Since the height difference between suction and discharge is small and can be neglected, and there is no heat exchange with the environment due to temperature difference,

$$\frac{dQ}{dt} - \frac{dW}{dt} = \dot{m}\,(h_d - h_s) \tag{4.90}$$

$$h_d - h_s = c_p\,(T_d - T_s) + \left[(1 - \beta T_d)\frac{p_d}{\rho_s} - (1 - \beta T_s)\frac{p_s}{\rho_s}\right] \tag{4.91}$$

The authors make the real study making the following assumptions: as it happen before, the terms containing the volumetric expansion coefficient $\beta$ will be unified due to the small jump in temperature. Moreover, considering now that the system is receiving both heat and work: the heat will be make of the extra energy/power added due to irreversibilities; and the work will be the (negative) theoretical work deliver to the impeller to pump the fluid. Thus,

$$\frac{N_{theory}}{\eta_p}(1 - \eta_p) + N_{theory} = \dot{m}c_p\,(T_d - T_s) + \dot{m}\left[(1 - \beta T_s)\frac{p_d - p_s}{\rho_s}\right] \tag{4.92}$$

$$\dot{m}\frac{(p_d - p_s)}{\rho_s}\frac{(1 - \eta_p)}{\eta_p} = \dot{m}c_p\,(T_d - T_s) - \dot{m}\beta T_s\frac{p_d - p_s}{\rho_s} \tag{4.93}$$

$$T_d = T_s + \frac{p_d - p_s}{c_p\rho_s}\left(\frac{1 - \eta_p}{\eta_p} + \beta T_s\right) \tag{4.94}$$

Like in the case of the viscous dissipation, a coefficient $K_{\mu,p}$ is added, going from 0 to 1, to let us quantify how much of the energy irreversibly transformed into heat goes to the pump and how much to the fluid.

$$T_d = T_s + \frac{p_d - p_s}{c_p \rho_s}\left(\frac{1-\eta_p}{\eta_p} K_{\mu,p} + \beta T_s\right) \tag{4.95}$$

## 4.8.5 Selection of Centrifugal or Positive Displacement Pumps

An important issue to consider in the design of a liquid transport system through pipelines is to define the type of pumps to be applied, i.e., dynamic type with rotors or volumetric type—pistons, vanes, screws, etc. They are generally referred to as centrifugal pumps and positive displacement (D+) pumps, respectively.

Relying on the specific number of rotations ($n_S$) as the selection parameter does not solve the problem because the $n_S$ is obtained from the application of similarity in centrifugal pumps that had not taken viscosity into account. For positive displacement pumps there are no similarity criteria.

$$n_S = \omega \frac{Q^{0.5}}{(gH)^{0.75}} \tag{4.96}$$

As the viscosity increases centrifugal pumps tend to be expensive. A basic selection guideline could be:

**Important: Pumps and Viscosity**

- High viscosity → Positive displacement (D+).
- Low viscosity → Centrifuges (if the $n_S$ becomes too small, it must be subdivided into many stages and again they are not convenient from the economic point of view).

For pipelines both types of pumps can work well. In the D+, the higher the viscosity, there will be fewer leaks and therefore the volumetric efficiency improves but the mechanical one deteriorates due to the greater viscous friction.

As neither water nor oil are "pure" fluids, the fluid must be well filtered before it enters a type D+ pump due to its operating principle. Filtering well means separating foreign elements, this is, mud, sand, oxides and scale, etc., which is not always possible and then greater clearances must be left and then internal leaks lower the volumetric efficiency (Fig. 4.13).

For $\mu > 600$ cP, the D+ pumps are suitable, but there is an intermediate area where the D+ and the centrifuges both work and you have to look at the efficiencies:

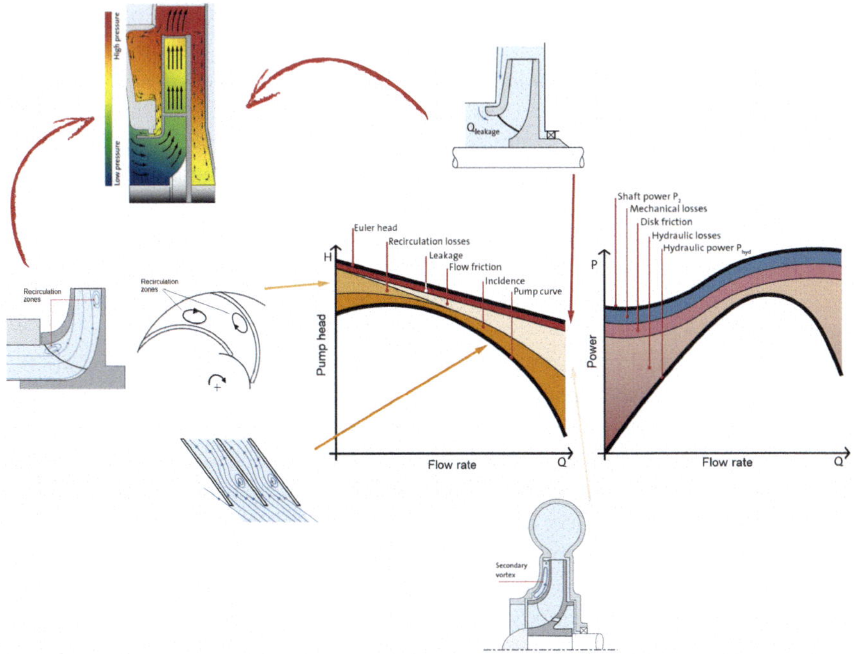


**Fig. 4.13** Scheme depicting the losses in a centrifugal pump, showing their location and general influence on the total loss as a function of the volumetric flow rate (Ayad Hassan 2016; Ludwig et al. 2003; Zhu et al. 2022)

**Important: Pumps and Efficiencies**

- D+ increases the efficiency with an increasing viscosity.
- Centrifuges decrease efficiency with increasing viscosity.

In short, the selection of the type of pumping equipment generally depends on four main parameters: height and flow rate required by the system, and viscosity and dirt of the fluid to be transported.

### *4.8.6 Waterhammer*

There is a very important dynamic aspect to consider in pipelines for the transport of liquids resulting from the consequential circumstances of relatively sudden changes in the flow regime, such as the closure of line valves or the shutdown of a pumping plant.

Waterhammer is caused by the transformation of the kinetic energy into pressure energy of the moving liquid due to a sudden variation in the flow that translates into a very significant local rise in pressure, due to the compressibility of the liquid. and the elasticity of the pipes, which propagates along the conduit creating standing waves (actually damped by the viscosity of the fluid).

It is shown that the propagation speed of the wave, i.e., speed of sound in the liquid, varies inversely with the compressibility. It should be noted that the compressibility is very different between water and hydrocarbons, since in the latter it is twice that of water.

In oil pipelines and multi-product lines for the transportation of hydrocarbons, wave propagation speeds fluctuate between 900 and 1,250 m/s, depending on the type of product and the thickness of the pipes. In long ducts of small or medium diameters, the damping of the wave is fast. On the other hand, D+ pumps, valve closures, and the free surfaces of the liquids in the tanks are wave reflection points, while centrifugal pumps reflect the wave to a much lesser extent.

When a pumping station stops, the high pressure at the discharge causes the check valves to close abruptly and consequently an interruption of the flow, which causes an increase in pressure in sections before the valve and a depression in subsequent sections to the valve.

This propagating wave adds to the operating pressure in the normal regime and high pressures occur in the pipes that could cause them to fail at some point. Some authors differentiate the phenomenon of waterhammer ("pressure surges" or "coups de bélier") from "water-hammer", but both have a similar theoretical treatment.

The primary reason for analyzing pressure fluctuations is to determine if any peak pressures exceed the maximum transient operating pressure (MAPO) at any point in the pipeline. The allowable transient MAPO is generally specified in standards as 110% of the design pressure. If the transient or pressure fluctuation is excessive, actions must be taken to reduce its magnitude.

# Appendix

## *Derivation of the General Formulation*

The system of equations to be optimized is,

$$\begin{aligned} c_T = {} & \kappa_{11}\frac{1+K_D}{2}D^2\frac{p}{2\cdot\sigma_{adm}}\pi g\rho_p L + (\kappa_{12}+\kappa_{13})\pi DL + \ldots \\ & + \kappa_{02}\kappa_2\frac{time}{\eta_{pump}\eta_m}\frac{\dot{m}}{\rho}\left(\rho g H + \frac{8f}{\pi^2}\frac{\dot{m}^2}{\rho}\frac{L}{K_D^5 D^5}\right) + \ldots \end{aligned}$$

$$+\left[\kappa_{03}+\kappa_3\frac{\dot{m}}{\eta_{pump}K_{SB}\rho}\left(\rho g\frac{H}{\chi}+\frac{8f}{\pi^2}\frac{\dot{m}^2}{\rho\chi}\frac{L}{K_D^5 D^5}\right)\right]\chi+\ldots$$

$$+K_4K_{04}K_D^2\frac{\pi D^2}{4}L \quad (4.97)$$

$$\phi(p_1,\chi,D)=(p_1-p_2)-\left(\rho g\frac{H}{\chi}+\frac{8f}{\pi^2}\frac{\dot{m}^2}{\rho K_D^5 D^5}\frac{L}{\chi}\right)=0 \quad (4.98)$$

$$p_1=(1-K_D)\,\sigma_{adm} \quad (4.99)$$

The total cost along with the condition function. The Lagrange Multiplier Method consist in solving the following system of equations,

$$\frac{\partial c_T(p_1,\chi,D)}{\partial D}+\lambda\frac{\partial\phi(p_1,\chi,D)}{\partial D}=0 \quad (4.100a)$$

$$\frac{\partial c_T(p_1,\chi,D)}{\partial\chi}+\lambda\frac{\partial\phi(p_1,\chi,D)}{\partial\chi}=0 \quad (4.100b)$$

$$\phi(p_1,\chi,D)=(p_1-p_2)-\left(\rho g\frac{H}{\chi}+\frac{8f}{\pi^2}\frac{\dot{m}^2}{\rho K_D^5 D^5}\frac{L}{\chi}\right)=0 \quad (4.100c)$$

The equations are then,

$$\left(\frac{1-K_D^2}{2}\kappa_{11}\pi\rho_p g+K_4K_{04}K_D^2\frac{\pi}{2}\right)D^7+(\kappa_{12}+\kappa_{13})\,\pi D^6-\ldots$$

$$\ldots-\left(\frac{\kappa_{02}\kappa_2 time}{\eta_{pump}\eta_m}+\frac{\kappa_3}{K_{SH}\eta_{pump}}\right)\frac{40f\dot{m}^3}{\pi^2\rho^2K_D^5}+\lambda\frac{40f\dot{m}^2}{\pi^2\rho K_D^5\chi}=0 \quad (4.101a)$$

$$\chi=\frac{1}{(p_1-p_2)}\left(\rho gH+\frac{8f}{\pi^2}\frac{\dot{m}^2}{\rho K_D^5}\frac{L}{D^5}\right) \quad (4.101b)$$

$$\kappa_{03}+\frac{\lambda}{\chi^2}\left(\rho gH+\frac{8f}{\pi^2}\frac{\dot{m}^2}{\rho}\frac{L}{K_D^5 D^5}\right)=0 \quad (4.101c)$$

From Eqs. 4.101c and 4.100c,

$$-\kappa_{03}\chi^2\cdot\left(\rho gH+\frac{8f}{\pi^2}\frac{\dot{m}^2}{\rho}\frac{L}{K_D^5 D^5}\right)^{-1}=\lambda\rightarrow-\kappa_{03}\chi^2\cdot[(p_1-p_2)\,\chi]^{-1}=\lambda$$

$$(4.102)$$

$$-\kappa_{03}\chi\cdot(p_1-p_2)^{-1}=\lambda \tag{4.103}$$

Replacing this in Eq. 4.104,

$$\left(\frac{1-K_D^2}{2}\kappa_{11}\pi\rho_p g+K_4K_{04}K_D^2\frac{\pi}{2}\right)D^7+(\kappa_{12}+\kappa_{13})\pi D^6-\ldots$$
$$\ldots-\left(\frac{\kappa_{02}\kappa_2 time}{\eta_{pump}\eta_m}+\frac{\kappa_3}{K_{SH}\eta_{pump}}\right)\frac{40f\dot{m}^3}{\pi^2\rho^2K_D^5}-\frac{\kappa_{03}}{\left[(1-K_D)\sigma_{adm}-p_2\right]}\frac{40f\dot{m}^2}{\pi^2\rho K_D^5}=0 \tag{4.104}$$

Which is the same equation presented in the main body of the chapter.

## References


Alkhomairi A, Jung BS. Comprehensive optimization of project cost for long supply pipelines. J Engin Res. 2021;9(3A), 2021. https://doi.org/10.36909/jer.v9i3A.8857.

Ayad Hassan AFS. Experimental and computational study of semi-open centrifugal pump. LAP LAMBERT Academic Publishing; 2016.

Babuk T, Mohitpour M, Jenkins A, Murray MA Pipeline cost reduction through effective project management and applied technology. 2005.

Brackett C, Zietlow D. Optimal design of a pump and piping system. In: 2011 ASEE annual conference and exposition. 2011. p. 22–1126.

Cookenboo L Jr. Costs of operating crude oil pipe lines. Rice Inst Pamphlet-Rice Univ Stud. 1954;41(1).

Gill K. Alaska oil pipeline - wikimedia commons cc by-sa 4.0. https://commons.wikimedia.org/wiki/File:Alaska_Oil_Pipeline_141.jpg. 2018. [Accessed 03 June 2024].

Grundfos Ecademy. Basic principles and pump types course - about pump curves. 2024. https://www.grundfos.com/solutions/learn/ecademy/all-courses/basic-principles-and-pump-types/about-pump-curves. [Accessed 04-06-2024].

Hafner M, Luciani G. The Palgrave handbook of international energy economics. Springer International Publishing; 2022. ISBN 9783030868840. https://books.google.nl/books?id=XdxxEAAAQBAJ.

Institute Hydraulic. Effects of liquid viscosity on rotordynamic (centrifugal and vertical) pump performance. ANSI-HI. 2010;9(6):7–2010.

Jokić A, Zavargó Z. Optimization of pipeline network for oil transport. Hungarian J Ind Chem. 2011;113–17.

Ludwig G, Meschkat S, Stoffel B. Design factors affecting pump efficiency. In: Energy efficiency in motor driven systems. Springer; 2003. p. 532–38.

McAllister EW, editor. 3 - pipe design. In: Pipeline rules of thumb handbook, 8th edn. Boston: Gulf Professional Publishing; 2014. p. 93–123. ISBN 978-0-12-387693-5. https://doi.org/10.1016/B978-0-12-387693-5.00003-6. https://www.sciencedirect.com/science/article/pii/B9780123876935000036.

Menon ES. Gas pipeline hydraulics. Taylor & Francis; 2005. ISBN 9780849327858. https://books.google.nl/books?id=uFpHngEACAAJ.

Menon ES, Menon PS. Working guide to pumps and pumping stations: calculations and simulations. Gulf Professional Publisher; 2010. ISBN 9781856178280. https://books.google.nl/books?id=YUyrPwAACAAJ.

Mohitpour M, Golshan H, Murray MA. Pipeline design and construction: a practical approach. ASME Press: Pipelines and Pressure Vessels; 2007.

Moran S. Pump sizing: bridging the gap between. The best of equipment series. 2016. p. 3.

Sery G, Kabamba B, Slatter P. Paste pumping with centrifugal pumps: evaluation of the hydraulic institute chart de-rating procedures. In: Paste 2006: proceedings of the ninth international seminar on paste and thickened tailings. Australian Centre for Geomechanics; 2006. p. 403–12

Sharma P, Shelza GSR, Virk RS. Cost optimization of pipeline systems using genetic algorithm. Int J Comput Engin Technol (IJCET). 2013;4(4):485–94.

Shashi Menon E, Pramila PE, Menon S. Liquid pipeline hydraulics: second edition. Trafford Publishing; 2013. ISBN 9781466977402. https://books.google.nl/books?id=e6VbRPl6PzEC.

Singh RP, Mahar PS. Optimal design of multidiameter, multioutlet pipelines. J Water Resourc Plan Manag. 200;129(3):226–33.

Tee KF, Khan LR, Chen HP, Alani AM. Reliability based life cycle cost optimization for underground pipeline networks. Tunnell Underground Space Technol. 2014;43:32–40. ISSN 0886-7798. https://doi.org/10.1016/j.tust.2014.04.007. https://www.sciencedirect.com/science/article/pii/S088677981400056X.

Zhu H, Zhu J, Zhang H-Q. Mechanistic modeling of gas effect on multi-stage electrical submersible pump (esp) performance with experimental validation. Chem Eng Sci. 2022;252: 117288.

# Chapter 5
# Transport of Various Liquid Substances in Sequence Through Pipeline Systems

> Any idiot can build a bridge that stands, but it takes an engineer to build a bridge that barely stands.
>
> *Unknown*

## 5.1 Introduction

Although the ideal would be to transport a single product at a time, in practice it is often necessary for different products to circulate through the same pipe over time. These lines, called multi-product pipelines, are configured and designed in this way. The design of a multi-product pipeline can also be optimized in order to find a pipe diameter fulfilling all the technical requirements while minimizing the costs involved.

In general, it is not convenient to install different pipes for different products in the same route, and it is better to use a single duct, since this is most cost-effective. As the internal volumes are generally very large and the pipeline cannot be filled with a single product and then they are pumped sequentially so that one displaces the other. For instance, multiple hydrocarbon products such as Diesel, Gasoline, Jet A-1 can be transported in a multi-product line from the refinery to the tank farm for further transportation. This is most notorious feature of a multi-product line, the transport of several products or commodities in what is called "batches". Nevertheless, this also brings associated one of the most difficult challenges in multi-product pipelines: the batch tracking and managing (see Fig. 5.1) (Cafaro et al. 2010; Cafaro and Cerdá 2010).

The central thing is to define the lengths that the products would occupy in the pipeline and maintain the proportion, and that proportion is given by the volumes. If a diameter is estimated for the pipeline, it does not matter if the sum of the individual volumes,transformed into individual lengths, is greater than the total length of the pipeline, because what matters is the proportion. Regarding the flow, it should be



P. Druetta et al., *Transport by Pipelines and Its Optimization*, Petroleum Engineering,
https://doi.org/10.1007/978-3-031-66566-0_5

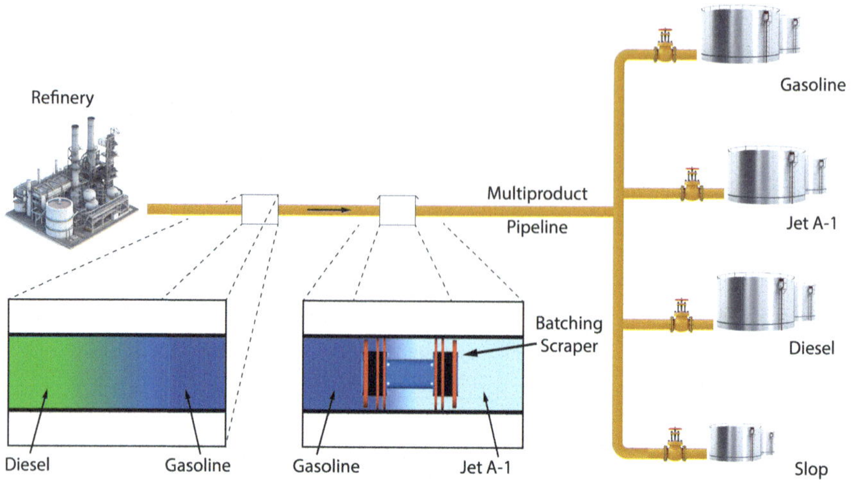


**Fig. 5.1** Scheme showing the transportation of a multi-product line with the two possible separation schemes: fluid mixing and scrapers

noted that it is basically a very slow transitory regime (conditions may take hours or days to vary) and then it can be studied as a permanent regime. In fact, it is a transitory regime with long times, that is why the quasi-static approach can be used.

For the purposes of the project, the volume of each product to be transported at a time is important, that is, the volume to be transported per batch of each product, size of the batches in volume and length, the sequence of the different products within the pipeline and the volume to be transported (e.g. per month, bi-monthly) of each product. These are very important topics in engineering research, involving the optimization and sequencing of the products to be transported. These include mathematical models applied to the production, transport and delivery in order to minimize the operational costs. The reason for this is associated with a topic that is developed in this chapter and it is related with the interface between the products, losing a certain amount of product due to the interface between different commodities.

## 5.2 Sequence of Pure Substances or Homogeneous Mixtures

In order to develop the optimization methodology in this instance for multi-product pipelines, the chapter will be dedicated to the consideration of the flows of different pure liquid substances or homogeneous mixtures of liquids that are miscible with each other (i.e. forming solutions) transported in batches, one after the other. In both cases, what is required is to be able to know the parameters of each pure fluid or of each mixture, in both cases as unique data—for each fluid in question- at the established transport temperatures and pressures (Tu et al. 2023).

Initially, it is required to obtain the greatest amount of data on the fluids and the flow (e.g. viscosity, density, vapor pressure) and their variations with respect to temperature and pressure. If laboratory tests are not available, it is highly recommended to perform test at least to determine correlations for the variation of density and viscosity, given the implication that these parameters have on the development of the optimal design.

In effect, as the calculation of the losses is very sensitive to the aforementioned parameters, the adopted values for the calculation play a major role in the design process, both dependent mainly on the temperature of the flow and somewhat less on the pressure. In the case of transporting very different liquids in sequence through the same pipe, it will be necessary to generate a "model" fluid that precisely makes it possible to carry out the optimization calculations with a certain simplicity.

## 5.3 Definitions of the Parameters of the Model Fluid from Single Substances or Homogeneous Mixtures in Sequence

For the optimization calculations, the properties of each fluid cannot be taken separately, but pseudo-mean values for the density and viscosity of the model fluid should be used. A similar approach is followed for the friction factor. Thus, for each product a series of initial conditions will be given, in terms of:

- Volumes: $V_1, V_2, V_3, \cdots, V_i$;
- Densities: $\rho_1, \rho_2, \rho_3, \cdots, \rho_i$;
- Viscosities: $\mu_1, \mu_2, \mu_3, \cdots, \mu_i$;
- Friction Factors: $f_1, f_2, f_3, \cdots, f_i$;

If the volumes $V_i$ are given in m$^3$/month (usually these quantities are incorporated into the pipeline in several batches per month until completing the volume per month required), then the total volume occupied by the products can be written in terms of the pipeline diameter,

$$\sum_i V_i = \sum_i \frac{\pi D_{int}^2}{4} L_i = \frac{\pi D_{int}^2}{4} \sum_i L_i, \qquad \sum_i L_i = L_{T,i} \tag{5.1}$$

Being that $L_{T,i}$ and $L_i$ have to do with the total length of the batches and not with the length $L$ of the pipe,

$$\sum_i m_i = \sum_i \rho_i V_i = \frac{\pi D_{int}^2}{\rho_m L_t} = \sum_i \frac{\pi D_{int}^2}{4} \rho_i L_i = \frac{\pi D_{int}^2}{4} \sum_i L_i \quad \rightarrow \quad \rho_m = \frac{\sum_i \rho_i L_i}{L_{T,i}} \tag{5.2}$$

This, seen from the point of view of the pipeline, renders the same result when analyzing the masses contained therein,

$$\rho_m A_p L_p = \sum_i \rho_i A_p L_i \quad \rightarrow \quad \rho_m = \frac{\sum_i \rho_i L_i}{L_{p,i}} \tag{5.3}$$

However, the main question is how to know the length that each batch occupies. It then makes sense to express the length of each product as a function of the mass flow rates inside the pipe,

$$\sum_i \dot{m}_i t = \dot{m}_m t \tag{5.4}$$

where $t$ is a random time during which the system is studied. In other words, considering the previous equations, it is the same to write the length proportional to the flow, taking the length of each product $L_i$ proportional to each flow rate, obtaining the following expression,

$$L_i = \frac{\dot{m}_i}{\rho_i A_p} t \quad \rightarrow \quad \rho_m = \frac{\sum_i L_i \rho_i}{\sum_i L_i} \tag{5.5}$$

Using the volumetric flow rates, this can also be written as,

$$L_i = \frac{Q_i}{A_p} t \quad \rightarrow \quad L = \sum_i \frac{Q_i}{A_p} t \quad \rightarrow \quad \frac{L_i}{L} = \frac{\frac{Q_i}{A_p} t}{\sum_i \frac{Q_i}{A_p} t} = \sum_i Q_i \quad \rightarrow \quad L_i = L \frac{Q_i}{\sum_i Q_i} \tag{5.6}$$

And the total time to fill in the pipeline is,

$$t = \frac{L_T}{\sum_i Q_i} \tag{5.7}$$

Each product requires a flow rate, occupying a length, being its main characteristics for the purposes of flow, and then the average values of density and viscosity will be:

$$\rho_m = \frac{\sum_i \rho_i L_i}{\sum_i L_i} \tag{5.8a}$$

$$\mu_m = \frac{\sum_i \mu_i L_i}{\sum_i L_i} \tag{5.8b}$$

Considering also that for each product the viscosity $\mu_i$ and the friction factor $f_i$ are related to the pressure losses according to the following expression,

$$\Delta p_i = f_i \rho_i \frac{v_i^2}{2} \frac{L_i}{D_{int}} \quad \rightarrow \quad \Delta p_T = \sum_i \Delta p_i \tag{5.9}$$

Accepting that the velocity $v_i$ and the diameter $D_{int}$ are the same for all the products (because the mass flow is not constant, but the volume flow is),

$$\Delta p_T = f_m \rho_m \frac{v^2}{2} \frac{L}{D_{int}} \quad \text{and} \quad f_m = \frac{\sum_i f_i \rho_i L_i}{\sum_i rho_i L_i} \tag{5.10}$$

Assuming the same speeds for all products, one can ensure the total flow. In general, multi-product lines are operated by constant speed, and therefore with slow pressure fluctuations. If one assumes a laminar regime, the friction factor is obtained from the analytical solution to the differential equations $f = 64/Re$ and then,

$$\Delta p_T = \frac{32V}{D_{int}^2} \mu_m \sum_i L_i \tag{5.11}$$

while in transition and turbulent regimes,

$$\Delta p_T = \frac{8 f_m}{\pi^2 g} \frac{Q_2}{D_{int}^5} \sum_i L_i \tag{5.12}$$

Subsequently, for the verification of the system by segments, as the situation changes over time, that is, the position of the batches of different products in the pipeline, each one with its own length, it is necessary to analyze and verify different situations. In the multi-product pumping plants, the pumps are going to pump different products, variables in time. In any case, for the optimization calculation, a "pseudo-average" efficiency must be found, the pumping power being,

$$N = \frac{Q \Delta p}{\eta_p} = Q \sum_i \frac{\Delta p_i}{\eta_{p,i}} \tag{5.13}$$

Eliminating the parameters that remain constant and solving for the global efficiency $\eta_{p,m}$ (not to be confused with the efficiency $\eta_m$ of the pump driver in the energy cost equation), it results:

$$\eta_{p,m} = \frac{f_m \rho_m L}{\sum_i \frac{f_i \rho_i}{\eta_{p,i}} L_i} \tag{5.14}$$

Where $\eta_{p,i}$ is the maximum value of the pump efficiency for each product that passes through it, corrected for a different viscosity than the test one, if applicable.

Finally, it will be necessary to consider the cost of the product based on the cost of each one of them, since they could be several commodities that are very different in characteristics and price.

$$\kappa_4 = \frac{\sum_i \kappa_{4,i} L_i}{\sum_i L_i} \tag{5.15}$$

The rest of the approach to obtain the optimal diameter is similar to the case developed for oil pipelines, based on taking the pseudo-mean values of a "model" fluid.

## 5.4 Contamination of Batches in Multi-product Pipelines

Contamination in a pipeline is defined as the mixing phenomenon that is produced by the contact of two different successive batches and the product of the mixture is called contaminated or interface, as presented in Fig. 5.2. It is recognized that the consequences are not serious within a crude pipeline where each shipment can withstand light pollution without great damage. It is not the same for fine or finished products (e.g., naphtha's of different types, JP1, etc.)

A contamination is characterized by the variation curve of the density of a product as a function of the volume that passes. In the case of finished or refined products (e.g. gasoline, LPG, JP1, gas-oil) whose contamination is inconvenient, they must be separated with separator plugs or scrapers, commonly known by the industry as "pigs" (Mohitpour et al. 2007).

The latter has the drawback that when the separator arrives at a pumping plant, it must either be stored in a suitable reception pig trap, and send a different one from the launch trap, or design some device that automatically allows the separator to pass from the low pressure zone, i.e. pump suction, to the high pressure zone, this is the pump discharge. In any case, whatever the scheme chosen, a counter-flow transport (with trucks or rail) of the separators must be established, from the reception traps to the launch ones.

An initial and essential precaution consists in studying the sequence of the batches so that the consequences of the inevitable contamination are reduced to a minimum and the contamination can be directed towards the less refined product. For example, in a Premium naphtha—Super naphtha interface, the contamination of the Premium naphtha could be carried to the Super naphtha tank.

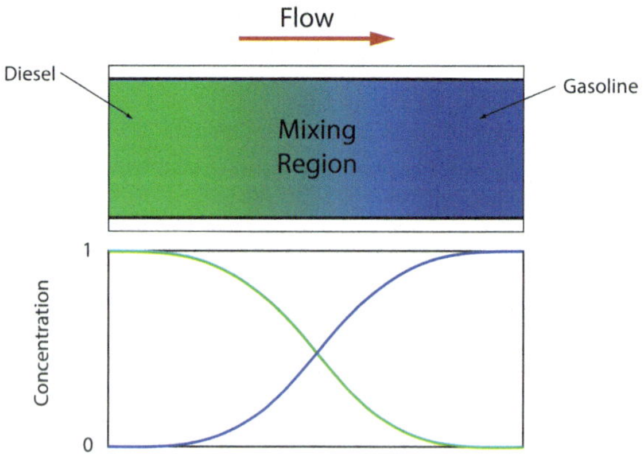


**Fig. 5.2** Detailed cut of a multi-product line, showing the interface between Diesel and Gasoline and the concentration profile in the area

**Fig. 5.3** Photo of a group of slop tanks in a refinery (Image reproduced with permission from Menno Gazendam)

As stated, the loading sequence of the multi-product lines must be studied so that contamination is reduced to a minimum and that it harms the less noble product. However, there are cases in the sequence of products whose characteristics make them completely incompatible (e.g., octane number, vapor pressure of gasoline and other fuels, ignition point). In such cases the contaminated volume must be withdrawn to a tank, known as *slop tank* (Fig. 5.3), from where it is then derived for further treatment at the distillery.

### 5.4.1 Product Separation in Multi-product Pipelines

There are mainly two different techniques when transporting products in multi-product lines, namely: pig batching and interface batching. Both have their advantages and disadvantages, from technical and economic points of view, which will be discussed hereinafter (Mostafaei et al. 2021).

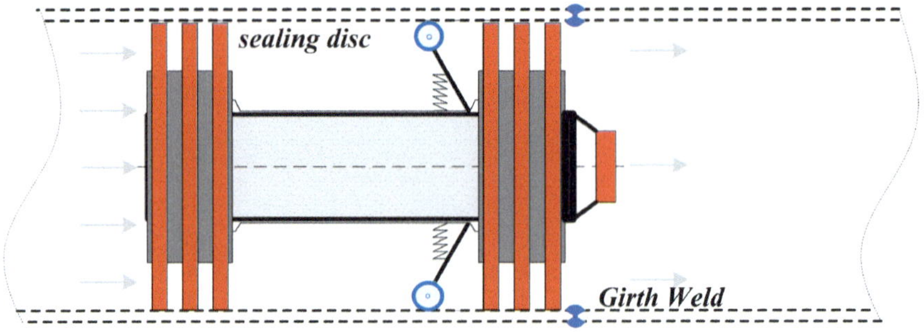


**Fig. 5.4** Scheme showing a typical pig/scraper in pipeline operation (Zhang et al. 2017)

### Pig Batching

As mentioned before, this techniques requires the use of special devices called scrapers, or colloquially in the industry, pigs (Fig. 5.4). This names comes from the fact that first models of scrapers, when passing through the pipeline, made a sound similar to a pig squealing. The main advantage of this is that the products are not mixing significantly, minimizing the contamination and possible reprocessing. Moreover, since the mixing is not as important, the order of the batches can be planned using a more simplified approach, without a deep consideration of the fluids' properties (Fig. 5.5).

However, its main disadvantage comes from the costs and logistics to operate these scrapers. Firstly, they require launching and receiving stations. Secondly, these scrapers must travel in a way to be determined, back to the launching plant where they can be reused. This creates an extra expense in the operational costs. Finally, the whole facility should be designed to be compatible with pigging operations, namely: all ball valves in the line should be full bore in order to let the pig to pass; tee's and other fittings, such as elbows and bends, should have a minimum curvature radius in order to let the pig to pass and not block the pipe. For instance, for bends it is required a minimum radius of curvature between 3 and 5 times the pipe diameter.

### Interface Batching

This techniques does not use any mechanical device to separate the different fluids and, as expected, in the interface a mixture/blending will be formed, which will lead to the contamination of (one of) the products. The amount of fluid contaminated can be calculated, as it will be shown in Sect. 5.4.2 and, obviously, should be kept to a minimum in order to reduce possible reprocessing costs (Vlot 2017).

In this case, the sequence of products and its planning become critical in the success of the operations. Usually, Diesel-oil is place between different products, since Diesel is the least sensitive to contamination. In comparison with the pig batching, it

**Fig. 5.5** A pipeline inspection gauge, commonly known in the industry as a pig, which was used to clean the inside of oil pipelines. This is on display near Fairbanks, Alaska, as part of the Trans-Alaska Pipeline, photography taken by Barrison (2007)

is not required here any launch/receiving facilities, nor any extra costs for transportation. However, since a large amount of product will be contaminated, it is deemed necessary to have a number of tanks, called *slop tanks*, to which the contaminated length can be diverted. These tanks are generally smaller than the main ones and several can be found in the tank farm, depending on the difference among the products transported.

Numerous works have been developed to improve knowledge of the factors that increase pollution or the blending interface volume, the results of which can be summarized in a series of factors that influence it, namely:

- Velocity and type of regime (laminar or turbulent): higher velocities render lowers interface volumes;
- Density differences: smaller differences render smaller interface volumes;
- Viscosity of each product: smaller differences render smaller interface volumes, products with a large difference in viscosity should be batched one after the other;
- Shutdowns and incidents in pumping;
- Passing through the pumping facilities.

An important question raised within this technique is: what needs to be done with the contaminated product/interface? As mentioned before, the blending causes the contamination of (one of) the products, causing a degradation of its properties as a commodity, requiring in most cases, a reprocessing. This should be carefully analyzed, whether the contaminated interface can go to one of the products or should be diverted to the slop tanks.

**Interface Reprocessing**

There are several techniques used with the contaminated products. These products cannot be simply discarded, since this would not be following a sustainable approach. Approximately, from 5 to 15% of the total length of all the batches (not the pipeline). Thus, this represent a significant cost of commodities waiting to be reprocessed. The costs associated with this reprocessing should also be considered in the operational costs analyses. The most commonly used techniques are:

- Transmix: the contaminated products are transported to a (new) refinery, in which they are again refined and sold as uncontaminated products;
- Downgrading (Regrading): the contaminated products are blended with the lower quality product, downgrading it even more with this addition.
- Blending: this involves mixing the products found in the slop tanks with a series of additives or ancillary compounds in order to obtain a products, which can be sold according to a set of (inter-)national regulations.

**Conclusions**

Summing up, there are mainly two techniques to handle the transport of multiple products via a pipeline. Both approach the problem from a different perspective, with advantages and disadvantages. Taking into consideration the brief study done here, it is considered that the best technique that will or should be used in the future is the pig batching. Though it represents some additional costs in facilities, it handles in a much better way the interface among the products, reducing noticeably the contamination and therefore, further reprocessing costs, which represents a sustainable approach in pipeline engineering.

## *5.4.2 Length and Volume of the Interface*

Continuing with the analysis of the interfaces in pipelines without pigging operations, it is important to establish the length of the blended/contaminated product. There are several techniques, which will be presented here. The importance to study and determine this will not be visible during the optimization process, since one is taking average values, but it will play a major role during the quasi-static segment analysis, in which the viscosities and densities of all the products in the pipeline should be considered in order to calculate accurately the pressure drops between pumping stations.

The first step consists in calculating the interface's density and viscosity, which will be considered constant throughout the whole length of the contaminated product. The subscript $b$ is used to indicate the properties of the blend.

$$\frac{1}{\rho_b} = \frac{1}{2}\left(\frac{1}{\rho_1} + \frac{1}{\rho_2}\right) \tag{5.16}$$

In order to calculate the viscosity of the blend, several correlations have been proposed based on the kinematic viscosity and volume fractions. One of the most used ones to calculate the viscosity of the blending of two products is the formula developed by Kendall and Monroe (1917), which states,

$$\nu_b^{1/3} = x_1 \nu_1^{1/3} + x_2 \nu_2^{1/3} \tag{5.17}$$

where $\nu_i$ is the kinematic viscosity and the $x_i$ are the mass fractions of the two fluids. Refutas (2000) proposed a new formulation based on two different alternatives: the mass and the volumetric expressions. The first step using the Refutas equation is to calculate a factor denominated Viscosity Blending Number (VBN) for each of the components in the blending.

$$VBN_i = 14.534 \ln\left(\ln\left(\nu_i + 0.8\right)\right) + 10.975 \tag{5.18}$$

where $\nu_i$ is the kinematic viscosity expressed in mm$^2$/s or cSt. Once this numbers have been calculated, the Viscosity Blending Number of the mixture is calculated,

$$VBN_b = \sum_i x_i VBN_i \tag{5.19}$$

As final step, with the first formula, knowing the VBN$_b$, the viscosity of the blending is calculates according to,

$$\nu_b = e^{\left(e^{\left(\frac{VBN_b - 10.975}{14.534}\right)}\right)} - 0.8 \tag{5.20}$$

Other formulas have also been used in industry, with average results when they have to predict the viscosity of blendings, for instance, Lederer (1937), Grunberg and Nissan (1949) and Oswal and Desai (2004).

With the properties of the blend properly defined, the length of the interface can be calculated. There have been many correlations proposed with the years, many of them depending on the Reynolds numbers, the properties of the blend and/or fluids in contact, the length of the pipeline, the friction factor and the inner diameter of the pipeline. Several will be presented in this chapter, summarizing the results in a series of graphs to determine its influence, and that of the parameters involved, in the length of the interface.

Several methods have been developed for calculating the length of the interface between two products in a pipeline. Due to the complex nature of modelling interface mixing, solutions from these methods lie in a wide range (this will be further explored in subsequent sections). A selection of these methods, commonly cited for quantification of interface mixing are listed below.

### Smith and Schulze

This formula, published in 1948, does not consider directly the properties of the fluids or the blend, but it makes the length of the interface as a function of the length of the pipeline and the operative conditions, presented in the Reynolds number (Smith et al. 1948).

$$L_b = L^{0.62}\left(0.55 + \frac{1055}{Re^{0.87}}\right) \tag{5.21}$$

### Taylor

Following the same approach as those presented before, Taylor introduced in 1954 a different correlation expressing also the length of the interface as a function of the pipeline physical properties and the operation parameters (Taylor 1954).

$$L_b = 6.59998 Re^{-0.0625}\sqrt{D_{int}L} \tag{5.22}$$

### Sjenitzer

This formula was presented in 1958 and calculated the interface as a function of the physical properties of the pipeline, and the operating conditions, specified with the Reynolds number (Sjenitzer 1958).

$$L_b = 245.999 D_{int}^{0.43} L^{0.57} Re^{-0.45} \tag{5.23}$$

### Austin and Palfrey

A commonly found equation in oil and gas industry to calculate the length of the blend was presented by Austin and Palfrey in 1964 (Austin and Palfrey 1963). The first step consists in calculating the Reynolds number selecting the proper formulation.

$$L_b = \begin{cases} \text{If } Re > 10000\exp\left(2.7539\sqrt{D_{int}}\right) \rightarrow 11.75 Re^{-0.1}\sqrt{D_{int}L} \\ \text{If } Re \leq 10000\exp\left(2.7539\sqrt{D_{int}}\right) \rightarrow 18419.9 Re^{-0.9}\sqrt{D_{int}L}\exp\left(2.19168\sqrt{D_{int}}\right) \end{cases} \tag{5.24}$$

### Udoetok and Nguyen

Finally, the experimental correlation presented by Udoetok and Nguyen in 2009 is one of the latest models introduced to calculate the length of an interface (Udoetok and Nguyen 2009). The first step consists in calculating a parameter $n$, which is function of the friction factor,

$$n = \frac{1}{\sqrt{f}} \tag{5.25}$$

With this value the length of the interface can be calculated introduced a constant $\omega$ according to the following expression,

$$L_b = \left\{1 - \left[1 - \left(\frac{2n^2}{(n+1)(2n+1)}\right)^n \omega^n\right]^2\right\} L \tag{5.26}$$

On their paper, they suggested that $\omega = 0.585$ based on field survey and experimental information used to create the correlation, although they strongly recommend to adapt this value to every situation and pipeline in particular.

The techniques described previously can be easily compared and studied in Figs. 5.6, 5.7, 5.8 and 5.9. The first three figures depict the interface length as a function of the inlet Reynolds number for the different formulations presented and the pipeline length. Figure 5.9, on the other hand, shows the dependence of the interface length as a function of the inlet Reynolds number.

## 5.5 Application of the Lagrange Method—Cost Functions

The methodology explained in Chap. 4, defining the cost in a pipeline for incompressible fluids, followed by the optimization calculation and selection of commercial parameters, can also be applied to multi-product pipelines. In fact, the fluids being transported are also incompressible, and the design codes of the pipeline wall applicable to single product pipelines are the same than for multi-product pipelines. Hence the cost functions will be exactly the same as those in Chap. 4, with the particularity that that fluid parameters, such as density and viscosity should be replaced for the average values explained at the beginning of this Chapter. Also the cost function of the fluid, $\kappa_4$, must be taken as the average of the commodities being transported.

Therefore, the procedure explained before is valid also for this type of pipelines and the same approach and pseudo-codes can be used. A significant difference is the calculation by segments when the optimization is done, which is explained in the following section.

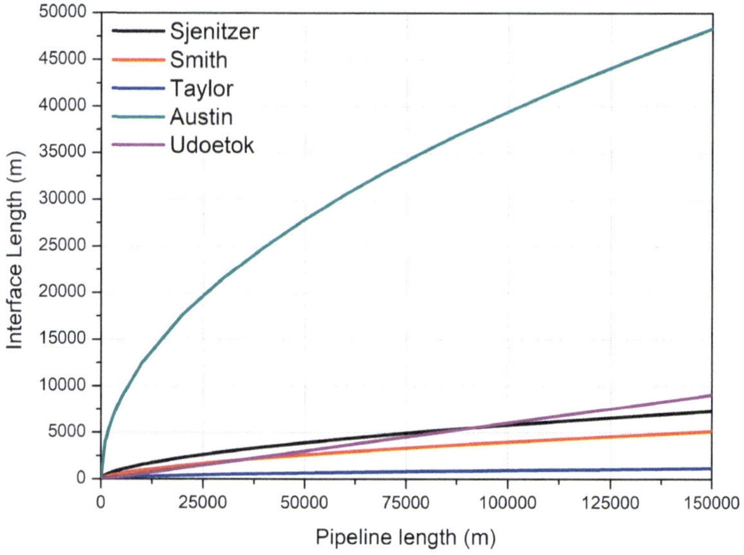


**Fig. 5.6** Interface length as a function of the pipeline length for the different formulations, considering a Reynolds number = 1000

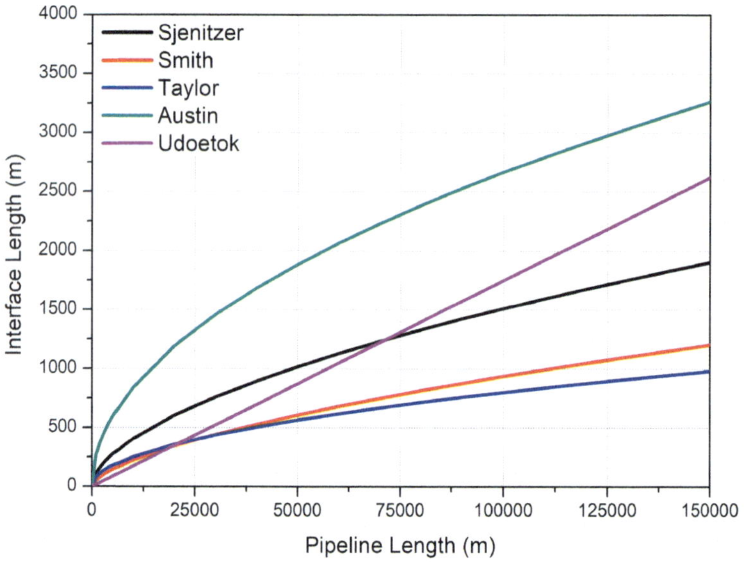


**Fig. 5.7** Interface length as a function of the pipeline length for the different formulations, considering a Reynolds number = 20000

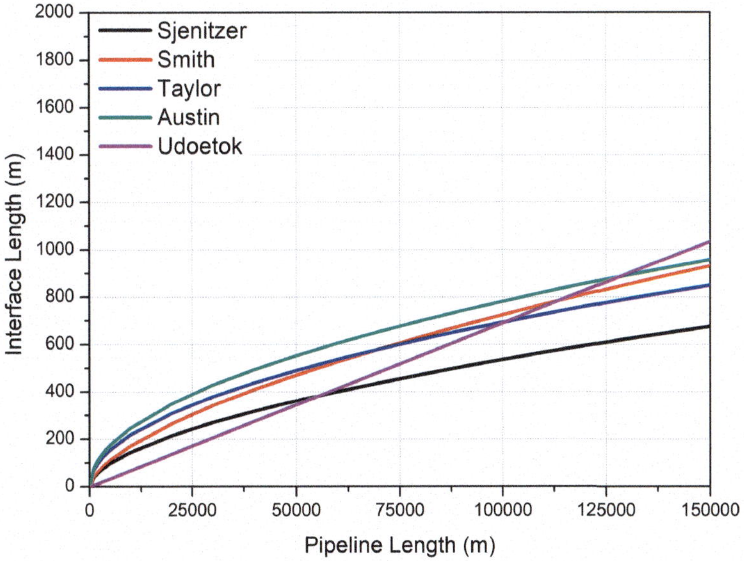


**Fig. 5.8** Interface length as a function of the pipeline length for the different formulations, considering a Reynolds number = 200000

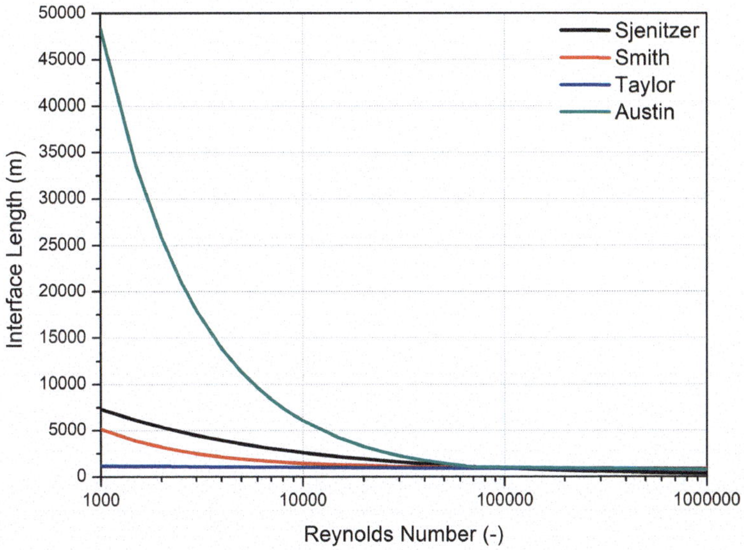


**Fig. 5.9** Interface length as a function of the Reynolds number for some formulations presented in this chapter

## 5.6 Calculation by Segments in Pipelines for Liquid Transportation

In the case of multi-product lines, with greater reason, the calculation of the optimum diameter is an approximate calculation to solve the economic problem when considering a "fluid model of weighted average characteristics", and after that, the fluid dynamic response of the system must be verified accurately as time passes.

To do this, a defined sequence of products must be generated within the pipeline, with their respective lengths for each batch, and the successive positions of the "batches" with respect to the progressive pipeline as time elapses. All of this studied at defined time intervals that allow a return to the initial configuration, i.e., having completed a complete cycle.

In this way, for each fluid in particular, the pressure changes as it circulates, and consequently the density and viscosity change, as well as the temperature, this for each "batch" starting from its initial conditions. Thus, the study must be carried out by sections—within each individual product- and performing the "jump" of characteristics, i.e., pressure, density and temperature, in the section (or as it is called in the industry, the *chainage* or *progressive mileage*) that is treated according to the temperature and pressure found based on the product studied.

Being and $\rho(p, T)$ and $\mu(p, T)$, the Reynolds changes and therefore the friction factor also changes section by section. In addition, in this case, since the speed remains constant at the interfaces, the mass flow rate defined by does not remain constant; afterwards, already inside the "batch", if the density changes, the speed also changes, and the length of each shipment should be adjusted more accurately with the fluid velocity.

In short, within each "batch" the treatment of the calculation by sections is identical to that explained for single-product pipelines in Chap. 4, taking the precaution of modifying the base characteristics for the next product in the planned sequence, once the interface, or the separator plug, has been reached. This is at the interface between two different products, the same speed is a known value, as well as the same pressure and the same temperature, which will determine the density and viscosity conditions to start the calculation in small sections for the product that follows, until reaching the next pump.

Lastly, it should be noted that this procedure should be repeated for regular periods of time (e.g., every 4 h) until the cycle mentioned in the second paragraph is completed. In this way, the maximum and minimum power requirements in each pumping plant could be distinguished according to the location of the batches within the pipeline. In other words, starting from $t = 0$ with a defined format of sequence of products in the line and occupied lengths, and then for $t = \Delta t$ (e.g., $t = 4$ h) the interfaces of each product will have moved within the duct a distance $\Delta L = v \cdot \Delta t$. A third arrangement of products follows for $t = 2\Delta t$ (i.e., $t = 8$ h), continuing with the previous example and so on until after $t = n\Delta t$ finding that the arrangement of the products returns to that of $t = 0$, which confirms that the cycle has been completed.

It is possible that in some of these dispositions of the products, the set pressure does not allow reaching the next pump with sufficient pressure, and in such a case the entire cycle must be analyzed to see if modifying the distance between the pumping stations, or in some case increase the discharge pressure of any plant to the extent that the thickness of the pipe allows it, it is possible to solve the problem.

In addition, it will be necessary to consider the conjunction between the pressure and the altimetry of the pipeline. In fact, at specific points of the pipeline, away from the pumps and at "high" points, the internal pressure remaining in the line must be verified. This is due to the fact that said pressure drops below the vapor pressure of the product at the temperature it is flowing at that point and a vapor bubble is produced that restricts the passage of the liquid. This verification will need to be carried out for the most volatile product and for different batch configurations.

Once the pumps have been relocated, the power needs of each pumping station can be verified in each case, depending on each of the pumped products.

### *5.6.1 Summary of Formulas for the Calculation by Sections*

The pipe deforms elastically and can expand with temperature, changing the diameter. There is heat transfer from the medium to the fluid, or vice versa. Friction generates additional heat by a viscous dissipation mechanism that modifies the temperature of the fluid, only a part of the heat due to friction, and the enthalpy variation due to fluid expansion is considered (Li et al. 2021).

It is assumed that there is no heat input in the $\Delta x$, the segment itself, and it is fully introduced in the section $n + 1$ as if it were supplied all of a sudden, making the temperature change, and with the new pressure $p_{n+4}$ and temperature $T_{n+1}$ the new diameter is calculated, along with the new viscosity, density, and velocity.

The length $\Delta x$ should be such that it is a section "sensitive" to the change in density and viscosity of the product in question due to changes in temperature.

This calculation must be repeated until reaching the section where the product changes and then with the pressure and temperature the viscosity and density of the next product will be defined, to continue the sequence of calculations until completing the length of the duct and products circulating through it.

Once this first detailed calculation is finished, the configuration must be changed, allowing the chosen period to elapse and the calculation will be restarted by segments with different lengths for each product. This calculation should be repeated for as many configurations until the cycle is complete and the product configuration is equal to the one at the start of the detailed calculations.

**Algorithm 5.1:** Calculation by segments for multi-product pipelines

**Data**: $n_{products}$, $i$, $\dot{m}$, $L$, $p_1$, $p_2$, $H$, $\Delta x$, $\rho_i(p, T)$, $\mu_i(p, T)$, $c_{i,p}$, $D_{p,T}$, $\chi$, sequence, interface management (pig/no pig)
**Result**: Pump stations located, temperature profile known, power for each station
Calculate number of segments $\rightarrow n_s = \frac{L}{\Delta x}$;
Calculate length of every product batch;
Calculate time step $\rightarrow \Delta t = \frac{t_{total}}{n_{steps}}$;
Calculate entry parameters to the pipeline;
Calculate length and properties of interfaces;
**for** $t = 0$ **to** $t_{total}$ **do**
  **for** $i = 1$ **to** $n_s$ **do**
    Get new pressure with Eq. 4.45;
    Calculate overall heat transfer coefficient (See Chap. 2), $U_i$;
    Get new temperature at the end of the segment with Eq. 4.66;
    Calculate the new diameter with Eq. 2.75, $D_{i+1}$;
    Calculate new density and viscosity with Eqs. 1.9 and 1.10, respectively, $\rho_{i+1}$ and $\mu_{i+1}$;
    Calculate new velocity with Eq. 4.67;
    Calculate new Reynolds and friction factor;
    **if** $p_{i+1} \leq p_2$ **then**
      Locate a pumping station in this coordinate;
      $p_{i+1} = p_1$;
      $T_{i+1} =$ Eq. $4.66 + \Delta T_{pump}$;
      Calculate real power for the pump;
    **else**
      No need to locate a pump after this segment
    **if** *there is an interface* **then**
      **else if** *no pigs are used* **then**
        Use properties of blend, i.e., $\rho_b$ and $\mu_b$;
      Change properties of the fluid, new $\rho$ and $\mu$;
  $t = t + \Delta t$;
  Calculate new interface chainages $\rightarrow x_{t+\Delta t} = x_t + v\Delta t$;
**Result**: Arrival pressure at every station for every fluid configuration, pumps power as function of time, discharge pressures at every station as a function of time

### Crossing an Intermediate Pump

Since the flow rate is fixed, if the pressure drop in the pump changes, the operating point on the *H-Q* curve of the pump changes and therefore also the efficiency $\eta$ and with this the temperature of the fluid discharged by the pump changes. As the operating point is not exactly known at the beginning, the efficiency must be preselected (corrected by the viscosity) and only after the problem has been resolved, re-verify the passage through the pump.

It can be seen that in the intermediate pumping stations the arrival pressures are highly fluctuating (the discharge ones are fixed and equal to $p_1$) depending on the configuration in question and therefore also the powers at stake for each configuration, and in each plant, will be variables between a minimum and a maximum.

### 5.6.2 *Verification of Centrifugal Pumps in Multi-product Pipelines*

A significant issue—and one that should not be ignored- is the situation of the multi-product pumping plants, since in each of them the required power will fluctuate over time between a maximum and a minimum. According to the sequence of products, which it will necessarily demand to study the complete cycle of foreseen “batching”, generally repetitive, to adequately select the pumps and the power of their motors, and the way to control their rotation speed, generally with the application of variable frequency drivers (VFD’s).

It should be noted that in general it is a matter of moving in the *H*-*Q* curves of each pump while maintaining $Q =$ constant, and therefore H, n, $\eta$ and N variables, which would enable the arrangement of pumping equipment in series within the same plant to absorb extreme fluctuations. This situation requires the corresponding and successive performance correction depending on the fluid and operating point, in which a given pump is operating at the instant considered. Regarding the influence of the transported fluids and their viscosity, see the corresponding points in the Chap. 2.

Centrifugal pumps in a pipeline are generally selected to give one head and flow rate for a single type of product. The speed of the pump is constant for that particular product. When operation with a lower or higher flow rate is desired, the rotation speed of the pump is reduced or increased, it being necessary to provide it with a variable speed motor. Otherwise the operating points could be varied through the regulation of a valve, but this introduces losses of power and performance.

In the case of a multiple pipeline where two or more products must be pumped, it will be necessary to select the pump for a Head of Pumped Flow condition that can handle all the products. The height required to pump the heaviest liquid (dense and/or viscous) would be the highest value, considering each pumped product separately.

In a real operation the pump has to move more than one product in the pipeline at the same time. For example, the pump could be propelling gasoline and downstream the pipeline could be transporting diesel and kerosene. The required pump head in meters of liquid would be the greater of the lighter product being pumped by the pump and the line that was filled with the heavier product. The pump selection will be made on this basis (Cafaro et al. 2010).

**Example: Multi-product Line**

As an introduction to the topic, consider the following case:

- Pipe NPS 16" - $D_{ext} = 0.406$ m
- thickness = 8.74 mm
- Material (grade) API 5L X52 (L360)
- Length = 322 km
- Maximum Allowable Operating Pressure (MAOP) = 99.3 bar
- Design flow rate = 100,000 bbl/day = 15,900 m$^3$/day
- Altitude of pumping station A is $H_1 = 152.4$ m
- Altitude of pumping station B is $H_2 = 457.2$ m
- Fluid: diesel oil
- Density = 870 kg/m$^3$
- Kinematic viscosity = 5.9 cSt

Using one of the applicable formulas, the line pressure drop results in approximately 0.5 bar/km, and then the total pressure loss would be around 160.5 bar. The static elevation height to be overcome is equivalent to 305 meters, which represents 26 bar. Thus, the total pressure required at station A to pump the 15,900 $m_3$/day of diesel oil will be 186.5 bar.

In order for the product to reach plant B with a pressure higher than its vapor pressure, the discharge pressure of plant A must be increased to 190 bar, thus ensuring a delivery pressure of 3.5 bar at plant B, and above the vapor pressure of diesel oil.

Since the pressure limitation of the line is 99.3 bar, a pressure of 190 bar is not admissible and then it will be necessary to incorporate at least one intermediate pumping station between A and B and the discharge pressure of plant A and plant middle floor of,

$$p = \frac{\Delta p_T + p_v}{2} + p_v = 98.25\,\text{bar} \tag{5.27}$$

Knowing the altimetry of the pipeline route, the intermediate station can be found graphically. It must be taken into account that each pumping station operates at a differential head of $98.5 - 3.5 = 95$ bar and a suction pressure of 3.5 bar. Suction pressure at plant A can be provided by a booster pump which supplies product from the storage tanks to the main pump. In each pumping station the differential head will be $\Delta p = 95$ bar.

When selecting the main pumps, this pressure is converted to meters of height of diesel oil, that is, 1094 meters. Thus, the pumps of each station have to develop a differential head of 1094 m and a flow of 662.5 m$^3$/hour. For greater flexibility, two pumps could be installed in series, each of which will develop a head of 547 m and

662.5 $m^3/h$, trying to make this point as close as possible to the point of maximum performance of the selected pump.

If the required differential head in the station is calculated, caused by different flow values, the characteristic curve of the system between plant A and the intermediate pumping station is obtained, superimposed on the pump curve is the operating point for the system.

If it is necessary to operate the pipeline at an operating point lower than 15,900 $m^3$/day, the head required by the system will be less, while the pumps generate a higher head. Consequently, the discharge pressure of the pumps must be regulated to an equivalent $\Delta p$ value, which would mean lower efficiency and loss of power. Another way to reach the new operating point would be to reduce the rotation speed of the pump, by means of a frequency inverter if it is an electric pump; on the contrary, for operating values higher than the design values, the pump should rotate faster.

**Example: Evaluation of Flow Rates and Batches**

Let there be a requirement to transport 500,000 ton/month of oil of three different characteristics according to the following information:

- Length = 300 km
- Number of pumping plants (with header) = 3
- Product 1
  - Mass flow rate = 150,000 ton/month
  - Density = 910 kg/$m^3$
  - Delivered in batches of 15,000 ton each.
- Product 2
  - Mass flow rate = 120,000 ton/month
  - Density = 880 kg/$m^3$
  - Delivered in batches of 20,000 ton each.
- Product 3
  - Mass flow rate = 230,000 ton/month
  - Density = 850 kg/$m^3$
  - Delivered in batches of 46,000 ton each.

Equivalent volumes to transport are then equal to $Q_i = \frac{\dot{m}_i}{\rho_i}$

- Product 1 $= 164,800$ $m^3$/month
- Product 2 $= 136,363$ $m^3$/month
- Product 3 $= 270,588$ $m^3$/month

The total volume to be transported through the pipeline is then 571,751 $m^3$/month = 794.1 $m^3$/day = 0.2201 $m^3/s$

The next step is to estimate the diameter of the pipeline, which is necessary to calculate the length of the batches. Assuming a transport velocity of 0.9 m/s, an area of 0.244 $m^2$, which is representative of a diameter of 0.558 m corresponding to a pipe of NPS 22" (D = 0.5588 m). The lengths are then,

- One batch of product 1 (16,680 $m^3$) would occupy 67.24 km of pipeline (pink color).
- One batch of product 2 (22,737 $m^3$) would occupy 97.73 km of pipeline (blue color).
- One batch of product 3 (54,117 $m^3$) would occupy 220.81 km of pipeline (yellow color).

Assuming that the plants are separated 100 km each, the following figures show the sequence of batches every 8 h until the cycle repeats after approximately 92.6 h (almost 4 days). For the study by segments, approximately 24 different formats should be analyzed in intervals every 4 h to return to the initial format. It is also observed that the pumping plants are working with different products as time goes by (Fig. 5.10).

It should be noted that in the case of similar products (e.g., oils with different characteristics), it is generally avoided to introduce batching separators (pigs) to

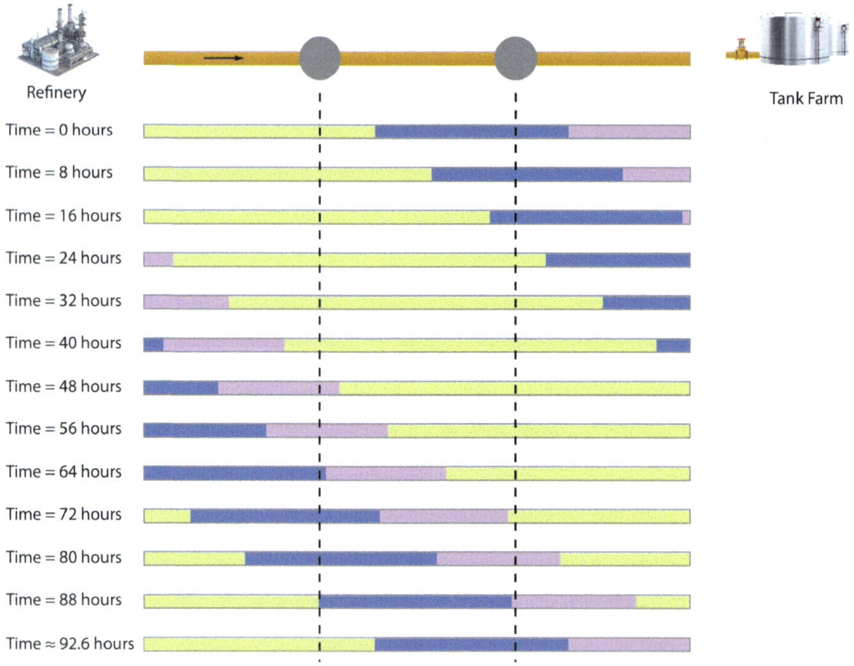


**Fig. 5.10** Scheme showing the variability of the fluids present in a multi-product pipeline as a function of time

identify the beginning and end of each one by monitoring the density and viscosity together with the length from the batch sent and its position on the line, eventually sending the "contaminated" portion at the interface to the slop tanks.

In the case of very dissimilar products (e.g., automotive gasoline, JP1 kerosene, LPG, gas oil) and in general "finished" products, it is essential to separate with one (or sometimes two "scrapers" with adequate separation) to avoid all types of contamination between products. This operation causes inconveniences when crossing through the intermediate pumping stations (continuous and sequential extraction and introduction of the separator caps). In addition, all the logistics dedicated to returning the scrapers back to the header station must be foreseen, by means of cargo vehicles that transport them from the receiving traps to the throwing traps.

## References


Austin JE, Palfrey JR. Mixing of miscible but dissimilar liquids in serial flow in a pipeline. Proc Inst Mech Engin. 1963;178(1):377–89.

Barrison H. Pipeline pigging operations. 2007. https://commons.wikimedia.org/wiki/File:PipelinePIG.jpg. [Accessed 17 Mar 2024].

Cafaro VG, Cafaro DC, Méndez CA, Cerdá J. Oil-derivatives pipeline logistics using discrete-event simulation. In: Proceedings of the 2010 winter simulation conference. IEEE; 2010. p. 2101–13.

Cafaro DC, Cerdá J. Operational scheduling of refined products pipeline networks with simultaneous batch injections. Comput Chem Engin. 2010;34(10):1687–704.

Grunberg L, Nissan AH. Mixture law for viscosity. Nature. 1949;164(4175):799–800.

Kendall J, Monroe KP. The viscosity of liquids. iii. ideal solutions of solids in liquids. J Am Chem Soc. 1917;39(9):1802–06.

Lederer EL. Viscosity of binary mixtures. Nature. 1937;139(3505):27–8.

Li J, Su Y, Yu B, Wang P, Sun D. Influences of hydrogen blending on the joule-thomson coefficient of natural gas. ACS Omega. 2021;6(26):16722–35. https://doi.org/10.1021/acsomega.1c00248

Maples RE. Petroleum refinery process economics. PennWell Corporation; 2000. ISBN 9780878147793. https://books.google.nl/books?id=bLtTAAAAMAAJ.

Mohitpour M, Golshan H, Murray MA. Pipeline design and construction: a practical approach. ASME Press: Pipelines and Pressure Vessels; 2007.

Mostafaei H, Castro PM, Oliveira F, Harjunkoski I. Efficient formulation for transportation scheduling of single refinery multiproduct pipelines. Eur J Oper Res. 2021;293(2):731–47. ISSN 0377-2217. https://doi.org/10.1016/j.ejor.2020.12.034. https://www.sciencedirect.com/science/article/pii/S0377221720310778.

Oswal SL, Desai JS, Ijardar SP. Studies of viscosities of dilute solutions of alkylamine in non-electrolyte solvents: I. aliphatic and aromatic hydrocarbons. Thermochimica Acta. 2004;423(1–2):29–41.

Sjenitzer F. How much do products mix in a pipeline? Pipeline Engin. 1958;30:31–4.

Smith SS, Schulze RK, et al. Interfacial mixing characteristics of products in products pipe line-part 1. Petroleum Engin. 1948;20(8):330–7.

Taylor GI. The dispersion of matter in turbulent flow through a pipe. Proc R Soc London. Ser A. Math Phys Sci. 1954;223(1155):446–68.

Tu R, Liao Q, Huang L, Jiao Y, Wei X, Liang Y. Pipeline sharing: remaining capacity estimation of multiproduct pipelines. Chem Engin Res Design. 2023;191:338–52. ISSN 0263-8762. https://doi.org/10.1016/j.cherd.2023.01.028. https://www.sciencedirect.com/science/article/pii/S0263876223000291.

Udoetok ES, Nguyen AN. A disc pig model for estimating the mixing volumes between product batches in multi-product pipelines. J Pipeline Engin. 2009;8(3).
Vlot SJ. Batch scheduling of multi-product pipeline networks. Master of science in systems and control. Delft University of Technology. 2017.
Zhang H, Zhang S, Liu S, Wang Y. Collisional vibration of pigs (pipeline inspection gauges) passing through girth welds in pipelines. J Natural Gas Sci Engin. 2017;37:15–28.

# Chapter 6
# Transport of Compressible Fluids or Gases Through Pipeline Systems

> Engineering is the art of directing the great sources of power in nature for the use and convenience of man.
>
> *Thomas Tredgold*

## 6.1 Introduction

The problem of long-distance transportation of gases through pipelines is relatively modern, beginning historically in cities with the distribution of manufactured gas at very low pressure (200 mm of water column or less, above atmospheric pressure) distributed through cast iron pipes by means of large diameter, mainly for public lighting. Indeed, gas began to be used for lighting cities until it was replaced by electricity. When it stopped being used for lighting, new uses for gas arose, mainly for cooking, heating water, and heating buildings, demonstrating that they were useful, safe, and efficient applications that brought well-being and comfort to societies (Mohitpour et al. 2007).

In the early days of gas, three milestones in its production and use must be highlighted, namely:

- The Scotsman William Murdock, who in 1792 managed to light his house with gas produced from coal.
- The Frenchman Lebon, who in 1801 was trying to apply the gas obtained from the distillation of wood for similar purposes.
- The German Frederick Albert Winsor, who in 1804 presented gas lighting experiments in London.

 
P. Druetta et al., *Transport by Pipelines and Its Optimization*, Petroleum Engineering,
https://doi.org/10.1007/978-3-031-66566-0_6

**Fig. 6.1** View of a NPS 26 (DN 650) natural gas pipeline at main line valve station, near Dampier, Australia (Dillon 2008)

It was Winsor who in 1812 created the "Gas Light and Coke Company"[1] in London, the first gas company in the world to provide public lighting with gas produced through coal.

As soon as it was technically possible, the construction of a network of gas pipelines from Texas and Louisiana to the large industrial centers near the Great Lakes area began in the US between 1925 and 1930. A little later in the Argentine Republic, the first long distance gas pipeline—almost 1000 km—between Comodoro Rivadavia and Buenos Aires was built in 1949, which began the replacement of manufactured gas by natural gas in the City of Buenos Aires. Somewhat later, in the 1960's, the transportation of natural gas appeared in Western Europe, growing very rapidly, derived from the presence of networks for manufactured gas in the main industrial areas and local discoveries of natural gas (Fig. 6.1).

The transport of gases through pipelines is also applied to other gaseous products such as $CO_2$, hydrogen and nitrogen, as well as for superheated water vapor, although in these cases in general for shorter journeys. Some aspects related to $CO_2$ are mentioned below. The rest of the chapter will be dedicated fundamentally to the transportation of high pressure natural gas.

[1] Also known as the Westmister Gas Light and Coke Company, it was a company that produced and supplied coal gas and coke in London, parent and whence descends the British Gas plc.

To develop the optimization methodology in this instance, the chapter will be dedicated to the consideration of the flows of pure gases or homogeneous mixtures of them; natural gas is precisely a homogeneous mixture of hydrocarbon and non-hydrocarbon gases. In both cases, what is required is to be able to know the parameters of the pure gas or the mixture of gases. In effect, it is initially necessary to obtain the greatest amount of data on the fluid, basically its molar composition. If they were not available, a chromatography will allow to know the same.

## 6.2 Definitions of the Parameters of the Ideal Gas from Single Real Gaseous Substances or Homogeneous Mixtures

### *6.2.1 Perfect Gas Equation*

In order to explain the behavior of all gases, a simplified gas model was developed that gave rise to the theory of perfect gases, which, within certain limits of pressure and temperature, satisfactorily explains the behavior of real gases.

In general, when referring to pressure and temperature in the case of perfect gases, absolute pressures and absolute temperatures must be considered, which must be taken into account when applying the formulas.

Perfect gases are those that comply with the laws of Gay-Lussac and Boyle and Mariotte. A real gas will be in perfect gas conditions and will comply with sufficient approximation with the aforementioned laws, when it is in a state far from those corresponding to the liquid-vapor equilibrium.

#### Boyle-Mariotte Law

The law of Boyle and Mariotte or isothermal transformations states that at constant temperature, the volume varies in inverse proportion to the absolute pressure.

$$p_1 \cdot V_1 = p_2 \cdot V_2 = \cdots = p_i \cdot V_i = C \tag{6.1}$$

Being $p$ the absolute pressure of the initial and final state, and $V$ the initial and final volume.

#### Gay-Lussac Law

When a transformation is carried out in a closed and non-deformable enclosure, it is true that, maintaining the volume constant, the absolute pressure of a certain mass of gas varies in direct relation to the absolute temperature; that is, if the gas increases its temperature, the pressure also increases and on the contrary, when it cools its pressure decreases.

$$\frac{p_1}{T_1} = \frac{p_2}{T_2} = \cdots = \frac{p_i}{T_i} = C \tag{6.2}$$

Being $p$ the absolute pressure of the initial and final state and $T$ the absolute temperature of the initial and final state.

**Charles and Gay-Lussac Law**

Charles, Gay-Lussac law or transformations at constant pressure states that at constant pressure, the volumes occupied by a given mass of gas are directly proportional to the absolute temperatures, that is, at constant pressure, when a mass of gas increases its temperature, it also increases its volume and, conversely, when it cools, it reduces its volume.

$$\frac{V_1}{T_1} = \frac{V_2}{T_2} = \cdots = \frac{V_i}{T_i} = C \tag{6.3}$$

### 6.2.2 General Equation of Perfect Gases

The General Equation of Perfect Gases, or Equation of State of Perfect Gases, determines the state of a given mass of a perfect gas, based on its three state variables: absolute pressure, volume and absolute temperature, without involving the path followed to pass from the initial state to final state.

$$\frac{p_1 \cdot V_1}{T_1} = \frac{p_2 \cdot V_2}{T_2} = \cdots = \frac{p_i \cdot V_i}{T_i} = C \tag{6.4}$$

or also,

$$\frac{p \cdot V}{T} = mR_p \tag{6.5}$$

being $m$ the gaseous mass considered and $R_p$ the particular constant of the gas. It can also be used as,

$$\frac{p}{\rho} = RT \tag{6.6}$$

or also,

$$p \cdot V = n \cdot R \cdot T \tag{6.7}$$

In which $n$ represents the number of moles and $R$ is the universal constant, resulting in $R_p = \frac{R}{M_w}$, where $M_w$ is the molecular weight. The value of the universal constant is indicated below.

$$R = 8.3145 \frac{J}{mol \cdot K} = N_a \cdot k_B \tag{6.8}$$

where $N_a$ is the Avogadro's number and $k_B$ is the Boltzmann constant.

### 6.2.3 *Perfect Gas Mixture*

For the mixture of perfect gases, the concepts of partial pressure and partial volume are used, applying Dalton's and Amagat's laws.

**Dalton's Law**

It is stated by saying that in a mixture of gases the total pressure that the mixture supports is equal to the sum of the partial pressures corresponding to each component, understanding by partial pressure that to which each component would be subjected if only the volume of the mixture were occupied and at the same temperature at which it is.

$$p_{p,i} = x_i \cdot p \tag{6.9}$$

In other words, the partial pressure of a component will be equal to the product of its mole fraction by the total pressure.

**Amagat's Law**

It is stated by saying that in a mixture of gases the total volume that the mixture occupies is equal to the sum of the partial volumes corresponding to each component, understanding by partial volume that which each component of the mixture would occupy if it were only subjected to the total pressure that exists in the mixture and at the same temperature.

$$V_{p,i} = V_i \cdot p \tag{6.10}$$

### Molecular Weight of the Mixture

The molecular weight of a mixture of gases is defined as,

$$M_w = \frac{m}{n} \tag{6.11}$$

Where $m$ is the mass of the mixture and $n$ the total number of moles of the same. Likewise, the molecular weight of the mixture $M_w$ can be obtained from,

$$M_w = \sum_i x_i M_i \tag{6.12}$$

### Particular Constant of the Mixture

As defined before, the particular constant can be calculated according to,

$$R_p = \frac{R}{M_w} \tag{6.13}$$

With this, the following equation of state can be used for the mixture:

$$\frac{p \cdot V}{T} = m R_{m,p} \tag{6.14}$$

The true behavior of real gases, for many states, is quite far from that expressed by the perfect gas laws. This fact has motivated the need to find other equations of state, explaining the behavior of real substances as a function of the state parameters.

## *6.2.4 Equation of State for Real Gases*

Several correlations of EoS have been developed in order to explain the real behavior of (mixture of) gases. One of the classes of these EoS are known as *cubic equations of state*.

### Van der Waals Equation of State

One of the first classic formulas[2] proposed to represent the thermoelastic behavior of real gases is,

$$\left(p + \frac{a}{V^2}\right)(V - b) = RT \tag{6.15}$$

[2] Johannes Diderik van der Waals was a Dutch physicist (1837–1923), who received the Nobel Prize in 1910 for his work on the equation of state for gases and liquids

Being $a$ and $b$ characteristics of the gas, calculated from the critical properties of the gas/mixture.

**Redlich-Kwong Equation of State**

It was introduced in 1949 and it represented the first improvement with respect to the Van der Waals EoS. The equation is given as,

$$p = \frac{RT}{V_m - b} - \frac{a}{\sqrt{T} V_m (V_m + b)} \tag{6.16}$$

where $a$ and $b$ are also function of the critical properties of the gas. An equivalent form of this equation is the one mentioned in Chap. 1 as the Law of Corresponding States (see Eq. 1.12). The compressibility factor is then,

$$Z_c = \frac{V_m}{V_m - b} - \frac{a}{RT^{3/2} (V_m + b)} \tag{6.17}$$

**Peng-Robinson Equation of State**

This EoS was presented in 1976 in order to improve the accuracy near the critical point. The equation is as follows,

$$p = \frac{RT}{V_m - b} - \frac{a\alpha}{V_m^2 + 2bV_m - b^2} \tag{6.18}$$

where $a$, $b$ and $\alpha$ are a function of the critical pressure and temperature, and the accentric factor $\omega$. In this case, the compressibility factor is found as the solution of the following cubic equation,

$$Z_c^3 - (1 - B)Z_c^2 + (A - 2B - 3B^2)Z_c - (AB - B^2 - B^3) = 0 \tag{6.19}$$

where,

$$A = \frac{\alpha a p}{R^2 T^2} \tag{6.20a}$$

$$B = \frac{bp}{RT} \tag{6.20b}$$

**Table 6.1** Chemical composition of the natural gas used in the example

| Component | Mole fraction (–) |
|---|---|
| Methane (C1) | 0.9156 |
| Ethane (C2) | 0.0320 |
| Propane (C3) | 0.0136 |
| Isobutane (iC4) | 0.0023 |
| Normal-butane (nC4) | 0.0041 |
| Isopentane (iC5) | 0.0010 |
| Normal-pentane (nC5) | 0.0012 |
| Hexane and above (C6) | 0.0014 |
| Nitrogen ($N_2$) | 0.0078 |
| Carbon Dioxide ($CO_2$) | 0.0210 |
| Oxigen ($O_2$) | 0.0000 |
| Hydrogen ($H_2$) | 0.0000 |

**Example: Properties of Natural Gas Mixture**

Natural gas is a mixture of light hydrocarbons that are in the gaseous state under normal conditions of ambient temperature and pressure, in the reservoirs. Composed mostly of methane, some ethane, and small amounts of propane and butane, it may also contain carbon dioxide, oxygen, and sulfur compounds; but, to be used, these last components must be registered in very limited quantities. For the gas presented in the table below, calculate all the characteristic parameters (Table 6.1).

The first part of this problem consist in calculating the mixture's molecular weight, which can be done according to the following table (Table 6.2),

With the molecular weight found, one can now calculate the mixture's particular constant and the relative density,

$$\mathcal{R}_p = \frac{R}{M_w} = \frac{8.3145\frac{J}{mol\cdot K}}{0.01804\frac{kg}{mol}} = 460.887\frac{J}{kg\cdot K} \tag{6.21}$$

$$\rho_{rel} = \frac{M_w}{M_{w,air}} = \frac{18.04\frac{g}{mol}}{28.96\frac{g}{mol}} = 0.6229 \tag{6.22}$$

The last step consists in calculating the critical pressure and temperature, based on Dalton's law (Table 6.3).

**Table 6.2** Mixture's molecular weight calculation as a function of the chemical composition

| Component | Mole fraction (–) | Molecular weight (g/mol) | Mixture's molecular weight (g/mol) |
|---|---|---|---|
| Methane (C1) | 0.9156 | 16.043 | 14.69 |
| Ethane (C2) | 0.0320 | 30.070 | 0.96 |
| Propane (C3) | 0.0136 | 44.097 | 0.60 |
| Isobutane (iC4) | 0.0023 | 58.124 | 0.13 |
| Normal-butane (nC4) | 0.0041 | 58.124 | 0.24 |
| Isopentane (iC5) | 0.0010 | 72.151 | 0.07 |
| Normal-pentane (nC5) | 0.0012 | 72.151 | 0.09 |
| Hexane and above (C6) | 0.0014 | 86.178 | 0.12 |
| Nitrogen ($N_2$) | 0.0078 | 28.013 | 0.22 |
| Carbon Dioxide ($CO_2$) | 0.0210 | 44.010 | 0.92 |
| Oxigen ($O_2$) | 0.0000 | 31.999 | 0.00 |
| Hydrogen ($H_2$) | 0.0000 | 2.016 | 0.00 |
| Total | 1 | – | 18.04 |

**Table 6.3** Mixture's critical pressure and temperatures calculation

| Component | Mole fraction (–) | Critical pressure (kPa) | Mixture's critical pressure (kPa) | Critical temperature (K) | Mixture's critical temperature (K) |
|---|---|---|---|---|---|
| Methane (C1) | 0.9156 | 4604 | 4215.42 | 190.55 | 174.47 |
| Ethane (C2) | 0.0320 | 4880 | 156.16 | 305.43 | 9.77 |
| Propane (C3) | 0.0136 | 4249 | 57.79 | 369.82 | 5.03 |
| Isobutane (iC4) | 0.0023 | 3648 | 8.39 | 408.13 | 0.94 |
| Normal-butane (nC4) | 0.0041 | 3797 | 15.57 | 425.16 | 1.74 |
| Isopentane (iC5) | 0.0010 | 3381 | 3.38 | 460.39 | 0.46 |
| Normal-pentane (nC5) | 0.0012 | 3369 | 4.04 | 469.39 | 0.56 |
| Hexane and above (C6) | 0.0014 | 3012 | 4.22 | 507.4 | 0.71 |
| Nitrogen ($N_2$) | 0.0078 | 3399 | 26.51 | 126.1 | 0.98 |
| Carbon Dioxide ($CO_2$) | 0.0210 | 7382 | 155.02 | 304.19 | 6.39 |
| Oxigen ($O_2$) | 0.0000 | 5081 | 0.00 | 154.60 | 0.00 |
| Hydrogen ($H_2$) | 0.0000 | 1300 | 0.00 | 33.20 | 0.00 |
| Total | 1 | – | 4646.50 | – | 201.06 |

**Table 6.4** Chemical composition of the Dutch natural gas used in the example

| Component | Mole fraction (–) |
|---|---|
| Methane (C1) | 0.85047 |
| Ethane (C2) | 0.03067 |
| Propane (C3) | 0.00610 |
| Isobutane (iC4) | 0.00092 |
| Normal-butane (nC4) | 0.00132 |
| Isopentane (iC5) | 0.00036 |
| Normal-pentane (nC5) | 0.00045 |
| Hexane and above (C6) | 0.00214 |
| Nitrogen ($N_2$) | 0.09561 |
| Carbon Dioxide ($CO_2$) | 0.01196 |
| Oxigen ($O_2$) | 0.0000 |
| Hydrogen ($H_2$) | 0.0000 |

**Example: Properties of a Dutch Natural Gas Mixture**
Repeat the previous example but use now a commonly found mixture in the Dutch gas fields in the north of the country (Groningen gas field) (Table 6.4).

## 6.3 Application of the Optimization Method to an Indeterminate Compressible and Isothermal Flow

Regarding the specific issue of the optimization of a gas pipeline project, it has already been mentioned that any project for the transmission of fluid substances over long distances must comply with basic requirements that link the facilities with their basic design parameters, namely: flow rate, maximum pressure, pressure minimum and route through which the pipeline will be laid.

With these imposed bases and with the respective mathematical and calculation formulation, it is ensured that the project is convenient to transport the required flow, although the key objective is to transport the fluid at a **minimum cost**, which—as already said—will be called the *optimal project*. The fundamental objective of this book is to introduce a calculation methodology to obtain the optimal project, which is strictly related to the continuous supply of fluid to consumers at the lowest cost. The fact of looking for a minimum cost, forces a detailed evaluation of the costs of all the components involved in the system.

With the theoretical and experimental tools of fluid mechanics, resistance of materials and heat transfer from the thermodynamic perspective, cost functions corresponding to all the elements of the installation (e.g. line pipe, compression plants, accessories, necessary energy, initial charge of product) can be established. Next, with appropriate mathematical methods, the variables are optimized, in such a way that the installation results in a minimum cost.

It was seen that in order to optimize the variables in the case of liquid transport, as a first approximation it is advisable to consider the characteristics of the fluid and the temperature throughout the pipeline as constant. The case of gases is different, since because they are compressible fluids, the variations in viscosity, and especially in density, must be taken into account in the calculations, but in this case the optimization will be considered with the simplification of assuming the flow of gas circulating at constant temperature, that is, an isothermal flow (Fig. 6.2a and b).

Once the optimal variables have been determined, it is advisable to carry out a more precise calculation, considering all the variables and variations in play. In this last case, given the complexity that an integral calculation would require, the calculation scheme is facilitated by making an approximate calculation by a finite number of segments, for pre-defined short sections of the pipe, thus obtaining for any section along the pipe, all the parameters that define the conditions of the compressible fluid (i.e., pressure, density, velocity, temperature, Mach number) (Levenspiel 1977).

Although the development of the methodology refers to the so-called gas pipelines, a name that is generally applied to the transport of natural gas through pipelines, nothing prevents that with some minimum readjustment of the deduced expressions, they can be applied to systems that transport other types of gases, for example: nitrogen, $CO_2$, hydrogen, etc.

All long-distance gas transportation pipelines are generally made up of the line pipeline, a head compressor station, and successive intermediate compressor stations. In a similar way to what was done for pipelines, and denominating $L$ the length of the entire pipe line and $\chi$ the number of compressor stations (including the header), then the pipe—if $\chi$ is greater than 1—It will be divided into $\chi$ sections. In each section, the pipe will have a maximum pressure $p_1$ at the beginning, denominated the *compressor discharge pressure*, and a minimum pressure $p_2$ at the end of the pipe, the suction piping, due to the pressure loss caused by the flow.

For the purposes of this optimization calculation, it is assumed that a steady state isothermal gas flow is established inside the pipe throughout its entire length (Guangda et al. 2014). The equation corresponding to this type of isothermal permanent flow[3] is the following:

$$\dot{m}^2 = \frac{\pi^2 D_{int}^4}{16RT} \frac{\left(p_1^2 - p_2^2\right)}{\left[\frac{f}{D_{int}}\left(\frac{L}{\chi}\right) + 2\ln\frac{p_1}{p_2}\right]} \tag{6.23}$$

[3] See book Flow of Fluids in valves, accessories and pipes—edited by the Engineering Division of Crane Co. from the USA.

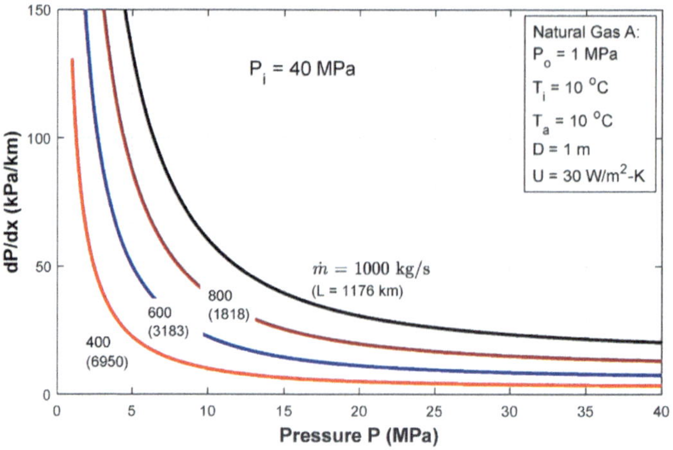


(a) Pressure drop per unit length in a natural gas pipeline with an inlet pressure of 40 MPa as a function of the mass flow rate. The linear (quasi-incompressible) behavior is clearly observed at the beginning of the pipeline[158].

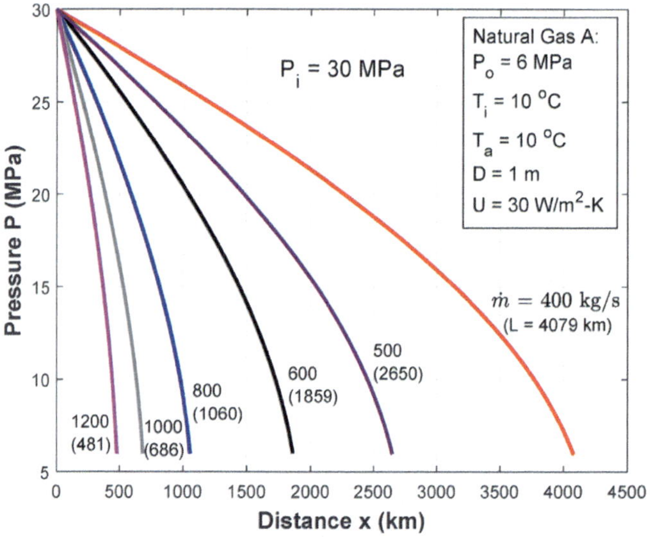


(b) Pressure drop as a function of the distance in a natural gas pipeline with an inlet pressure of 30 MPa, for different mass flow rates[158].

**Fig. 6.2** Pressure drop in compressible flow pipelines

The previous equation can be simplified for long pipes, since the logarithmic term is very small compared to the length one. This represents the mass flow rate of the transported gas[4],

$$\dot{m}^2 = \frac{\pi^2 D_{int}^5 \chi}{16RTfL}\left(p_1^2 - p_2^2\right) = \frac{\pi^2 D_{int}^5 p_1^2 \chi}{16RTfL}\left[1 - \left(\frac{p_1}{p_2}\right)^{-2}\right] \quad (6.24)$$

Being,

- $\dot{m}$ is the mass flow rate [kg/s].
- $D_{int}$ is the internal diameter [m].
- $\chi$ is the number of compression stations, including header.
- $R$ is the gas particular constant, corrected with the compressibility factor.
- $T$ is the equivalent gas temperature [K].
- $f$ is Moody's friction factor. At fully developed turbulent regimes, it only depends on the relative rugosity.
- $L$ is the pipeline's total length [m].
- $p_1$ is the maximum absolute pressure at the beginning of the pipeline section [Pa].
- $p_2$ is the minimum absolute pressure at the end of the pipeline section [Pa].

For these reasons, the mass flow rate of a certain gas that can be transported through a long pipe is a function of the diameter; the number of compression plants $\chi$, the pressure $p_1$ and the compression ratio $\left(\frac{p_1}{p_2}\right)$.

Then a single equation with 4 unknowns is obtained, which admits infinite solutions. The problem is to select—among the infinite solutions—the one that minimizes the cost of the entire installation, that is, the optimal solution. For this objective, the cost functions of all the elements that intervene in the system must be created.

As previously stated, the costs of the system will be divided into four large groups: the cost of the pipe, $c_1$; the cost of the energy used in the compression plants, $c_2$; the cost of the compression plants, $c_3$; and the cost of the initial filling gas of the pipeline $c_4$.

## 6.3.1 Determination of the Gas Equivalent Temperature

The equivalent temperature of the transported gas, to apply in the isothermal formula, can be determined by the equation[5]:

[4] Some authors affect the flow expression with a drag coefficient, also known as drag factor, which takes into account the effect of the number and severity of situations that necessarily increase the effect of roughness (e.g., welding seams, terrain curves, imperfect presentations between plumbing). The suggested range of values is from 0.93 to 0.985.

[5] 8th International Gas Conference (Stockholm 1961) "The calculations of pipelines and prerequisites for choosing optimum gas transmission conditions" authors: A.V.Alexandrov, B.V. Barabash, I.E. Khodanovich.

$$T_{ave} = T_t + \frac{T_1 - T_t}{\frac{62.6UDL}{Q\rho}}\left(1 - e^{-\frac{62.6UDL}{Q\rho c_p}}\right) \tag{6.25}$$

where,

- 62.6 is a conversion constant, derived from $24\,\frac{hours}{day} \cdot \frac{\pi}{1.204\,\frac{kg}{m^2}}$ (density of the air at standard conditions).
- $T_1$ is the gas temperature at the beginning of the section in [°C].
- $T_t$ is the terrain temperature at the pipeline depth in [°C].
- $U$ is the overall heat transfer coefficient of the system in $[\frac{kcal}{m^2 \cdot h \cdot {}^\circ C}]$.
- $c_p$ is the gas calorific capacity in $[\frac{kcal}{kg {}^\circ C}]$.
- $Q$ is the gas volumetric flow rate in [m$^3$/day].
- $\rho_r$ is the relative density of the gas to the air.
- $D$ is the pipe's external diameter in [m].
- $L$ is the section length in [km].

Using the SI system of units,

$$T_{ave} = T_t + \frac{T_1 - T_t}{\frac{U\pi DL}{\dot{m}c_p}}\left(1 - e^{-\frac{U\pi DL}{\dot{m}c_p}}\right) \tag{6.26}$$

where,

- $T_1$ is the gas temperature at the beginning of the section in [°C].
- $T_t$ is the terrain temperature at the pipeline depth in [°C].
- $U$ is the overall heat transfer coefficient of the system in $[\frac{kcal}{m^2 \cdot h \cdot {}^\circ C}]$.
- $c_p$ is the gas calorific capacity in $[\frac{kcal}{kg {}^\circ C}]$.
- $\dot{m}$ is the gas mass flow rate in [kg/s].
- $D$ is the pipe's external diameter in [m].
- $L$ is the section length in [km].

## 6.4 Lagrange Method—Cost Functions

### *6.4.1 Cost of Installed Pipe*

As previously stated, the cost of the installed pipe can be made up of two parts, one function of the weight of the pipe and another function of the exterior surface.

$$c_1 = c_{11} + c_{12} \tag{6.27}$$

Where $c_{11}$ is a function of the weight of the pipe material and a constant of proportionality, obtained from market values, and $c_{12}$ represents the coating and installation costs based on the external surface of the pipe, which for practical purposes will be assumed linear functions, valid within a certain range of diameters (Taghavi 2013). Both costs were discussed and analyzed in Chap. 4 and they do not suffer substantial modifications but those described hereunder.

The expression used to relate the material properties, the thickness and the design pressure come in this case from gas-related standards, such as ASME B31.8.

$$t = \frac{pD}{2FET\sigma_{SMYS}} \tag{6.28}$$

Where $\sigma_{SMYS}$ is the specified minimum yield strength of the material; $F$ is the design factor, with values ranging between 0.8 and 0.4 depending on the location of the line and the the population density of the adjacent area[6]; $E$ is the longitudinal welding safety factor (e.g. for seamless pipes, $E = 1$); and $T$ is the temperature factor.

### Absolute and Relative Pressures

It should be mentioned that the definition of the pressures for fluid and thermodynamic calculations (compressor condition and power equation) must be done in absolute pressures, while in the Barlow thickness formula—as in general in all expressions for mechanical calculation of casings and containers to pressure—the pressures are expressed in relative pressure values, and then a correction would have to be made in the formulas, but its influence is so small in this instance, that for simplicity it will remain in absolute pressure.

However, considering that the buried pipe will support the stress exerted by the ground, which will surely be greater than atmospheric pressure, when calculating the thickness with $p_1$ absolute, one is over-evaluating the pipe (i.e. incorporating a small over-thickness), completely acceptable within the precision limits of the optimization calculation.

### Extra Thickness Due to Corrosion

If it is expected that the gas can cause significant corrosion, e.g., acid gases basically due to the presence of sulfur compounds,—$H_2S$, mercaptans and organic sulphides—or $CO_2$, generally corrosive in the presence of water. It is advisable to define an additional thickness to that resulting from the pressure calculation (for example, it is usually adopted an over-thickness between 1.6 and 3.2 mm). It is clear that this extra thickness should not be included in the calculation of the resistance (stresses)

[6] The layout class unit is a surface that extends 200 m on each side of the longitudinal axis of a continuous 1600 m section of pipeline (see e.g. NAG-100 Section 5, or also ASME B31.8.

that is, of $c_{11}$, but it should undoubtedly be considered when calculating the cost of the $c_{12}$ installation. The weight added by the extra-thickness results in,

$$W = \pi D t_c L \rho_p g \tag{6.29}$$

and its associated cost,

$$c_{14} = \kappa_{11} t_c \rho_p g \pi D L = \kappa_{14} \pi D L \tag{6.30}$$

The final cost $c_1$ is then,

$$c_1 = \kappa_{11} \pi D_m t L g \rho_p + (\kappa_{12} + \kappa_{13} + \kappa_{14}) \pi D_{ext} L \tag{6.31}$$

$$c_1 = \kappa_{11} \frac{1 + K_D}{2} D_{ext}^2 \frac{p}{2 \cdot \sigma_{adm}} \pi g \rho_p L + (\kappa_{12} + \kappa_{13} + \kappa_{14}) \pi D_{ext} L \tag{6.32}$$

**Important: Extra Thickness**

This excess thickness due to corrosion must be considered for the selection of the D/t, since at the time of installation it eventually helps to generate the necessary strength to install the pipe. However, it should not be considered for verification by the "limited maximum pressure" formulas set forth below in Sect. 6.4.9. However, in the calculation of commercial alternatives, the total cost must include the extra-thickness due to corrosion.

**Important: Extra Thickness**

In the trunk gas pipelines of the Argentine Republic, the quality of the gas within the specification established by ENARGAS avoids any need to adopt extra-thickness due to corrosion, unlike the case of regional lines or other interconnecting gas pipelines that could carry sour gases until they are "sweetened" or suitably treated gas streams. In particular, the current exploitation of gas from the Vaca Muerta Field or other non-specification gases should be careful in this regard.

### 6.4.2 Cost of the Energy Consumed by the Compression Plants

The expression used to calculate the power necessary for the compression of the gas between two determined pressures, assuming an isentropic compression process and affecting it by a performance $\eta_c$, which depends on the type of compressor[7], of such process is the following:

$$N_c = \dot{m}\frac{k}{k-1}\frac{R_pT}{\eta_c}\left[\left(\frac{p_1}{p_2}\right)^{\frac{k-1}{k}} - 1\right]\chi\left(\frac{Z_{c,ave}}{Z_{c,2}}\right) \tag{6.33}$$

where,

- $N_c$ is the total compression power required in [W].
- $k$ is the heat capacity ratio.
- $\dot{m}$ is the mass flow rate in [kg/s].
- $R_p$ is the gas particular constant in [J/kg K].
- $T$ is the gas temperature at the suction in [K].
- $p_1$ is the discharge pressure in [Pa].
- $p_2$ is the suction pressure in [Pa].
- $\chi$ is the amount of compression plants, including the header.
- $Z_c$ is the compressibility factor.

The compressors are driven by motor machines—which can be either reciprocating gas engines, gas turbines, or electric motors—which have their own performance $\eta_m$, understood as the relationship between the heat or electrical energy delivered to the machine and the power available on the output shaft to drive the compressor. Following the same approach as described in Chap. 4, the final cost of the energy is then,

$$c_2 = \frac{\kappa_{02}\kappa_2 time}{\eta_c\eta_m}\frac{k}{k-1}\dot{m}RT\left[\left(\frac{p_1}{p_2}\right)^{\frac{k-1}{k}} - 1\right]\chi Z_{c,m} \tag{6.34}$$

### 6.4.3 Cost of Compression Plants

This cost is a non-linear function of the power installed in the compression plants; being that the compression power is basically a direct function of the pressure ratio $\frac{p_1}{p_2}$. To simplify the calculations, a valid linear function will be assumed within a certain range of powers.

$$c_3 = (\kappa_{03} + \kappa_3 N_c)\,\chi \tag{6.35}$$

[7] As approximation, for reciprocating compressors, $\eta_c = 0.85$ and for centrifugal compressors, $\eta_c = 0.83$.

- $N_c$ is the total compression power required in [W].
- $\chi$ is the amount of compression plants, including the header.
- $\kappa_{03}$ Independent costs of the compression power, representing mostly ancillary costs, such as control rooms, maintenance workshop in [EUR].
- $\kappa_3$ market constant corresponding to the unitary cost of power, for the range of power estimated in [EUR/W].

$$c_3 = \kappa_{03}\chi + \kappa_3 \dot{m} \frac{k}{k-1} \frac{R_p T}{\eta_c} \left[ \left( \frac{p_1}{p_2} \right)^{\frac{k-1}{k}} - 1 \right] \chi Z_{c,ave} \tag{6.36}$$

Since the power to be applied to each plant is not originally known, it is initially tested with approximate values of $\kappa_{03}$ and $\kappa_3$. The verification of the cost per plant can be done with the following, approximate, theoretical-practical expression obtained as a simple regression of market data, once the power involved in each plant is known:

$$K_{03} = \frac{C_0}{2} (N)^{n_{c3}} \tag{6.37a}$$

$$K_3 = \frac{C_0}{2} \frac{1}{(N)^{n_{c3}}} \tag{6.37b}$$

The coefficient $n_{c3}$ depends on several economical and market parameters, with its value going from 0.4 to 0.5. On the other hand, the value $C_0$ is taken equal to 9000 EUR/W.

### 6.4.4 Cost of the Gas to Fill the Pipeline

As previously explained, the mass cost of the gas for filling the pipeline is a function of the internal volume of the pipeline, i.e., its length and internal diameter, and also the transport pressure. As said pressure varies in each section between compressors from $p_1$ to $p_2$, for calculation purposes, the average pressure corresponding to the isothermal flow in long pipes is considered according to the following expression:

$$p_m = \frac{2}{3} \left( \frac{p_1^3 - p_2^3}{p_1^2 - p_2^2} \right) = \frac{2}{3} \frac{1 - \left( \frac{p_1}{p_2} \right)^{-3}}{1 - \left( \frac{p_1}{p_2} \right)^{-2}} \tag{6.38}$$

In order to calculate the mass inside the pipeline the average density is used, based on the Equation of State. This renders,

$$\rho_m = \frac{2}{3}\frac{p_1}{RT}\frac{1-\left(\frac{p_1}{p_2}\right)^{-3}}{1-\left(\frac{p_1}{p_2}\right)^{-2}} \tag{6.39}$$

Considering the compression ratios commonly used in gas compression in trunk pipelines (in general, they range from 1.3 to 2), said simplified cost function results in an average compression ratio of 1.5. Thus, the average pressure can be approximated to $\frac{8}{9}$, and the accumulated mass can then be estimated with:

$$m = \rho_m V_{int} = \rho_m \frac{\pi D^2 K_D^2}{4} L \tag{6.40}$$

The approximate cost of the initial fill gas, also called the initial linepack in industry jargon, will be,

$$c_4 = \kappa_{04}\kappa_4 K_D^2 \frac{2}{9}\frac{p_1}{RT}\pi D^2 L \tag{6.41}$$

Being:

- $\kappa_4$ is the unit cost[8] of gas as it enters the pipeline (treated and conditioned) in [EUR/kg].
- $\kappa_{04}$ is the residual value coefficient of the cost $c_4$. Indeed, at the end of the operational life, and prior to the abandonment of the pipeline, the initial filling gas must be extracted[9], recovering part of that initial cost, which must be evaluated with expressions of financial mathematics, setting an appropriate discount rate "i", as explained in Chap. 4.

### 6.4.5 *Total Cost of the System*

The Total Cost $c_T$ at present value of the gas transportation system, considering the installation of the pipeline and its plants, and the operation and maintenance of the system, is obtained from the sum of the costs calculated above,

$$c_T = c_1 + c_2 + c_3 + c_4 \tag{6.42}$$

The goal is then to achieve the optimization by obtaining the minimum of the cost function, that is, the minimum cost of installation and operation of the gas pipeline

[8] Generally the cost of the natural gas is expressed in EUR/MMBTU, and BTU, or British Thermal Unit, is an unit of energy equivalent to approximately 1055 J.

[9] Technical and environmental standards generally require that the abandonment of a gas pipeline be preceded by complete emptying and inerting.

system, conditioned to the pressure loss function for the isothermal compressible flow at steady state.

$$\dot{m}^2 = \frac{\pi^2 K_D^5 D^5 p_1^2 \chi}{16RTfL}\left[1 - \left(\frac{p_1}{p_2}\right)^{-2}\right] \tag{6.43}$$

Once the total cost function $c_T = f(D, p_1, \frac{p_1}{p_2}, \chi)$, as the sum of the four costs in which the system has been evaluated, it is a matter of optimizing said total cost (i.e. search for the minimum cost) obtaining the variables diameter, maximum operating pressure, compression ratio in the compressors, and the number of compression plants in the line. All of them conditioned to the aforementioned gas dynamics equation (steady-state isothermal compressible flow).

To obtain the values of the four variables that make the minimum total cost, the Lagrange Multipliers Method will be applied, as shown below,

$$c_T = f(D, p_1, \frac{p_1}{p_2}, \chi) \tag{6.44a}$$

$$\varphi = \dot{m}^2 - \frac{\pi^2 K_D^5 D^5 p_1^2 \chi}{16RTfL}\left[1 - \left(\frac{p_1}{p_2}\right)^{-2}\right] = 0 \tag{6.44b}$$

The total cost function for a compressible gas pipeline is then,

$$c_T = \left[\kappa_{11}\frac{1+K_D}{2}\frac{\pi g \rho_p}{2\cdot\sigma_{adm}} + \kappa_{04}\kappa_4 K_D^2 \frac{2}{9}\frac{\pi}{RT}\right] p_1 D^2 L + (\kappa_{12}+\kappa_{13}+\kappa_{14})\pi DL + \kappa_{03}\chi + \ldots$$
$$+ \frac{k}{k-1}\dot{m}R_p T\left[\left(\frac{p_1}{p_2}\right)^{\frac{k-1}{k}} - 1\right]\chi Z_{c,m}\left(\frac{\kappa_{02}\kappa_2 time}{\eta_c \eta_m} + \frac{\kappa_3}{\eta_c}\right) \tag{6.45}$$

## 6.4.6 *Variable Optimization for Gas Pipelines*

In the case of gas transport systems, the unknowns of the system are: the outside diameter of the pipeline, $D$; the maximum operating pressure of the line, $p_1$; the compression ratio in the plants, $\frac{p_1}{p_2}$; and the number of compression plants, including the header, $\chi$. It must be recalled that the outer diameter has been chosen for optimization purposes to reduce the number of variables in the equations. With the outer diameter and the discharge pressure, it is feasible to calculate the (theoretical) thickness and the theoretical inner diameter, similarly the inner diameter can be inferred with the coefficient $K_D$ and the outer diameter.

As stated above, in order to find the minimum of the summation function of the costs, $c_T$, conditioned on the function $\phi$, the Lagrange Multipliers method will be applied, for which the following system of partial derivative equations is set, being $\lambda$ the Lagrange multiplier.

$$\frac{\partial c_T(p_1, \chi, D, \frac{p_1}{p_2})}{\partial D} + \lambda \frac{\partial \phi(p_1, \chi, D, \frac{p_1}{p_2})}{\partial D} = 0 \tag{6.46a}$$

$$\frac{\partial c_T(p_1, \chi, D, \frac{p_1}{p_2})}{\partial p_1} + \lambda \frac{\partial \phi(p_1, \chi, D, \frac{p_1}{p_2})}{\partial p_1} = 0 \tag{6.46b}$$

$$\frac{\partial c_T(p_1, \chi, D, \frac{p_1}{p_2})}{\partial \chi} + \lambda \frac{\partial \phi(p_1, \chi, D, \frac{p_1}{p_2})}{\partial \chi} = 0 \tag{6.46c}$$

$$\frac{\partial c_T(p_1, \chi, D, \frac{p_1}{p_2})}{\partial \frac{p_1}{p_2}} + \lambda \frac{\partial \phi(p_1, \chi, D, \frac{p_1}{p_2})}{\partial \frac{p_1}{p_2}} = 0 \tag{6.46d}$$

$$\phi(p_1, \chi, D, \frac{p_1}{p_2}) = \dot{m}^2 - \frac{\pi^2 K_D^5 D^5 p_1^2 \chi}{16RTfL}\left[1 - \left(\frac{p_1}{p_2}\right)^{-2}\right] = 0 \tag{6.46e}$$

Calculating the partial derivatives with respect to each unknown, 4 equations will have been obtained with the application of the Lagrange method, which together with the condition equation make up a determined system of 5 equations with 5 unknowns, the four previously mentioned plus the Lagrange multiplier, $\lambda$. Solving the system of equations and eliminating the Lagrange variable, the following expressions are obtained (see the appendix at the end of this chapter for its derivation):

$$\left(\frac{\kappa_{02}\kappa_2 time}{\eta_c\eta_m} + \frac{\kappa_3}{\eta_c}\right)\dot{m}RT\left\{\frac{1}{2}\left(\frac{p_1}{p_2}\right)^{\frac{k-1}{k}}\left[\left(\frac{p_1}{p_2}\right)^2 - \frac{3k-1}{k-1}\right] + \frac{k}{k-1}\right\} - \kappa_{03} = 0 \tag{6.47a}$$

$$D = \left\langle\left\{\frac{k}{k-1}\left(\frac{\kappa_{02}\kappa_2 time}{\eta_c\eta_m} + \frac{\kappa_3}{\eta_c}\right)\dot{m}RT\left[\left(\frac{p_1}{p_2}\right)^{\frac{k-1}{k}} - 1\right] + \kappa_{03}\right\} \times \ldots \right.$$
$$\left. \frac{\left(\kappa_{11}\frac{1+K_D}{8\sigma}g\rho_p + \frac{\kappa_{04}\kappa_4 K_D^2}{9RT}\right)^2}{\left[\pi\left(\kappa_{12}+\kappa_{13}+\kappa_{14}\right)\right]^3}\frac{16\dot{m}^2RTf}{K_D^5\left[1-\left(\frac{p_1}{p_2}\right)^{-2}\right]}\right\rangle^{\frac{1}{4}} \tag{6.47b}$$

$$p_1 = \frac{\kappa_{12}+\kappa_{13}+\kappa_{14}}{D\left(\kappa_{11}\frac{1+K_D}{8}\frac{g\rho_p}{\sigma} + \frac{\kappa_{04}\kappa_4 K_D^2}{9RT}\right)} \tag{6.47c}$$

$$\chi = \frac{16\dot{m}^2}{\pi^2 K_D^5} \frac{RTfL}{D_5 p_1^2 \left[1 - \left(\frac{p_1}{p_2}\right)^{-2}\right]} \tag{6.47d}$$

The solution process begins with Eq. 6.47a, whence the compression ratio is calculated, and then with this pressure ratio from Eq. 6.47b the outer diameter $D$ is reckoned. Successively from Eq. 6.47c the maximum pressure in the pipeline; and finally, from Eq. 6.47d, the number of compressor stations, including the header, is computed.

During the solving cycle, the $K_D$, the friction factor Æ, the cost constants $\kappa_i$ and $\kappa_{0i}$, some o them depending on the power and/or diameter, must also be verified.

In this way it is possible to determine the four variables sought that make the total cost of the project a minimum, and for this reason they are called optimal or theoretical. Knowing the optimal variables calculated, the thickness of the pipe is calculated using the Barlow's formula, or any other from the code being considered in the calculations.

The results obtained obey a mathematical optimization and the approximation to commercial or regulatory values poses *a posteriori* a new optimization in terms of commonly used or standardized variables (e.g. commercial external diameters of the pipes), discrete variables, or dependent on current regulations (e.g. limits to maximum operating pressures), normally achieved through a series of detailed calculations of the gas pipeline being projected. In the case of pipe diameter, within the standardized series, according to API 5L Standard, for instance, the diameter closest to the calculated optimum should be selected, whether greater or less (see Chap. 4). If the theoretical optimum diameter is greater than the pipes possible to manufacture within the normal series, the project is subdivided into two pipes in parallel, also known as *loops*, with half the flow rate each, and the optimal variables are recalculated.

Algorithm 6.1 shows graphically for a gas transportation system through pipes, the logical calculation sequence for the minimum cost method and for the case of setting a maximum operating pressure, or what is the same, the fixing a specific "diameter/thickness" ratio.

## Cost Constants

Based on local costs in 2020[10], the indicative values of the cost constants are indicated to apply to the resolution of the system of optimization equations (Cerniauskas et al. 2020).

[10] Important: the estimated values for the beginning of 2020. The effects of the Coronavirus pandemic, the increase in hydrocarbon prices with the Russian invasion of Ukraine in 2022, the volatility of the dollar and high inflation do not allow us to confirm or reject them until the effects of the crisis are evaluated by the local and international market.

**Algorithm 6.1:** Solution of optimization for compressible pipelines

**Data**: $\dot{m}, L, k, R, T, H, \kappa_{ii}, \sigma_{adm}, \rho, \rho_p, \eta_i, time$
**Result**: Optimum diameter, discharge pressure, compression ratio, number of pumping stations
Estimate friction factor, $K_D$ and Diameter;
Calculate $\kappa_{11}, \kappa_{12}$;
Get $\frac{p_1}{p_2}$ from Eq. 6.47a;
Get $D$ from Eq. 6.47b;
Get $p_1$ from Eq. 6.47c;
Get $\chi$ from Eq. 6.47d;
Calculate $N_{comp}, \kappa_{03}, \kappa_3$;
Recalculate $f, \kappa_{11}, \kappa_{12}, K_D$ **if** *all these do not change* **then**
  Proceed to next verification;
**else**
  Take the calculate values and start again the optimization;
**if** *the maximum pressure* $p_1$ *is limited* **then**
  Maximum Pressure Calculation - See Sect. 6.4.9;
  Adopt $p_1$ maximum;
  Get $\frac{p_1}{p_2}$ from Eq. 6.49a;
  Get $D$ from Eq. 6.49b;
  Get $\chi$ from Eq. 6.49c;
  Calculate the friction factor $f, \kappa_{11}, \kappa_{12}, K_D, N_{comp}, \kappa_{03}, \kappa_3$; **if** $f$ *did not change* **then**
    Proceed to next verification;
  **else**
    Recalculate the parameters and start over the Maximum Pressure Calculation;
**else**
  Export all the results;
  Adopt commercial, integer and technical values for all the parameters;
  Set the 4 different alternatives based on $D$ and $\chi$ and choose the optimum alternative;

- $\kappa_{11}$ is the market constant corresponding to the cost of material equal to 0.368 [EUR/N].
- $\kappa_{12}$ is the market constant corresponding to the cost of the thermal insulating and/or anticorrosive coating per surface unit equal to 13 [EUR/m$^2$].
- $\kappa_{13}$ is the market constant corresponding to the cost of the pipe installation per surface unit equal to 240 [EUR/m$^2$].
- $\kappa_2$ is the constant corresponding to the cost of energy (see detailed calculation in Sect. 6.4.9). Bear in mind that the displayed values obtained from the specific consumption have already incorporated the $\eta_m$ and equal to 0.01015 [EUR/kW-h].
- $\kappa_3$ is the market constant corresponding to the average power unit cost of the compression stations (engine units + compressor mounted and running) for the resulting power range. The verification of the cost per plant can be done with the following approximate theoretical-practical expression obtained as a simple regression of market data, once the power (in Watts) involved in each plant is known equal to $\kappa_3 = C'nN^{n-1}$ [EUR/W].

- $\kappa_{03}$ is the figure independent to a certain extent from power, which mainly represents the cost of the land and auxiliary facilities indirectly related to the installed power, equal to $\kappa_{03} = C'(1 - n)N^n$ [EUR].
- For the two previous formulas for calculating and verifying these constant, it is possible to estimate $n = 0.48$ and $C' = 26000$.

### 6.4.7 Practical Adjustment Methodology of the Optimization Variables

Since the variables $D$ and $p_1$, and with them the theoretical thickness, are obtained from the theoretical formulation, standard commercial or manufacturing dimensions must be selected, which can be called practical or commercial adjustment of the results. In the same way, the result of $\chi$ must be operated, that is, the number of plants, which therefore must be a discrete number. Therefore, to practically solve the design of a Gas Pipeline by the optimization method, the following logical sequence must be followed:

1. Correctly solve the complete equations for obtaining $p_1$, $D$, $\chi$, $\frac{p_1}{p_2}$, achieving the verification of the adopted relationship $K_D$, the friction factor and the cost constants of the compression plants. This is achieved starting with a trial of the compression ratio in the first equation to equalize it to zero and with that compression ratio go to the Eq. 6.47b and obtain the Theoretical Optimum Diameter, and successively the Maximum Pressure and the number of compression stations.
2. Select, with respect to the theoretical optimal diameter $D$, the largest and smallest commercial diameter or. for example, if the $D$ is 31.2", select 30" (DN 750) and 32" (DN 800).
3. Select, regarding the number of compression stations $\chi$, a non-integer, the largest and smallest integer number. For example, if $\chi_{opt} = 6.4$ compressors, select 6 and 7.
4. Define several projects selecting the one with the largest diameter with the least number of compression plants and vice versa, and the largest diameter with the largest number of plants and vice versa, that is, developing at least four alternatives to evaluate technical performance and total cost.
5. Calculate for the commercial alternatives defined in the previous point the pressure losses of a Sect. $L/\chi$, applying the isothermal flow equation at a steady state, for each case of $D$ and $\chi$. In this way, the compression ratio will be obtained in each case, using the maximum pressure obtained from the resistant calculation for the selected thicknesses, also these are approximate from the commercial tables.
6. For all cases with each commercial values, with the initial cost expressions, i.e., without applying the diameter factor in $c_1$, nor the simplification of 2/9 in $c_4$, since the thickness and the real average pressure of the section respectively, calculate the total cost of each alternative and select as optimal the one that

results in the lowest cost or, in the case of comparable solutions in terms of cost, the one that meets some additional requirement that makes it more suitable for the purposes sought.

7. Finally, solve the selected optimal alternative in segments (see Sect. 6.5) and adjust the location of the compression stations in order to have similar powers in all the compression plants (i.e., achieving similar compression ratios in all compression plants).

**Important: Pressure Drop at Compressor Stations**

To calculate the pressure ratio that the compressor will have to face, it is necessary to consider that, since the compressor is far from the main duct (> 150 m—see figure) with the addition of filtering systems, headers, air-coolers, it will be necessary incorporate losses in suction and discharge (see Fig. 6.3). The allowable losses in the suction pipes must be < 1 bar and in the discharge pipes < 0.5 bar. Thus, the compression ratio for power calculation will be:

$$\frac{p_d}{p_s} = \frac{p_1 + \Delta p_s}{p_2 - \Delta p_d} \tag{6.48}$$

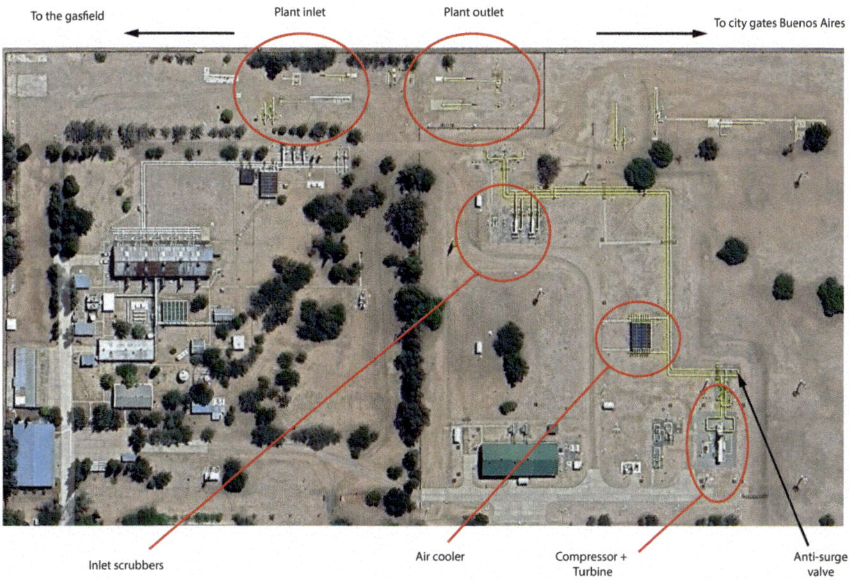


**Fig. 6.3** Satellite view of the compression plant Dean Funes (Córdoba, Argentina), part of the Northern Gas Pipeline

### 6.4.8 Final Adjustment of Projected System Variables

It should be mentioned that the calculation of the optimum diameter is an approximate calculation to solve the economic problem, and after that the fluid dynamic response of the system must be verified exactly.

Once the variables have been defined according to the eventual practical conditions for the commercial project that is selected, a detailed calculation, for small sections or pipe segments, of the gas pipeline must be undertaken, through an adequate programming of the finite difference type. Normally a formulation to permanent regime considering the heat transfer and later verification in transient behavior, to verify the parameters of the project and accurately define the operating conditions and limits for the different components of the projected duct. For this purpose, a "black-box" type software package may be used, such as those applied by the industry, e.g., PIPESIM®, TGNET®, OLGA®, ASPEN®, etc., that will allow the verification of the arrival and suction pressures and temperatures of the intermediate plants; and the discharge/suction pressure ratios in compression plants. However, if you do not have a licensed program, it is possible to organize a detailed calculation program for small sections or segments with the guidelines that will be explained below.

In case of deciding to organize a program for the resolution by sections of any long pipe, short lengths—segments—of the order 1000 or 2000 m can be used, depending on the type of flow, and in each section a constant speed can be assumed, and therefore hence the friction factor, or the Mach number, or the temperature, but never the pressure. The pressure variations due to viscous friction generate heat, a part of which is reincorporated into the flow. With the calculation of the pressure at the end of the section in question, $p_{n+1}$, you can calculate the heat transferred to or from the outside to the flow and find the temperature at the end of the section, $T_{n+1}$, and with these values obtain the new values for density and viscosity. If divergences are found with the optimization calculation, basically an inlet pressure very different from that set in the optimization calculation, said calculation should be reconsidered, for example: setting a more suitable equivalent temperature in the isothermal flow formula. Later in this chapter, the calculation method by segments is developed in detail (Popan 1977).

### 6.4.9 Limitation of the Maximum Operating Pressure

If, for reasons of safety standards or technological limitations[11], the optimum pressure, or its equivalent $t/D$, cannot be reached, the highest possible will be adopted. If

[11] It is common for countries to set maximum operating pressures for their high-pressure gas pipeline networks, although in general they do not do so formally. For example, in Argentina the limit is set at 100 bars, in Peru at 140 bars, in the Netherlands 80 bars, and for other European countries see Report of Working Committee 4 "Transmission of Gases"—21st World Gas Conference—June 6–9, 2000 "Nice" France.

this maximum pressure differs greatly from the optimum value, a new optimization must be carried out, calculating the discharge pressure according to Eq. 4.35, and looking for the optimal values of the diameter, the compression ratio and the number of compression plants, resulting in the following:

$$\left(\frac{\kappa_{02}\kappa_2 time}{\eta_c\eta_m}+\frac{\kappa_3}{\eta_c}\right)\dot{m}RT\left\{\frac{1}{2}\left(\frac{p_1}{p_2}\right)^{\frac{k-1}{k}}\left[\left(\frac{p_1}{p_2}\right)^2-\frac{3k-1}{k-1}\right]+\frac{k}{k-1}\right\}-\kappa_{03}=0 \tag{6.49a}$$

$$\left[\kappa_{11}\frac{1+K_D}{10}\frac{\pi p_1 g\rho_p}{\sigma_{adm}}+\frac{4}{45}\kappa_{04}\kappa_4 K_D^2\frac{\pi p_1}{RT}\right]D^7+\left(\frac{\kappa_{12}+\kappa_{13}+\kappa_{14}}{5}\right)\pi D^6-\ldots$$
$$-\left(\frac{\kappa_{02}\kappa_2 time}{\eta_c\eta_m}+\frac{\kappa_3}{\eta_c}\right)\dot{m}RT\left(\frac{p_1}{p_2}\right)^{3-\frac{1}{k}}\frac{8\dot{m}^2RTf}{\pi^2K_D^5p_1^2}=0 \tag{6.49b}$$

$$\chi=\frac{16\dot{m}^2}{\pi^2K_D^5}\frac{RTfL}{D^5p_1^2\left[1-\left(\frac{p_1}{p_2}\right)^{-2}\right]} \tag{6.49c}$$

To solve the system, the new compression ratio is obtained, which if the market constants have not been changed is identical to the previous one; then, the outer diameter is obtained, and from the last equation the number of compression plants is calculated, which, if the market constants have not been changed, is identical to the previous one. Then, the calculation of the commercial alternatives is carried out in a similar manner to the previous calculation.

### Pressure Ratings According to International Standards

The ASME B16.5 standard or its equivalent in Europe, EN 1092-1, determines the pressure-temperature ratings, materials, dimensions, tolerances, marking, tests, of flanges for pipes and fittings. ASME B16.5 flanges are rated Class 150, 300, 400, 600, 900, 1500, and 2500 pounds. The classes, also known as "rating" or "series", of the flanges correspond to service pressures at specified service temperatures. The European EN 1092-1 has a similar approach using what is known as nominal pressure (PN), with the following categories indicating the maximum pressure in bars at room temperature: PN 2.5, PN 6, PN 10, PN 16, PN 25, PN 40, PN 63, PN 100, PN 160, PN 250, PN 320 and PN 400.

**Table 6.5** Maximum allowable operating pressure (in bars) for different low-, medium- and high-pressure ASME B16.5 classes

| ASME B16.5 rating | 150 | 300 | 400 | 600 | 900 |
|---|---|---|---|---|---|
| MAPO (–28/+38 °C) | 19.6 | 51.1 | 68.1 | 102.1 | 153.2 |
| MAPO (+50 °C) | 19.2 | 50.1 | 66.8 | 100.2 | 150.4 |
| MAPO (+100 °C) | 17.7 | 46.6 | 62.1 | 93.2 | 139.8 |

The pressure-temperature ratings are the maximum permissible gauge working pressures expressed in bar (see the following summarized table of ASME 16.5 for material 1.1—Carbon Steel materials in the group C-Mn-Si). It can be seen from the table that the ratings are variable according to the service temperature, and it can be interpolated between them, for each Class (Table 6.5).

This table gives a certain hint on how the operating companies and regulators set the pressure *steps* of the different types of pipe service: trunk lines 140 bars (900#) or 70/100 bars (600#); HP branches 25/45 bars (300#), and MP branches 10/19 bars (150#).

**Example: Calculation of the cost coefficient dependent on time**

Given that the cost coefficients of the power consumed in the compression machines ($\kappa_2$) and of the initial filling gas ($\kappa_4$), are directly related to the quality of the natural gas, basically the calorific value, a calculation of said costs based on the prices of use in the industry normally measured in units of heat energy of the imperial system (millions of BTU). It should be noted that due to the synergy between the transported product (natural gas) and the fuel of thermal engines and turbines, it is not common to use other motorization supplied with electricity or liquid fuels.

- Higher Calorific Value = 9300 kCal/m$^3$ STD
- Lower Calorific Value = 8900 kCal/m$^3$ STD
- Approximate cost at the gas field = 3 to 5 €/MMBTU

This values are for the standard Argentinian natural gas, values may differ for different counties. Therefore, the cost related to the energy of compression can be calculated as,

- If the cost is 3 €/MMBTU $\Rightarrow \kappa_2 = 0.0103$ €/kW-h
- If the cost is 5 €/MMBTU $\Rightarrow \kappa_2 = 0.0171$ €/kW-h

The average density of the gas in standard conditions is 0.767 kg/m$^3$. This renders the following costs parameters for the cost of the fluid inside the pipeline,

- If the cost is 3 €/MMBTU $\Rightarrow \kappa_4 = 0.145$ €/kg
- If the cost is 5 €/MMBTU $\Rightarrow \kappa_4 = 0.241$ €/kg

Verification with the specific consumption of reciprocating compressors.

For a specific consumption of 4.4 STD m$^3$/day/HP = 5.95 STD m$^3$/day/kW = 0.248 STD m$^3$/kW-h

- If the cost is 3 €/MMBTU $\Rightarrow \kappa_2 = 0.0275$ €/kW-h
- If the cost is 5 €/MMBTU $\Rightarrow \kappa_2 = 0.0459$ €/kW-h

For a specific consumption of 5.66 STD m$^3$/day/HP = 7.65 STD m$^3$/day/kW = 0.319 STD m$^3$/kW-h

- If the cost is 3 €/MMBTU $\Rightarrow \kappa_2 = 0.0354$ €/kW-h
- If the cost is 5 €/MMBTU $\Rightarrow \kappa_2 = 0.0590$ €/kW-h

Considering a performance of the machine of 0.37 (37%) and 0.95 for the higher calorific values and of the compressor 0.85 (85%), the following values of $\kappa_2$ result,

- If the cost is 3 €/MMBTU $\Rightarrow \kappa_2 = 0.00822/0.01058$ €/kW-h
- If the cost is 5 €/MMBTU $\Rightarrow \kappa_2 = 0.01371/0.01763$ €/kW-h

**Verification of Specific Consumption for Centrifugal Compressors**

The first case will be to consider a turbine Saturn T-1000® (manufacturer Solar Turbines). For a specific consumption between 12 and 14 MMBTU/HP-h of in SI units, 0.4379 to 0.5108 5.66 STD m$^3$/kW-h (substantially higher than reciprocating compressors), and considering a thermal efficiency of 28%, it renders,

- If the cost is 3 €/MMBTU $\Rightarrow \kappa_2 = 0.01283$ €/kW-h
- If the cost is 5 €/MMBTU $\Rightarrow \kappa_2 = 0.2140$ €/kW-h

The second case will be the analysis of a turbine Frunze GPA-TS-5/71® (NK-12STE-8). For a specific consumption between 3200 and 3400 kCal/kW-h, it yields,

- If the cost is 3 €/MMBTU $\Rightarrow \kappa_2 = 0.008647/0.009186$ €/kW-h
- If the cost is 5 €/MMBTU $\Rightarrow \kappa_2 = 0.014412/0.015314$ €/kW-h

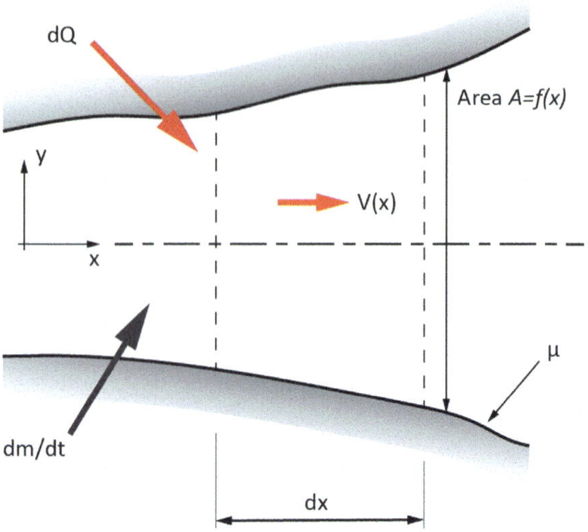


**Fig. 6.4** Differential analysis of the flow of a compressible fluid in a pipe, showing the different simplified flows that can take place

## 6.5 Calculation by Segments for Gas Pipelines

In reality, the isothermal hypothesis of the gas flow with which the optimization was solved is approximate, since in a gas pipeline the temperature of the fluid varies along the line, since the gas is discharged hot from the compressors due to the effect of the compression and it cools down in its transmission (Joule-Thomson effect), affecting the rest of the variables.

In effect, being $dT \neq 0$; $dQ \neq 0$; and $\mu \neq 0$ there is no possible analytical solution to solve system of differential equations. In this book, the technique used to solve the compressible flow is the Lorenzelli's unified method (see Appendix 6.8). Therefore, numerical techniques must be implemented in order to solve this system of nonlinear equations. In this case, a finite difference approximation is used to solve the problem, dividing the pipe in relatively small segments (Fig. 6.4).

Considering the differential equations of the one-dimensional compressible flow according to the method developed by Dr. Ing. Ezio Lorenzelli (Turin—Italy 1910, Buenos Aires—Argentina 2001), introducing the variations of the gas parameters due to height variations in the terrain (although at relatively low pressures—between 1.5 and 15 bars—its influence is negligible, it is not when transporting gas at high or very low pressures),

$$\frac{dz}{z} = M_{01}\frac{d\dot{m}}{\dot{m}} + M_{02}\frac{dA}{A} + M_{03}\frac{f}{2D}dx + M_{04}\frac{dQ}{c_pT} \tag{6.50a}$$

$$\frac{dp}{p} = M_{11}\frac{d\dot{m}}{\dot{m}} + M_{12}\frac{dA}{A} + M_{13}\frac{f}{2D}dx + M_{14}\frac{dQ}{c_pT} \tag{6.50b}$$

$$\frac{d\rho}{\rho} = M_{21}\frac{d\dot{m}}{\dot{m}} + M_{22}\frac{dA}{A} + M_{23}\frac{f}{2D}dx + M_{24}\frac{dQ}{c_pT} \tag{6.50c}$$

$$\frac{dV}{V} = M_{31}\frac{d\dot{m}}{\dot{m}} + M_{32}\frac{dA}{A} + M_{33}\frac{f}{2D}dx + M_{34}\frac{dQ}{c_pT} \tag{6.50d}$$

$$\frac{dT}{T} = M_{41}\frac{d\dot{m}}{\dot{m}} + M_{42}\frac{dA}{A} + M_{43}\frac{f}{2D}dx + M_{44}\frac{dQ}{c_pT} \tag{6.50e}$$

where $M_{ii}$ are the Lorenzelli coefficients. Being $z$ the Lorenzelli parameter,

$$z = \frac{(1+k)Ma^2}{1+kMa^2} \tag{6.51}$$

$$z_L = \lim_{Ma\to\infty} \frac{(1+k)Ma^2}{1+kMa^2} \tag{6.52}$$

Considering the pipe of constant section ($dA \approx 0$) and without extractions or injections in the considered section ($d\dot{m} = 0$), and for a small segment of pipe, a parameter will be maintained constant for each segment, choosing for this and in order to achieve greater precision and, in fact, a greater convergence, the constancy of the Lorenzelli parameter, or in other words what is ultimately keeping constant is the Mach number. The system to be solved is then:

$$\frac{dz}{z} = M_{03}\frac{f}{2D}dx + M_{04}\frac{dQ}{c_pT} \tag{6.53a}$$

$$\frac{dp}{p} = M_{13}\frac{f}{2D}dx + M_{14}\frac{dQ}{c_pT} \tag{6.53b}$$

$$\frac{d\rho}{\rho} = M_{23}\frac{f}{2D}dx + M_{24}\frac{dQ}{c_pT} \tag{6.53c}$$

$$\frac{dV}{V} = M_{33}\frac{f}{2D}dx + M_{34}\frac{dQ}{c_pT} \tag{6.53d}$$

$$\frac{dT}{T} = M_{43}\frac{f}{2D}dx + M_{44}\frac{dQ}{c_pT} \tag{6.53e}$$

Replacing the differentials by finite increments and also, as in the previous system, the Lorenzelli coefficients are functions of the Mach parameter ($M_{ii} = f(z, z_L)$), then everything remains as function of the Lorenzelli parameter such that:

$$\left(\frac{\Delta z}{z}\right)_n = \frac{z_{n+1} - z_n}{z_n} = \frac{z_{n+1}}{z_n} - 1 \rightarrow z_{n+1} = \left(1 + \frac{\Delta z}{z}\right) z_n \tag{6.54}$$

$$z_{n+1} = \left(1 + M_{03}\frac{f}{2D}\Delta x + M_{04}\frac{\Delta Q}{c_p T}\right)_n z_n \tag{6.55a}$$

$$p_{n+1} = \left(1 + M_{13}\frac{f}{2D}\Delta x + M_{14}\frac{\Delta Q}{c_p T}\right)_n p_n \tag{6.55b}$$

$$\rho_{n+1} = \left(1 + M_{23}\frac{f}{2D}\Delta x + M_{24}\frac{\Delta Q}{c_p T}\right)_n \rho_n \tag{6.55c}$$

$$V_{n+1} = \left(1 + M_{33}\frac{f}{2D}\Delta x + M_{34}\frac{\Delta Q}{c_p T}\right)_n V_n \tag{6.55d}$$

$$T_{n+1} = \left(1 + M_{43}\frac{f}{2D}\Delta x + M_{44}\frac{\Delta Q}{c_p T}\right)_n T_n \tag{6.55e}$$

Calculating for the following section $n + 1$, it is possible to calculate the pressure $p_{n+1}$ and temperature $T_{n+1}$, while the density and speed are obtained from the equation of state, in the following equations shown for an ideal gas,

$$\frac{p_{n+1}}{\rho_{n+1}} = RT_{n+1} \tag{6.56a}$$

$$V_{n+1} = Ma_{n+1}\sqrt{(kRT_{n+1})} \tag{6.56b}$$

Subsequently, $Ma_{n+1}$ and $z_{n+1}$ can be calculated, and what will allow us to calculate for the following section $n + 2$ the pressure and temperature, while the density and velocity are obtained as seen before.

In short, only three equations corresponding to the parameter are calculated using the finite difference approximation, i.e., the Lorenzelli parameter, the pressure, and the temperature. Moreover, considering a real-gas behavior, adopting the compressibility factor, function of $Z_c = f(p_{sr}; T_{sr})$, and from now on denominating $R$ to the particular constant of the gas already affected by it.

### 6.5.1 Correction for Altimetry and Viscous Dissipation

Based on the methodology developed by Lorenzelli,

$$\Delta\alpha_3 = \frac{f}{2D}\Delta x \tag{6.57a}$$

$$\Delta\alpha_4 = \frac{\Delta Q}{c_p T} \tag{6.57b}$$

Modifying both expressions, introducing terms that evaluate the influence of the altimetry, considering a positive sign for the elevations with respect to zero level, on said factors,

$$\Delta\alpha_3 = \frac{f}{2D}\Delta x + \frac{g}{V^2}\Delta h \tag{6.58}$$

Since $\Delta\alpha_3$ is a dimensionless function of the velocity $V$ through the friction factor, the altimetry correction is adimensionalized by dividing it by the square of the velocity. In the case of $\Delta\alpha_4$ also introducing a part of the heat generated by viscous friction,

$$\Delta\alpha_4 = K_\mu \frac{f}{2D}\frac{2 - z_L}{z_L - z} z\Delta x + \frac{Q}{c_p T} - \frac{g}{c_p T}\Delta h \tag{6.59}$$

where $K_\mu$ is a coefficient determining the amount of heat going to the fluid itself and the rest to the environment.

### 6.5.2 Final Formulation

Introducing the aforementioned corrections in the equations previously presented, it renders,

$$z_{n+1} = \left(1 + M_{03}\Delta\alpha'_3 + M_{04}\Delta\alpha'_4\right)_n z_n \tag{6.60a}$$

$$p_{n+1} = \left(1 + M_{13}\Delta\alpha'_3 + M_{14}\Delta\alpha'_4\right)_n p_n \tag{6.60b}$$

$$T_{n+1} = \left(1 + M_{43}\Delta\alpha'_3 + M_{44}\Delta\alpha'_4\right)_n T_n \tag{6.60c}$$

$$\rho_{n+1} = \frac{p_{n+1}}{Z_{c,n+1} R_{n+1} T_{n+1}} \tag{6.60d}$$

$$V_{n+1} = \frac{\dot{m}}{\rho_{n+1} A_{n+1}} \tag{6.60e}$$

### 6.5.3 Temperature Variation in the Compressor Stations

When, in the calculation by segments, an intermediate compressor is reached, the new section must be started considering the gas at the compressor discharge pressure and temperature conditions. For this, the discharge temperature must be estimated as follows.

The specific energy delivered by the compressor to the gas can be measured by the so-called *Isentropic Head*,

$$H_{ise} = \frac{k}{k-1}\frac{RZ_m T_2}{g}\left[\left(\frac{p_1}{p_2}\right)^{\frac{k-1}{k}} - 1\right] \tag{6.61}$$

In order to obtain the energy delivered to the gas, the isentropic hear should be multiplied by the mass flow rate and the isentropic performance, defined as,

$$\eta_{ise} = \frac{\left(\frac{p_1}{p_2}\right)^{\frac{k-1}{k}} - 1}{\frac{T_1}{T_2} - 1} = \frac{\left(\frac{p_1}{p_2}\right)^{\frac{k-1}{k}} - 1}{\left(\frac{p_1}{p_2}\right)^{\frac{k-1}{k\eta_{pol}}} - 1} \tag{6.62}$$

The polytropic performance can be obtained from measurements of pressure and temperatures at the nozzles of the compressor. The power can be calculated as,

$$N = \frac{\rho Q g H_{ise}}{\eta \eta_{ise}} \tag{6.63}$$

The discharge temperature of the gas at the exit of the compressor is obtained then as,

$$T_1 = T_2 + \frac{T_2}{\eta_{ise}}\left[\left(\frac{p_1}{p_2}\right)^{\frac{k-1}{k}} - 1\right] \tag{6.64}$$

Being the established premise to keep the Mach parameter constant for the segment, and knowing the initial data, namely: internal commercial diameter; absolute discharge pressure in the initial section; dynamic viscosity in the initial section; absolute temperature of the gas in the initial section; mass flow rate of gas to be transported; the properties for the next section are calculated according to the following algorithm,

## 6.6 Conclusions Relative to the Optimization

As seen, to focus on the problem, it is first necessary to carry out market research in order to obtain all the costs of the elements involved in the installation, i.e., materials, labor and equipment, and then develop adequate mathematical regressions to obtain the cost functions required by the method. Next, the equation that technically conditions the project must be considered, in this case the isothermal flow equation used in gas pipelines, in which all the variables to be determined are involved.

Subsequently, with the total cost function, the minimum cost conditioned to the flow equation is sought, using the Lagrange Multipliers Method, thus obtaining the optimized solutions. Then, the optimum values must be adjusted to practical values,

**Algorithm 6.2:** Calculation by segments

**Data**: $\dot{m}, L, p_1, \frac{p_1}{p_2}, H, \Delta x, \rho(p, T), \mu(p, T), p_{sc}, T_{sc}, c_p, \lambda, D_{p,T}, \chi$
**Result**: Compressor stations located, temperature profile known, power for each station
Calculate number of segments $\rightarrow n_s = \frac{L}{\Delta x}$;
Calculate entry parameters to the pipeline and dimensionless groups, i.e., Mach, Prandtl, Reynolds, Nusselt;
Calculate heat exchanged in the first segment;
**for** $i = 1$ **to** $n_s$ **do**
  Get new Lorenzelli parameter with Eq. 6.60a;
  Get new pressure with Eq. 6.60b;
  Get new temperature with Eq. 6.60c;
  Get the new Compressibility Factor $Z_c$;
  Get the new density with the equation of state;
  Get the new velocity with the mass conservation equation;
  Calculate the new Reynolds and friction factor;
  Calculate the new Mach number and the Lorenzelli parameter for $i = i + 1$;
  Calculate the heat exchanged for the segment in study;
  **if** $p_{i+1} \leq p_2$ **then**
    Locate a compressor station in this coordinate;
    $p_{i+1} = p_1$;
    $T_{i+1} =$ Eq. $4.66 + \Delta T_{compressor}$;
    Calculate real power for the compressor with adjusted compression ratio $\frac{p_1}{p_{i+1}}$;
  **else**
    No need to locate a compressor after this segment;

that is, diameter and maximum operating pressure; and finally execute the detailed verification of the flow by small segments.

The proposed method aims to meet the requirements of effective use of resources with rationality, and effective calculation possibilities, since most of the data used is usually available in companies in the sector, although some of them eventually require a limited investigation of market. Given the major importance of pipeline transportation systems within the gaseous hydrocarbon industry, the efforts made to achieve the greatest economic rationality will never be enough when projecting a new system.

This is not the first treatment given to the subject. Optimization methods in the hydrocarbons industry and in fluid transport in particular, are already long-standing. However, this new approach is that of an updated review in its mathematical language and fundamentally in the number of simultaneously optimized variables, with a scope that can be easily extended to different types of remote fluid transmission systems and adapted to local conditions in terms of units, norms, usual values, regulatory restrictions, etc.

## 6.7 Formation of Hydrates in Gas Pipelines

For many years, the freezing of a natural gas line was assumed to be the result of the formation of ice from the water present and that it occurred when the temperature of the adjacency of the pipeline was 0 °C. However, these assumptions were not verified in reality and the idea emerged that these freezing were caused by the formation of hydrates.

Hydrates are crystalline compounds formed by the chemical combination of natural gas and water that form under pressure at temperatures considerably above the freezing point of water. The first reliable records on the freezing suffered by the gas industry date from 1934. A gas hydrate extracted from a gas pipeline appears to be packed snow. The mass is very porous and light, and at atmospheric pressure large volumes of gas are given off. If the hydrate ignites under atmospheric conditions, it burns quietly and completely, leaving small residues of water.

The temperature of the gas together with its pressure and composition is the controlling factor in the possibility of formation of hydrates, so the information of the temperature of the gas in the pipeline is necessary to judge if there is a danger of formation of hydrates and the consequent blockages. The temperature of the gas in the pipeline can be known with certainty, but little can be practically done to change it. Methods to overcome difficulties during gas transportation operations are application of heat, use of alcohol, use of ammonia, and reduction of pipeline pressure.

The mentioned methods to remove hydrates are corrective and not preventive. The only completely satisfactory method of preventing hydrate formation in gas transmission lines is to dehydrate the injected gas to a dew point low enough to rule out hydrate formation at any point in the system.

## 6.8 Anti-surge Control

Pumping, or "surge", is a large fluctuation of pressure and flow that occurs when trying to operate at a pressure ratio greater than the maximum design and that causes the centrifugal compressor to operate unstable and with violent vibrations. that can lead to the destruction of its most delicate parts (seals, labyrinths, etc.).

The pumping limit can be reached from a stable operating point either by reducing the flow rate or by decreasing the suction pressure (Figs. 6.5 and 6.6). An anti-pumping system detects when conditions approach the pumping limit and maintains the unit's pressure ratio below the limit by recycling part of the discharge flow to the compressor suction, fictitiously increasing the suction flow.

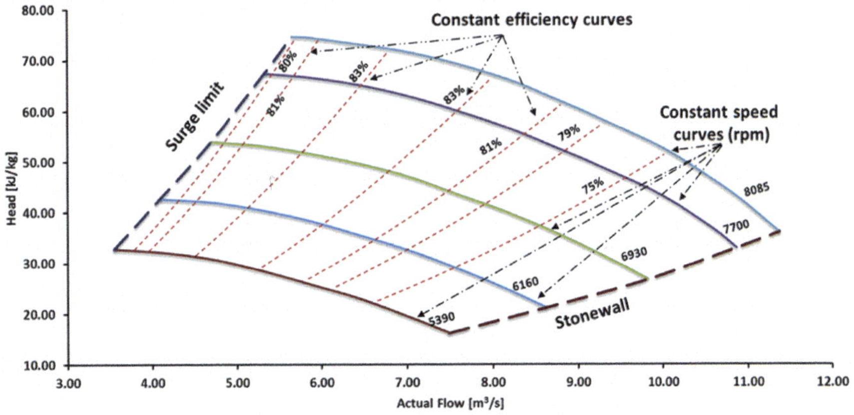


**Fig. 6.5** Typical centrifugal compressor map, showing the main lines limiting the compressor's working region (Behrooz 2016)

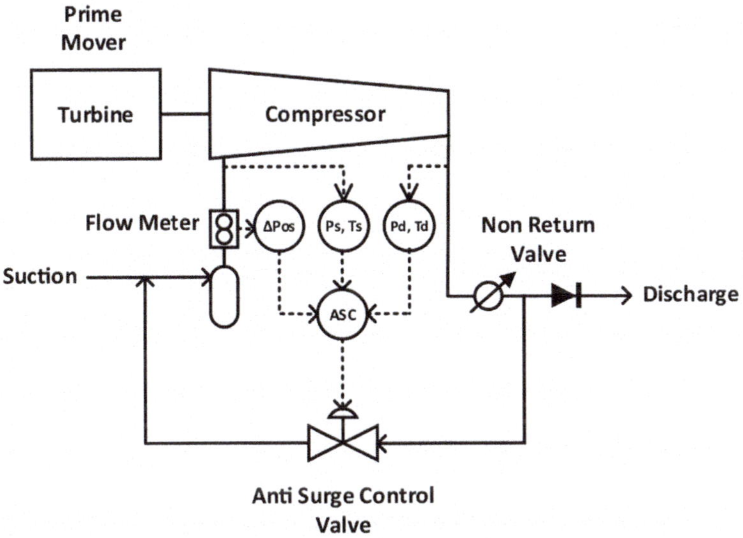


**Fig. 6.6** Anti-surge control system for a centrifugal compressor (Amin et al. 2021)

# Appendix

## *Derivation of the General Formulation*

This study contemplates the most general case, that is, the permanent uni-dimensional flow of a perfect gas with: mass input ($d\dot{m} \neq 0$), with heat input ($dQ \neq 0$), in a variable section pipe ($dA \neq 0$) and with viscous friction ($\mu \neq 0$).

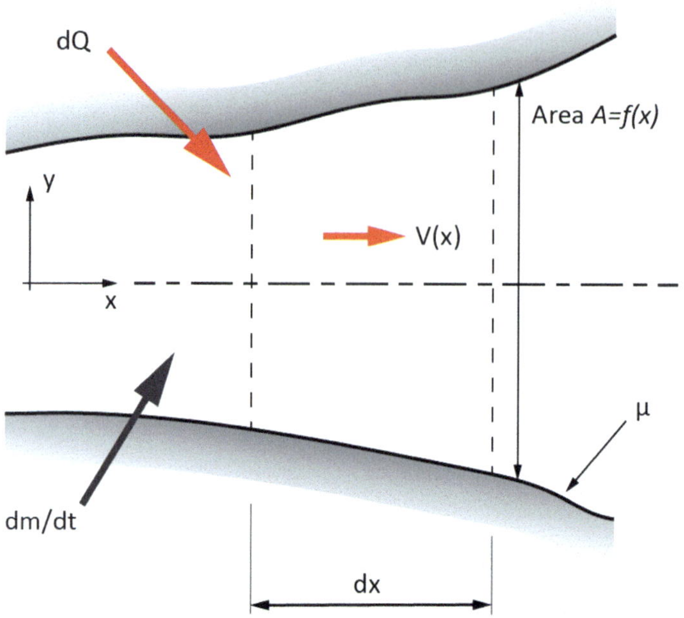


Figure 6.4 showing the different type of simple compressible flows that can be present in a flow in pipelines.

A differential analysis of the system is performed to a control volume, using Reynolds' Transport Theorem to obtain balances of the three fundamental parameters of classical mechanics: mass, momentum and energy. The appendix will only include a very simplified analysis of this balances, only to continue with the further derivation of the Lorenzelli's method.

## *Conservation of Mass*

The conservation of mass for permanent flow of a compressible flow is,

$$\dot{m} = \rho V A \rightarrow \frac{d\dot{m}}{\dot{m}} = \frac{d\rho}{\rho} + \frac{dV}{V} + \frac{dA}{A} \tag{6.65}$$

## *Conservation of Momentum*

Applying Euler's theorem to the reference volume that encompasses the fluid contained in the pipe of length dx. The total force on the volume due to pressures and velocities must be equal to the viscous resistance (due to steady state), and discarding the higher-order differentials,

$$dF = pA - (p + dp)(A + dA) + \left(p + \frac{dp}{2}\right)dA - [-v\dot{m} + (v + dv)(\dot{m} + d\dot{m})] = -Vd\dot{m} - \dot{m}dV - Adp \tag{6.66}$$

The viscous resistance is expressed as,

$$dF_{\mu} = f\frac{v^2}{2}\frac{dx}{D}\rho A \tag{6.67}$$

$$\frac{f}{2D}dx = -\frac{d\dot{m}}{\dot{m}} - \frac{dV}{V} - \frac{dp}{\rho V^2} \tag{6.68}$$

$$-\frac{f}{2D}dx - \frac{d\dot{m}}{\dot{m}} = \frac{dV}{V} + \frac{dp}{p}\frac{1}{kMa^2} \tag{6.69}$$

## *Conservation of Energy*

Considering the heat transmission in a segment of the differential control volume in steady-state conditions,

$$dQ = d\left[h + \frac{V^2}{2} + U_g\right] \tag{6.70}$$

Remembering that the entalphy for ideal gases can be written as,

$$h = u + \frac{p}{\rho} = c_pT = \frac{k}{k-1}\frac{p}{\rho} = \frac{c^2}{k-1} \tag{6.71}$$

Being $c = \sqrt{kRT}$ the speed of sound for an ideal gas. Discarding the potential term and dividing by the entalphy,

$$\frac{dQ}{c_pT} = \frac{dh}{h} + \frac{V^2dv(k-1)}{c^2V} \tag{6.72}$$

$$\frac{dh}{h} = \frac{dT}{T} = \frac{dp}{p} - \frac{d\rho}{\rho} \tag{6.73}$$

$$\frac{dQ}{c_pT} = \frac{dp}{p} - \frac{d\rho}{\rho} + (k-1)Ma^2\frac{dV}{V} \tag{6.74}$$

Therefore, the resulting system of equation is,

$$\frac{d\dot{m}}{\dot{m}} = \frac{d\rho}{\rho} + \frac{dV}{V} + \frac{dA}{A} \tag{6.75a}$$

$$-\frac{f}{2D}dx - \frac{d\dot{m}}{\dot{m}} = \frac{dV}{V} + \frac{dp}{p}\frac{1}{kMa^2} \tag{6.75b}$$

$$\frac{dQ}{c_pT} = \frac{dp}{p} - \frac{d\rho}{\rho} + (k-1)Ma^2\frac{dV}{V} \tag{6.75c}$$

It can be written the Lorenzelli parameter $z = f(Ma) = f(V, T)$ and then $\frac{dz}{z} = f\left(\frac{dV}{V}; \frac{dT}{T}\right)$. The Mach number varies between 0, known as tank or stagnation condition to infinite or limit condition, going through the subsonic flow for $Ma < 1$, by the sonic flow $Ma = 1$ (critical condition), and by supersonic flow $Ma > 1$. The upper limit makes it difficult to find the solution equations of the different flows and then, to solve the problem, a change of variable can be considered, called the Lorenzelli parameter, defined by the relation,

$$z = \frac{(1+k)Ma^2}{1+kMa^2} \tag{6.76}$$

when $Ma = 0 \rightarrow z = 0$ and $Ma \rightarrow \infty$,

$$z_L = 1 + \frac{1}{k} \tag{6.77}$$

Taking the differential of the equation,

$$\frac{dz}{z} = 2\left(1 - \frac{z}{z_L}\right)\frac{dMa}{Ma} = \left(1 - \frac{z}{z_L}\right)\left(2\frac{dV}{V} - \frac{dT}{T}\right) \tag{6.78}$$

Replacing this in the previous system of equations and rearranging,

$$\frac{d\dot{m}}{\dot{m}} - \frac{dA}{A} = 0\frac{dp}{p} + \frac{d\rho}{\rho} + \frac{dV}{V} \tag{6.79a}$$

$$-\frac{f}{2D}dx - \frac{d\dot{m}}{\dot{m}} = \left(\frac{z_L}{z} - 1\right)\frac{dp}{p} + \frac{d\rho}{\rho} + \frac{dV}{V} \tag{6.79b}$$

$$\frac{dQ}{c_pT} = \frac{dp}{p} - \frac{d\rho}{\rho} + \frac{z(2-z_L)}{z_L - z}\frac{dV}{V} \tag{6.79c}$$

Solving by determinants and everything as a function of the parameter Z and calling $d\alpha_1 = \frac{d\dot{m}}{\dot{m}}$, $d\alpha_2 = \frac{dA}{A}$, $d\alpha_3 = \frac{f}{2D}dx$ and $d\alpha_4 = \frac{dQ}{c_pT}$, with the system of equations the three unknowns, pressure, density and velocity, can be found. Using the equation of state, the temperature $\frac{dT}{T}$ can also be calculated. Thus, the final system renders,

$$\frac{dz}{z} = M_{01}\frac{d\dot{m}}{\dot{m}} + M_{02}\frac{dA}{A} + M_{03}\frac{f}{2D}dx + M_{04}\frac{dQ}{c_pT} \tag{6.80a}$$

$$\frac{dp}{p} = M_{11}\frac{d\dot{m}}{\dot{m}} + M_{12}\frac{dA}{A} + M_{13}\frac{f}{2D}dx + M_{14}\frac{dQ}{c_pT} \tag{6.80b}$$

$$\frac{d\rho}{\rho} = M_{21}\frac{d\dot{m}}{\dot{m}} + M_{22}\frac{dA}{A} + M_{23}\frac{f}{2D}dx + M_{24}\frac{dQ}{c_pT} \tag{6.80c}$$

$$\frac{dV}{V} = M_{31}\frac{d\dot{m}}{\dot{m}} + M_{32}\frac{dA}{A} + M_{33}\frac{f}{2D}dx + M_{34}\frac{dQ}{c_pT} \tag{6.80d}$$

$$\frac{dT}{T} = M_{41}\frac{d\dot{m}}{\dot{m}} + M_{42}\frac{dA}{A} + M_{43}\frac{f}{2D}dx + M_{44}\frac{dQ}{c_pT} \tag{6.80e}$$

The coefficients $M_{ij}$ are all functions of the Lorenzelli parameter. The $M_{1/2/3,i}$ parameters are obtained from the original system of equations. The $M_{4,i}$ are obtained from the equation of state. Finally, the $M_{0,i}$ are obtained according to,

$$M_{0,i} = \left(1 - \frac{z}{z_L}\right)(2M_{3,i} - M_{4,i}) \tag{6.81}$$

The solution to the system of equations can be applied to the following cases, also known as *pure* cases. This is when 3 out of 4 $d\alpha_i$ are null. The cases are then (Table 6.6),

- Elemental combustion flow with variable mass ($d\alpha_1 \neq 0$; $d\alpha_2 = d\alpha_3 = d\alpha_4 = 0$).
- Isentropic flow with variable area ($d\alpha_2 \neq 0$; $d\alpha_1 = d\alpha_3 = d\alpha_4 = 0$).
- Flow with viscous resistance (or Fanno flow) ($d\alpha_3 \neq 0$; $d\alpha_1 = d\alpha_2 = d\alpha_4 = 0$).
- Flow with heat addition (or Rayleigh flow) ($d\alpha_4 \neq 0$; $d\alpha_1 = d\alpha_2 = d\alpha_3 = 0$).

**Table 6.6** Table showing the different coefficients $M_{m,n}$ for the simple flows, allowing to calculating the different state/operational parameters

| Parameter | Combustion ($n = 1$) | Isentropic ($n = 2$) | Fanno ($n = 3$) | Rayleigh ($n = 4$) |
|---|---|---|---|---|
| Mach ($m = 0$) | $+\frac{2-z}{1-z}$ | $-\frac{(2-z)(z_L-z)}{(1-z)z_L}$ | $+\frac{(2-z)z}{(1-z)z_L}$ | $+\frac{z_L-z}{(1-z)z_L}$ |
| Pressure ($m = 1$) | $-\frac{(2-z)z}{(1-z)(z_L-z)}$ | $+\frac{z}{(1-z)z_L}$ | $-\frac{z_L-(z_L-1)z}{(1-z)(z_L-z)z_L}z$ | $-\frac{z}{(1-z)z_L}$ |
| Density ($m = 2$) | $-\frac{z}{1-z}$ | $+\frac{(z_L-1)z}{(1-z)z_L}$ | $-\frac{z}{(1-z)z_L}$ | $-\frac{z_L-z}{(1-z)z_L}$ |
| Velocity ($m = 3$) | $+\frac{1}{1-z}$ | $+\frac{(z_L-z)}{(1-z)z_L}$ | $+\frac{z}{(1-z)z_L}$ | $+\frac{z_L-z}{(1-z)z_L}$ |
| Temperature ($m = 4$) | $-\frac{z(2-z_L)}{(1-z)z_L}$ | $+\frac{(2-z_L)z}{(1-z)z_L}$ | $-\frac{(2-z_L)z^2}{(1-z)(z_L-z)z_L}$ | $-\frac{2z-z_L}{(1-z)z_L}$ |

When two of the four coefficients are different from zero and an extra condition is added (e.g. constant temperature, constant pressure, etc.), with the equation of the system corresponding to the conditioned parameter, it allows to relate two of the non-zero coefficients and in this way, integrals can be solved as in the case of pure flows. This leads to the *conditioned isothermal flow*. This is complemented with no mass addition ($d\alpha_1 = 0$) and constant area ($d\alpha_2 = 0$).

$$\frac{dT}{T} = 0 \tag{6.82}$$

$$d\alpha_4 = -\frac{M_{43}}{M_{44}} d\alpha_4 \tag{6.83}$$

$$\frac{dz}{z} = \left( M_{03} - M_{04}\frac{M_{43}}{M_{44}} \right) \frac{f}{2D} dx \tag{6.84a}$$

$$\frac{dz}{z} = \left( M_{04} - M_{03}\frac{M_{44}}{M_{43}} \right) \frac{dQ}{c_p T} \tag{6.84b}$$

And finally,

$$\frac{dy_n}{y_n} = \left( \frac{M_{n3} - M_{n4}\frac{M_{43}}{M_{44}}}{M_{03} - M_{04}\frac{M_{43}}{M_{44}}} \right) \frac{dz}{z} \tag{6.85}$$

## *Derivation of the equation of the Lagrange formulation*

The total cost for a compressible pipeline project is,

$$c_T = \left[ \kappa_{11} \frac{1 + K_D}{2} \frac{\pi g \rho_p}{2 \cdot \sigma_{adm}} + \kappa_{04}\kappa_4 K_D^2 \frac{2}{9} \frac{\pi}{RT} \right] p_1 D^2 L + (\kappa_{12} + \kappa_{13} + \kappa_{14}) \pi D L + \kappa_{03}\chi + \ldots$$
$$+ \frac{k}{k-1} \dot{m} R_p T \left[ \left( \frac{p_1}{p_2} \right)^{\frac{k-1}{k}} - 1 \right] \chi Z_{c,m} \left( \frac{\kappa_{02}\kappa_2 time}{\eta_c \eta_m} + \frac{\kappa_3}{\eta_c} \right) \tag{6.86}$$

Applying the Lagrange Multipliers method, the following system of equations can be written,

$$\frac{\partial c_T(p_1, \chi, D, \frac{p_1}{p_2})}{\partial D} + \lambda \frac{\partial \phi(p_1, \chi, D, \frac{p_1}{p_2})}{\partial D} = 0 \tag{6.87a}$$

$$\frac{\partial c_T(p_1, \chi, D, \frac{p_1}{p_2})}{\partial p_1} + \lambda \frac{\partial \phi(p_1, \chi, D, \frac{p_1}{p_2})}{\partial p_1} = 0 \tag{6.87b}$$

$$\frac{\partial c_T(p_1, \chi, D, \frac{p_1}{p_2})}{\partial \chi} + \lambda \frac{\partial \phi(p_1, \chi, D, \frac{p_1}{p_2})}{\partial \chi} = 0 \tag{6.87c}$$

$$\frac{\partial c_T(p_1, \chi, D, \frac{p_1}{p_2})}{\partial \frac{p_1}{p_2}} + \lambda \frac{\partial \phi(p_1, \chi, D, \frac{p_1}{p_2})}{\partial \frac{p_1}{p_2}} = 0 \tag{6.87d}$$

$$\phi(p_1, \chi, D, \frac{p_1}{p_2}) = \dot{m}^2 - \frac{\pi^2 K_D^5 D^5 p_1^2 \chi}{16RTfL} \left[ 1 - \left( \frac{p_1}{p_2} \right)^{-2} \right] = 0 \tag{6.87e}$$

The equations in the main body of the chapter, this is, Eqs. 6.47a–6.47d can be obtained using the following approach:

- Equation 6.47a is derived using Eqs. 6.87a and 6.87b.
- Equation 6.47b is derived using Eqs. 6.87a and 6.87c. The Eq. 6.47c should also be used to replace the value of the pressure $p_1$ in the equation.
- Equation 6.47c is derived using Eqs. 6.87c and 6.87d.
- Equation 6.47d is derived using the condition equation 6.87e.

# References


Amin AA, Maqsood MT, Mahmood-ul Hasan K. Surge protection of centrifugal compressors using advanced anti-surge control system. Measur Control. 2021;54(5–6):967–82.

Behrooz HA. Managing demand uncertainty in natural gas transmission networks. J Nat Gas Sci Engin. 2016;34:100–11.

Cerniauskas S, Junco AJC, Grube T, Robinius M, Stolten D. Options of natural gas pipeline reassignment for hydrogen: Cost assessment for a germany case study. Int J Hydrog Energy. 2020;45(21):12095–107. ISSN 0360-3199. https://doi.org/10.1016/j.ijhydene.2020.02.121. https://www.sciencedirect.com/science/article/pii/S0360319920307023.

Dillon G. File: main line valve number 7 2008. https://commons.wikimedia.org/wiki/File:MainLineValve7.jpg. [Accessed 27 Mar 2024].

Guangda ZHAO, Qiji S, Xiaohu W, Xu W. Natural gas transmission pipeline temperature drop calculation. Adv Pet Explor Dev. 2014;7:127–31.

Levenspiel O. The discharge of gases from a reservoir through a pipe. AIChE J. 1977;23(3):402–3.

Mohitpour M, Golshan H, Murray MA. Pipeline design and construction: a practical approach. ASME Press: Pipelines and Pressure Vessels; 2007.

Popan VA. How to calculate pressure drop in compressor station piping. Pipeline Gas J. 1977;204(6):29–33.

Prasad V, Almara LM, Wang G-X. Ultra-long-distance transport of supercritical natural gas (sng) at very-high mass flow rates via pipelines through land, underground, water bodies, and ocean. Gas Sci Engin. 2023;117: 205053.

Taghavi N. Eonomic investigation on the use of internal coating for natural gas trunk-lines. Chem Engin Res Design. 2013;91(9):1725–30. ISSN 0263-8762. https://doi.org/10.1016/j.cherd.2013.03.012. https://www.sciencedirect.com/science/article/pii/S0263876213001020.

# Chapter 7
# Transportation of Liquid and Gaseous Substances Simultaneously Through Pipelines

> Engineering is not merely knowing and being knowledgeable, like a walking encyclopedia; engineering is not merely analysis; engineering is not merely the possession of the capacity to get elegant solutions to non-existent engineering problems; engineering is practicing the art of the organizing forces of technological change ... Engineers operate at the interface between science and society.
>
> *Gordon S. Brown*

## 7.1 Introduction

There are numerous applications where economics dictate the movement of more than one phase in a pipeline. The liquid-gas type frequently occurs in oil field lines and is the flow of hydrocarbon liquids stirred with vapors or gases. This type of flow is also usually found in flow lines and collection gas pipelines.

In this chapter, only applications for liquid and gaseous hydrocarbons are developed. In general, the primary liquid phase is a liquid hydrocarbon, but in reality water must be considered as a single liquid or in combination with oil or condensate. Since these liquids are insoluble in each other, two liquid phases could be present. Some water is present in most applications but a small amount will have a minor effect on line performance.

Historically, many two-phase pipelines had operational problems with excessive pressure drops as a result of inadequate design. However, the development of two-phase lines in many situations were the correct choice, both operationally and due to the fact that they saved 20 to 25% compared to two single-phase lines.

Due to this potential economic attractiveness of two-phase lines, many researchers set their focus from 1950 on wards. Although a single method has not been achieved, continued experience has allowed engineers to develop guidelines for the limits of application of available methodological approaches. In short, a *magic* solution for all systems has not appeared to date.



P. Druetta et al., *Transport by Pipelines and Its Optimization*, Petroleum Engineering,
https://doi.org/10.1007/978-3-031-66566-0_7

Few gas pipelines in production operations are truly liquid-free and the presence of separators is the tacit recognition to this. As many other factors affect flow, a factor of 0.92 to 0.94 (and above) represents a “dry” pipeline since the amount of liquids does not affect flow calculations. The use of cleaning spheres or *pigs* generally begins when the factor approaches 0.88.

It is convenient to divide multiphase flow into two large categories: (a) horizontal lines, those where the length is very large; and (b) vertical lines, those where the elevation is important with respect to the length.

## 7.2 Fluid Mechanics

In the transport of mixtures, the gas tends to be located in the center and the liquid to wet the walls, but there are various degrees of dispersion of each phase in the other, which causes different types of flows.

It will be seen in the next point that researchers have detected between 4 to 10 types of flow and beyond the number, ranging from a range of continuous liquid phase with discrete gas bubbles to a continuous gas phase containing a sprayed liquid (i.e., fog). Emulsions or foams are special cases of flows (Barnea 1987; Poesio et al. 2009; Grassi et al. 2008; Babu 2013).

It should be noted that it influences the type of flow regime:

- The relative quantity of the phases and their velocities (relative and absolute) are of great importance.
- The size and inclination of the line segments.
- The pressure and temperature of the fluid in each segment.
- In distributed flows the precision of the calculations can reach 5 to 10% and in other regimes between 10 and 15%.

The total pressure drop, considering the gradients due to elevation changes, friction losses and acceleration changes, is generally expressed by:

$$\left(\frac{\Delta p}{\Delta L}\right)_{TOT} = C\left[\left(\frac{\Delta p}{\Delta L}\right)_{mu} + \left(\frac{\Delta p}{\Delta L}\right)_{pot} + \left(\frac{\Delta p}{\Delta L}\right)_{acc}\right] \quad (7.1)$$

In distributed flows the pressure drop is between the composite fluid and the pipe, although cannot be treated as a single phase flow. On the other hand, in segregated or intermittent flows, the pressure drop is the result of friction with the pipe and the interaction between the phases (Cao et al. 2023).

Most of the mathematical correlations used to solve multiphase flow, or calculation of head loss, are (semi-)empirical. The problem is to determine the type of flow for a given set of conditions and apply the correct correlation (López et al. 2016; Barnea 1987; Poesio et al. 2009; Krumrick et al. 2016; Cheng and Xia 2023). The type of flow regime depends on:

- gas and liquid flow rates;
- gas-liquid relationship;
- physical properties of gas and liquid;
- line diameter; and,
- interface energies (i.e., interfacial tension) and shear forces (i.e., viscosity) between the phases.

With respect to horizontal lines, these cannot be truly found in reality, but it is understood in these cases that the length is much greater than any change in elevation.

In most lines, multiphase flow occurs in the uphill and downhill portions as well as in the horizontal sections. If the energy of the liquid entering a low point is insufficient to carry it towards the next slope the liquid will accumulate at that low point.

The accumulation of the liquid will reduce the section and the speed of the gas will increase until it has enough energy to start moving the liquid. A part will be propelled by the gas upwards and a part will return to the low point, these fluctuations cause excessive pressure losses. For this reason it is usually desirable to maintain high flows to keep most of the liquids moving. Some researchers suggest 4 to 15 m/s as minimum velocities and recommend not oversizing the pipes.

## 7.2.1 Definitions Applicable to the Study of Multiphase Flows

### Superficial and Mixture Velocities

These velocities are defined as the respective volumetric flow rate of the phase divided by the total cross sectional area of the pipeline, considered as if that phase is flowing alone through the pipe.

$$v_{SL} = \frac{4Q_l}{\pi D_{int}^2} \tag{7.2a}$$

$$v_{SG} = \frac{4Q_g}{\pi D_{int}^2} \tag{7.2b}$$

It it important to notice that the superficial velocities are calculated using the total area, not the actual area occupied in the pipeline. Thus, the real velocity in the pipeline is always higher than the superficial velocities. Finally, the mixture velocity is simply the addition of the superficial velocities,

$$v_m = v_{SL} + vSG \tag{7.3}$$

## Slippage and Liquid Hold-Up

Slippage is a phenomenon that can be observed not only in pipelines, but almost everywhere two different fluids have an interface and flow with certain velocities. Their actual velocities in a specific place tend to be different, and usually the less dense fluid is flowing with a higher velocity, starting to slip on the higher dense fluid. Therefore, the inlet volume fractions are different than those present in different points of the pipeline (Ma et al. 2023).

**Liquid hold-up with slip** This parameter is commonly used in many multiphase correlations, as it will be shown in the following sections. As mentioned previously, when the slippage is considered, then the input and actual volume fractions tend to differ. The liquid hold-up with slip ($E_l$) is the fraction of the pipeline's cross sectional area that is occupied by liquid, defined as,

$$E_l = \frac{A_l}{A_p} \tag{7.4}$$

**Input Liquid Hold-Up** These hold-ups ($C_l$, $C_g$) are defined based on the volumetric flow rates of the two phases,

$$C_l = \frac{Q_l}{Q_l + Q_g} = \frac{v_{SL}}{v_m} \tag{7.5a}$$

$$C_g = \frac{Q_g}{Q_l + Q_g} = \frac{v_{SG}}{v_m} \tag{7.5b}$$

These allows defining then the actual velocities at every point of the pipeline,

$$v_l = \frac{v_{SL}}{E_l} \tag{7.6a}$$

$$v_g = \frac{v_{SG}}{1 - E_l} \tag{7.6b}$$

### *Densities and Viscosities*

In a two-phase flow, the concept of densities and viscosities becomes more complicated than in single-phase flows, since one needs to consider the properties of the fluids themselves, i.e., compressible or incompressible, but also the percentage of every fluid. Based on this, one can define two different groups,

**Mixture Parameters** This is a value representing the in-situ density of the two-phase flow define as,

$$\rho_m = \rho_l E_l + \rho_g E_g = \rho_l E_l + \rho_g (1 - E_l) \tag{7.7a}$$

$$\mu_m = \mu_l E_l + \mu_g E_g = \mu_l E_l + \mu_g (1 - E_l) \tag{7.7b}$$

**No-Slip Parameters** This is calculated under the assumption that there is no slip between the phases, i.e., both are moving at the same velocity in that specific point of the pipeline.

$$\rho_{ns} = \rho_l C_l + \rho_g C_g = \rho_l C_l + \rho_g (1 - C_l) \tag{7.8a}$$

$$\mu_{ns} = \mu_l C_l + \mu_g C_g = \mu_l C_l + \mu_g (1 - C_l) \tag{7.8b}$$

## 7.3 Types of Flows

### *7.3.1 Horizontal Flow*

Three major divisions can be made of the basic regimes/types of two-phase flow (Figs. 7.1, 7.2 and 7.3):

1. Distributed: a single continuous phase and the other distributed (e.g., bubbles, fog and foam);
2. Segregated: two phases are continuous with the gas phase moving faster (e.g., strata, waves, annular); and,
3. Intermittent: the continuity of both phases is interrupted and a series of inertial and kinetic forces are modified (e.g., plug, waves).

### *7.3.2 Vertical Flow*

In the case of vertical flows, the patterns found are similar to those found in horizontal pipelines with one major distinction: due to the effect of the gravity forces, now acting on the pipeline axis, there is no stratified flow pattern present (Figs. 7.4, 7.5, and 7.6).

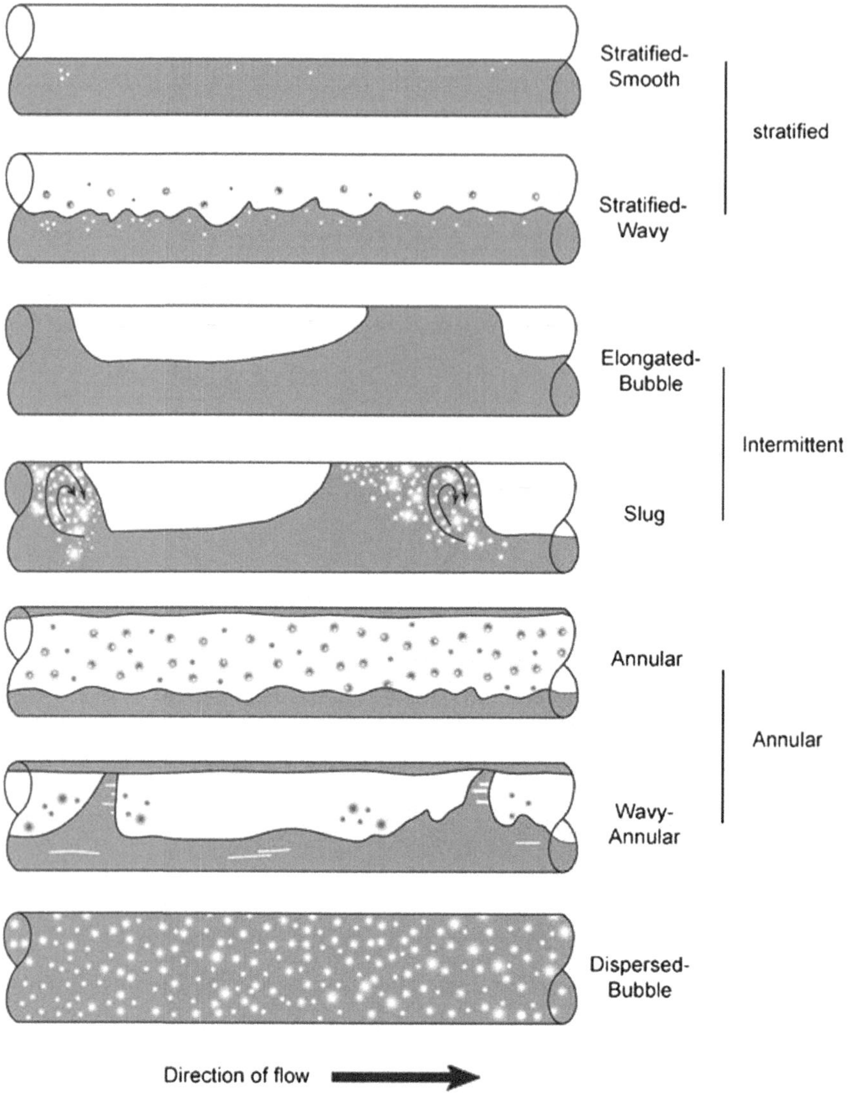


**Fig. 7.1** Different flow patterns that can be found in horizontal flow pipelines (Najmi et al. 2015)

## 7.4 Methodological Development of Two-Phase Flows

In general, the authors define a parameter $H_L$, or liquid hold-up, as the fraction of the pipe occupied by the flow of liquids and allows knowing the pseudo-average linear velocity of each phase. Consequently, the difference in speed between the phases, which is called slip speed, can be calculated. In general, as the gas moves faster than

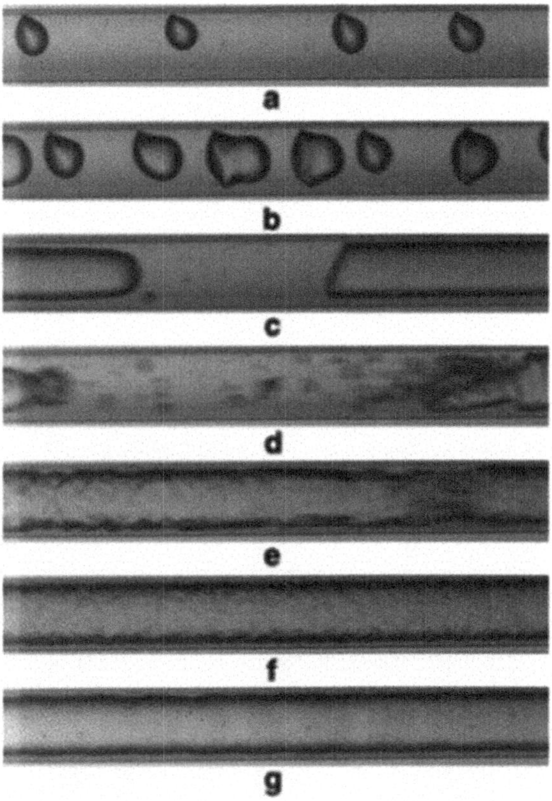


**Fig. 7.2** Experimental results showing different flow patterns found in horizontal pipelines (Revellin et al. 2006)

the liquid, a shear force is produced at the interface which causes an energy transfer between the phases.

In a given length of pipe various flow regimes can occur due to varying forces, elevations, and liquid-gas ratios; and the difference between correlations depends fundamentally on how each researcher treats the influence of each variable.

With these principles in mind, the most commonly used models will be considered, presented in the approximate order in which they were developed.

### *7.4.1 Lockhart-Martinelli*

This method, introduced in 1949 their authors,[1] was the first quantitative approach proposed, it has become a *classic* compared to other later works. It is a simplified method that proposes four combinations of *viscous* (or *laminar*) and *turbulent* flows and is based on two basic hypotheses:

[1] R.W Lockart and R.C. Martinelli "Chemical Engineering Progress" Vol 45 (1949).

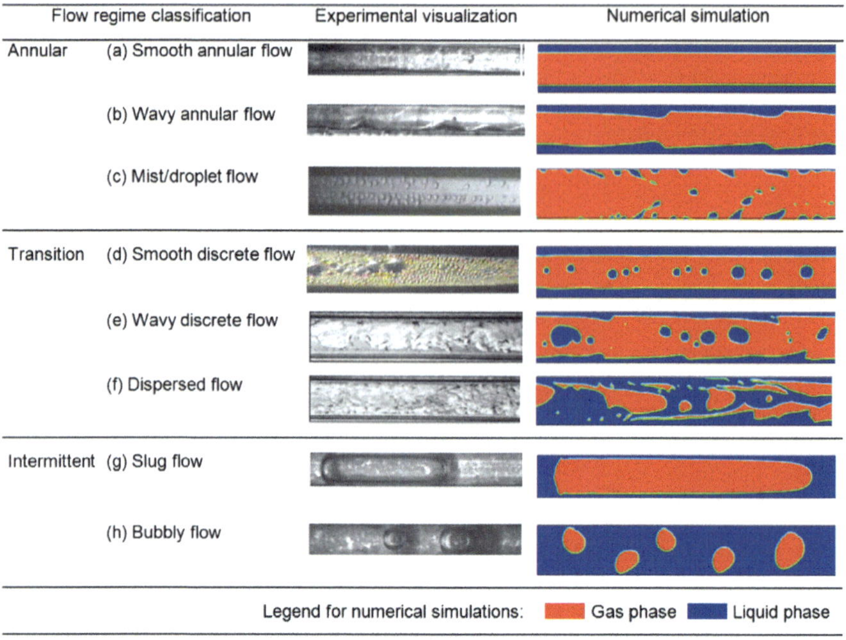


**Fig. 7.3** Table showing the different patterns found in horizontal flow with respective CFD simulations (Ganapathy et al. 2013)

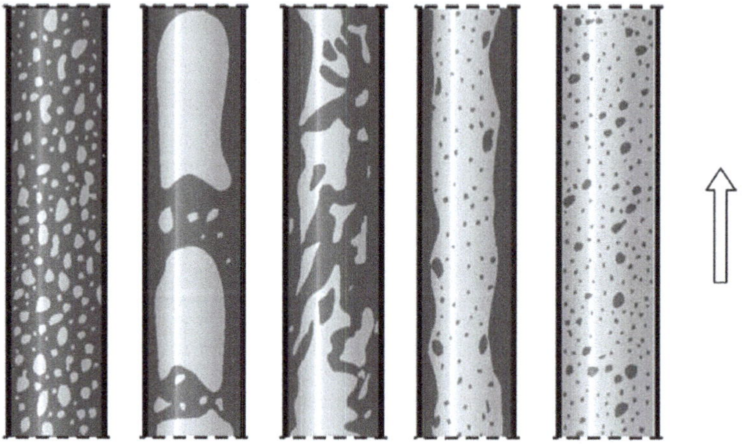

**Fig. 7.4** Flow patterns found in vertical pipelines, showing similarities with the patterns found in the horizontal pipes (Dziubinski et al. 2004)

- The pressure drop is the same for the gas and the liquid ($\Delta p_g = \Delta p_l$).
- The total volume of the pipe is equal to the sum of the volumes occupied by the gas and the liquid ($V_p = V_g + V_l$)

**Fig. 7.5** Experimental work showing the different flow patterns in vertical pipes (Schmid et al. 2022)

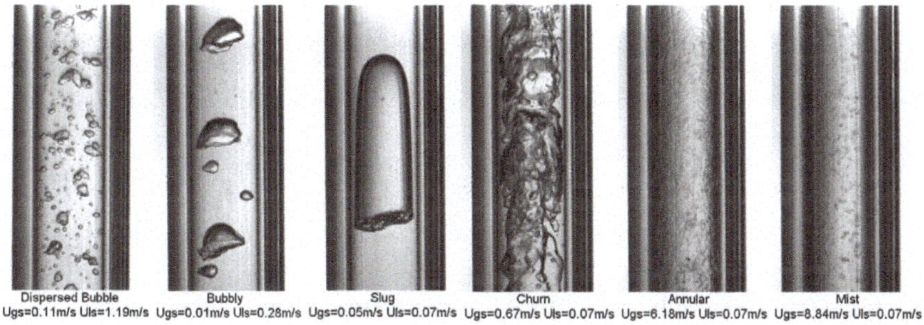


**Fig. 7.6** Images taken by Chen showing different experimental results of flow patterns in vertical lines (Chen et al. 2006)

The procedure proposed by Lockhart and Martinelli, with the Table 7.1 and Fig. 7.7 associated to the method, is:

1. Calculate the $\Delta p$ of each phase as if they flowed separately (inside the pipe as a single flow), $\Delta p_g$ and $\Delta p_l$.
2. Determine the correlation $\chi = \left(\frac{\Delta p_g}{\Delta p_l}\right)^{0.5}$.

**Table 7.1** Limits, based on the different flow regimes of gas and liquid, for the lines to be used in the Lockhard and Martinelli plot, Fig. 7.7

| Flow type | Reynolds number | | | | $C$ |
|---|---|---|---|---|---|
| | Liquid | $\phi_l$ | Gas | $\phi_g$ | |
| Liq. Turb.—Gas Turb. | >2000 | $L_{tt}$ | >2000 | $G_{tt}$ | 20 |
| Liq. Viscous—Gas Turb. | <1000 | $L_{vt}$ | >2000 | $G_{vt}$ | 12 |
| Liq. Turb.—Gas Viscous | >2000 | $L_{tv}$ | <1000 | $G_{tv}$ | 10 |
| Liq. Viscous—Gas Viscous | <1000 | $L_{vv}$ | <1000 | $G_{vv}$ | 5 |

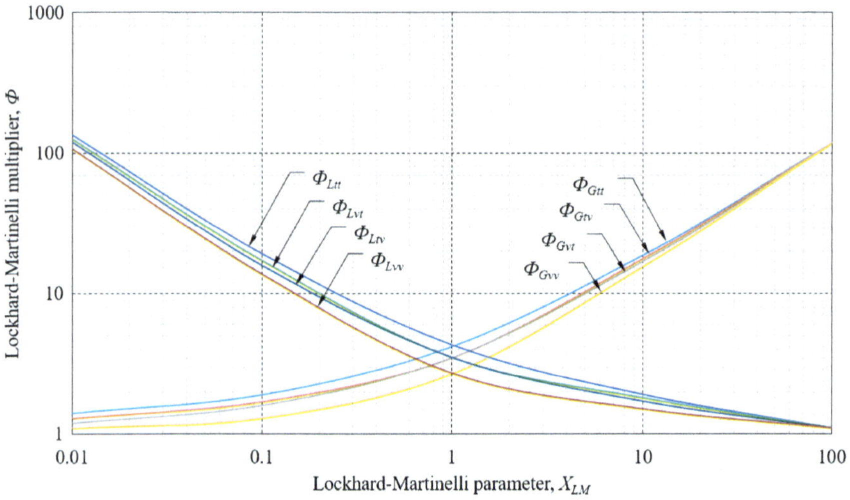


**Fig. 7.7** Lockhart and Martinelli plot showing the different lines limited by the values from Table 7.1 (Ligus et al. 2019)

3. Calculate the Reynolds of the liquid and gas as if they were individual flows and use the table and graph.
4. From the L&M graph, with known $\chi$, calculate $\phi_l$ and $\phi_g$.
5. Calculate $\Delta p_t = \phi_l^2 \Delta p_l = \phi_g^2 \Delta p_g$.

The factor $C$ can be used to calculate the Lockhart and Martinelli parameters,

$$\phi_l^2 = 1 + \frac{C}{\chi} + \frac{1}{+chi^2} \tag{7.9a}$$

$$\phi_g^2 = 1 + C\chi + chi^2 \tag{7.9b}$$

Though in this book it is consider that the parameter $C$ is a constant, it has been found in the literature that it is actually dependent on the size of the channel, which can be expressed as a function of the Reynolds and the Suratman (or Laplace) numbers (Kim and Kim 2023; Chisholm 1967; Sun and Mishima 2009).

The initial step, that is, the calculation of $\Delta p$ involves considering the diameter $D$ as if each fluid were through an independent pipe, using the known formulas a for isothermal and steady flow for liquids and for gases.

This method is successfully applied to absolutely horizontal lines containing two continuous phases and renders reasonably results for low and medium pressure two-phase pipelines.

### 7.4.2 Baker

The work of this researcher extended in 1958[2] the previous correlation and introduced different regimes, also known as flow regime map, and a correction for inclined flow into the Lockart and Martinelli method in the graph, being:

- G: gas mass velocity (by total pipe area) in kg/(s m$^2$).
- L: liquid mass velocity (by total pipe area) in kg/(s m$^2$).
- $\lambda = \left[\left(\frac{\rho_g}{\rho_{air}}\right)\left(\frac{\rho_l}{\rho_{water}}\right)\right]^{0.5}$
- $\psi = \left(\frac{\sigma_w}{\sigma}\right)\left[\frac{\mu_l}{\mu_w}\left(\frac{\rho_w}{\rho l}\right)^2\right]^{\frac{1}{3}}$
- $\rho_g$ is the gas density in lb/ft$^3$.
- $\rho_l$ is the liquid density in lb/ft$^3$.
- $\sigma$ is the surface tension of the liquid, in dynas/cm.
- $\mu_l$ is the absolute viscosity of the liquid in cP.

where the viscosity, density and surface tension of water, and the density of air are taken at room temperature. These dimensionless groups were introduced to consider variations in the physical properties of the media. The Baker pattern map gives reasonably good results for small pipes (NPS 2/DN 50 or smaller).

For each type of flow regime (Fig. 7.8), once the typology of the Baker flow map was found, equations were proposed when the liquid and the gas are in turbulent flow ($\phi_{G_{tt}}$) for 5 flow regimes: bubbles; plugs; stratified; wavy; and annular flow.

[2] O. Baker "Oil and Gas Journal" Vol. 59 Nov. 10. 1958.

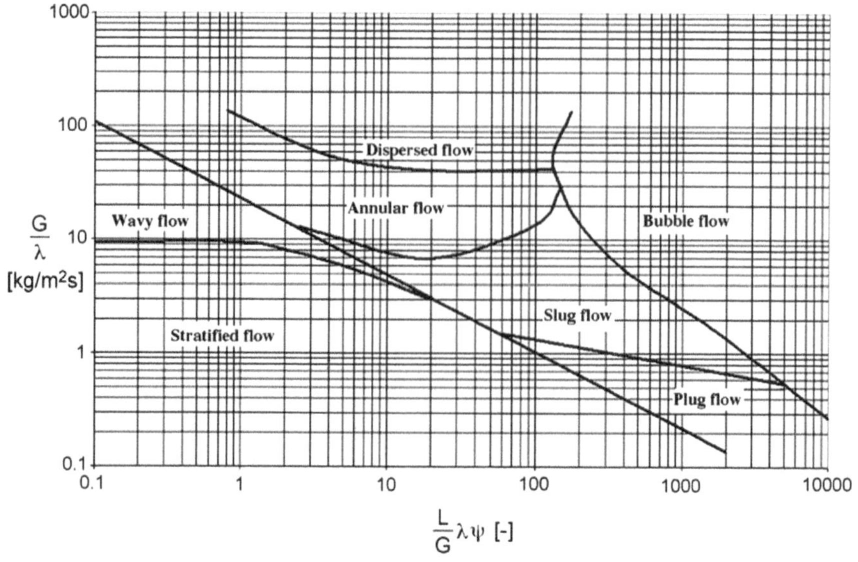


**Fig. 7.8** Baker flow pattern map, based mainly on the liquid and gas flow rates, showing the different patterns found (De Schepper et al. 2008)

The values of $\phi_{G_{tt}}$ obtained from the Baker equations are used in the Lockart and Martinelli calculations previously described for the horizontal lines. Later studies determined inconsistencies with respect to the flows of the regime map, although this does not necessarily condemn the method. Although as the method reflects the limitations of knowledge of those times, its use outside the tested ranges is not recommended.

### *7.4.3 Hewitt and Roberts*

Hewitt and Roberts created in 1969 (Hewitt 2012) a flow regime map applied to vertical flow of water/air or water/steam systems, which renders good results in small pipes for a broad range of pressures (Fig. 7.9). This methodology, one of the few applicable only to vertical flows, was originally created in Imperial Units, but has been adapted by other authors to the SI (e.g. Whalley, 1987). Due to its nature, it is evident that in the flow map no segregated or stratified flows can happen.

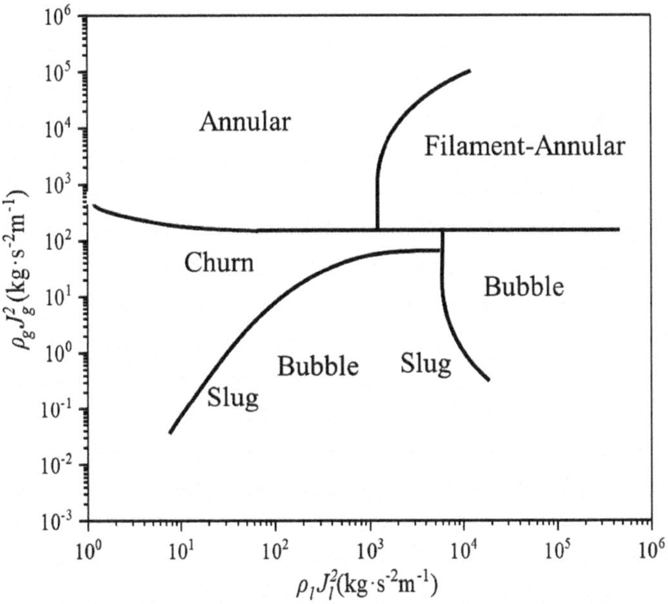


**Fig. 7.9** Hewitt and Roberts flow pattern map, expressing the flow type as a function of the liquid and gas mass flow rates (Zhang et al. 2024)

### *7.4.4 Flanigan*

Flanigan's work[3] done in parallel to Bakery is similar except for the fact that he uses the Panhandle An formula[4] as a reference. He used a tilt correction similar to Baker's that was modified by Campbell.[5]

This method was the result of field testing of lines up to 0.406 m (NPS 16" /DN 400). Subsequent tests confirmed the application of the method for efficiencies greater than 50%. Below this value the liquid loads are greater and no longer allow the concept of a simple correction in the gas flow due to the presence of liquids to be applied.

---

[3] O. Flanigan "Oil and Gas Journal" Vol 56, March 10, 1958.

[4] Equation for calculating pressure loss in gas pipelines based on the isothermal formula. There are several similar formulas Panhandle A, Panhandle B, and modified Panhandle. They take their name from a very important gas field in Texas.

[5] J. M Campbell "Gas Conditioning and Processing" Vol 1, Campbell Petroleum series, 1983.

### 7.4.5 Taitel and Dukler

The flow map developed by Taitel and Dukler's (1976) is one of the most widely used flow patterns for horizontal two-phase flow in industry. The characteristic of this technique compared to the previous ones is that it is based partly on a mechanistic (i.e. computational) study, hence its increased difficulty. It is also represented by a graph (Figs. 7.10 and 7.11), whose horizontal axis represents the term $\chi$ from Lockhart and Martinelli and there are two vertical coordinates: $K$ on the left-hand side; and $T$ or $F$ on the right-hand side (Triplett et al. 1999a, b). These last two are defined as follows:

$$F = \sqrt{\frac{\rho_l - \rho_g}{\rho_g}} \frac{u_{G^s}}{\sqrt{Dg\cos\alpha}} \tag{7.10}$$

$$K = FRe_{L^s}^{\frac{1}{2}} \tag{7.11}$$

$$T = \left[\frac{\left|\frac{dp}{dx}\right|_l^s}{\left(\rho_l - \rho_g\right) g\cos\alpha}\right]^{\frac{1}{2}} \tag{7.12}$$

The following relationship among the curves are observed:

- Curves A and B = F versus $\chi$
- Curve C = K versus $\chi$
- Curve D = T versus $\chi$

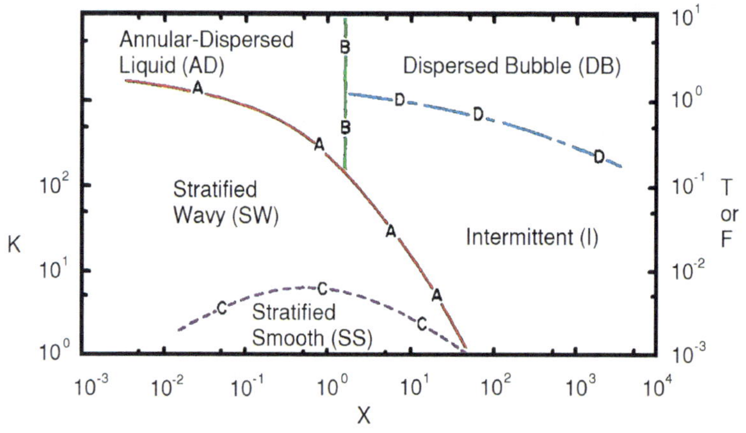


**Fig. 7.10** Taitel and Dukler flow pattern map, based mainly on the parameters $X$, $K$, $T$ or $F$ as shown in the Eqs. 7.10–7.12 (Taitel and Dukler 1976)

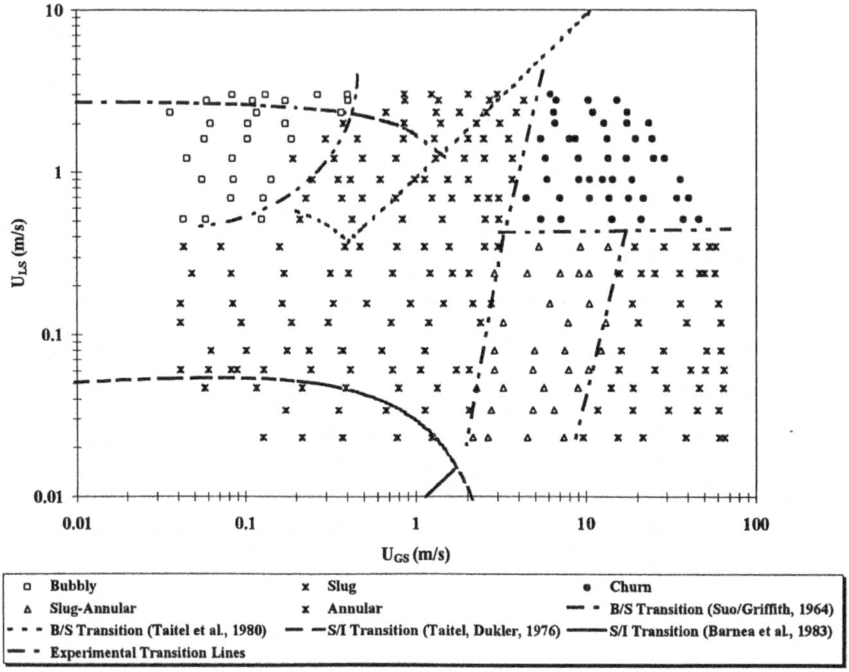


**Fig. 7.11** Experiments carried out by Tripplet showing the correspondence with the flow predicted by Taitel and Dukler (1999a)

## *7.4.6 Petalas and Aziz*

One of the most popular mechanistic models in oil and gas industry, the authors developed the model not based on a specific set of properties or experiments, but applied the conservation equations to the different possible patterns that can be observed. Petalas and Aziz used a large amount of experiments (circa 20000) and well measurements (circa 1800) to identify patterns at different directions and establish a relationship between the conservation principles and measurements (Petalas and Aziz 1996). The flow patterns considered are: stratified smooth; stratified wavy; Intermittent (e.g. slug, elongated bubble, plug); annular mist (w/dispersed bubbles); bubble; froth or churn. The flow pattern map also changes with the fluid and inclination, for instance, Fig. 7.12 shows the different patterns found in an oil/gas system in a vertical upward flow, such as the encountered in oil wells (Petalas and Aziz 2000).

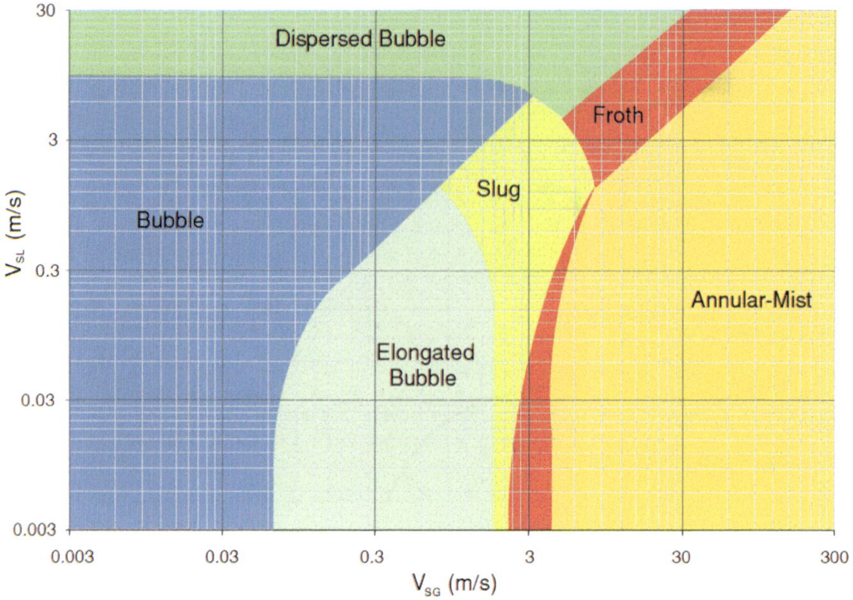


**Fig. 7.12** Petalas and Aziz flow map for oil and gas flow in a pipelines oriented 90° upwards. Image reproduced with permission of S&P Global Commodity Insights (https://www.spglobal.com/)

### *7.4.7 Summary of Different Techniques*

As presented during this Section, there are numerous different mathematical correlations, empirical or mechanistic, for vertical or horizontal flows, which have been developed during the years. All these models have certain limitations or range of applicability, and this must be carefully controlled before using formulas, which are meant to solve different type of flows. The following Table 7.2 summarizes all the techniques not mentioned previously, before introducing and explaining a certain method in particular, Beggs and Brill, taken as example for the design of two-phase pipelines.

## 7.5 Beggs and Brill

This method published in 1973[6] has had modifications and expansions during the course of years, and is widely used for both horizontal and vertical flows since the inclination angles range from 0° to 90° (Beggs and Brill 1973).

[6] Beggs, H.D. and Brill, J.P. (1973) A Study of Two-Phase Flow in Inclined Pipes. JPT, Trans., AIME, 255.

**Table 7.2** Different techniques used to calculate the pressure drop and predict the flow regime

| Author(s) | Year | DOI | Type of correlation | Flow direction |
|---|---|---|---|---|
| Beggs and Brill (1973) | 1973 | 10.2118/4007-PA | Empirical | All directions |
| Barnea Unified Model (1987) | 1987 | – | Empirical | All directions |
| Hagedorn and Brown (1965) | 1965 | 10.2118/940-PA | Empirical | All directions |
| Duns and Ros (1963) | 1963 | – | Empirical | Vertical |
| Aziz and Govier (1972) | 1972 | 10.2118/72-03-04 | Mechanistic | Vertical |
| Ozon et al. (1987) | 1987 | 10.2118/16535-MS | Mechanistic | All directions |
| Kaya et al. (2001) | 2001 | 10.2118/72998-PA | Mechanistic | All directions |
| Petalas and Aziz (2000) | 2000 | – | Mechanistic | All directions |
| Stanford drift flux (2005) | 2005 | 10.2118/89836-PA | Mechanistic | Up to 88° |

The studies applied many data from previous research and the testing apparatus was similar to that used by previous authors, except that the inclination could be varied. Air and water were the fluids originally used in the tests made with acrylic tubes of 25 and 38 mm (NPS 1" to 1½"—DN 25 to DN 40) in diameter.

The Froude number was applied to the basic correlations in a similar way to that done by Ros and by Guzhov, and a Dukler-like flow map now contains four regions: segregated, transitional, intermittent and distributed.

The total head loss was divided into three components (pressure gradients): friction, acceleration and gravity:

$$-\frac{dp}{dz} = \left(\frac{\partial p}{\partial z}\right)_{ele} + \left(\frac{\partial p}{\partial z}\right)_{acc} + \left(\frac{\partial p}{\partial z}\right)_{\mu} \tag{7.13}$$

$$\left(\frac{\partial p}{\partial z}\right)_{ele} = g\rho_{tp} \sin\alpha \tag{7.14a}$$

$$\left(\frac{\partial p}{\partial z}\right)_{\mu} = f_{tp}\rho_{ns}\frac{v_m^2}{2}\frac{1}{d} \tag{7.14b}$$

$$\left(\frac{\partial p}{\partial z}\right)_{acc} = \rho_{tp} v_m \frac{dv_m}{dz} = -\frac{\rho_{tp} v_m v_{sg}}{p}\frac{dp}{dz} \tag{7.14c}$$

where $\alpha$ is the inclination with respect to the horizontal plane, and $\rho_{tp}$ is the density of the two phase flow, expressed as,

$$\rho_{tp} = \rho_l H_l + \rho_g (1 - H_l) \tag{7.15}$$

the mixture velocity is calculate as $v_m = v_{sl} + v_{sg}$. Finally, the no-slip density in two-phase flow may have a different value from the instantaneous or *in-situ* density due to the slippage velocity between the two phases. It can be calculated as,

$$\rho_{tp} = \rho_l \lambda + \rho_g (1 - \lambda) \tag{7.16}$$

where $\lambda$ is the liquid content, equal to $\frac{q_l}{q_l + q_g}$. Combining all these terms one reaches the final differential version of the pressure drop,

$$-\frac{dp}{dz} = g\left[\rho_l H_l + \rho_g (1 - H_l)\right] \sin\alpha + f_{tp}\rho_{ns}\frac{v_m^2}{2}\frac{1}{d} - \frac{\left[\rho_l H_l + \rho_g (1 - H_l)\right] v_m v_{sg}}{p}\frac{dp}{dz} \tag{7.17}$$

$$-\frac{dp}{dz} = \frac{g\left[\rho_l H_l + \rho_g (1 - H_l)\right] \sin\alpha + f_{tp}\rho_{ns}\frac{v_m^2}{2}\frac{1}{d}}{1 - \frac{\left[\rho_l H_l + \rho_g (1 - H_l)\right] v_m v_{sg}}{p}} \tag{7.18}$$

**Flow Pattern**

The first step in this methodology consists in identifying what type of flow is present with the present operating conditions in-situ. This is done with the fluids properties, the pipeline diameter and the flow rates. This renders the type of flow, i.e., segregated, intermittent or distributed.

This methodology then uses a flow pattern plot based on the initial liquid hold-up ($C_l$) and the Froude number of the mixture, calculated as,

$$Fr_m = \frac{v_m^2}{gD} \tag{7.19}$$

The original map is shown in Fig. 7.13, dividing the types of flow in three types: distributed (bubble and mist), segregated (stratified, wavy, and annular) and intermittent (plug and slug). The map was later revised into the version which contains straight lines instead of the curves, including then a new section, a transition between intermittent and segregated.

It is noteworthy mentioning that the flow pattern map in this method is one for horizontal flows. If the pipe is oriented not in horizontal position, then the flow map cannot determine the actual pattern in the pipeline. The procedure is then based on

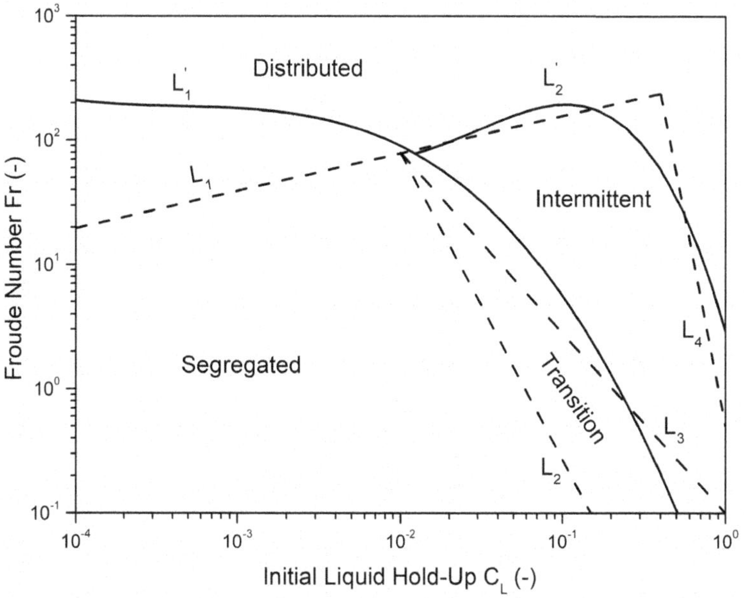


**Fig. 7.13** Beggs and Brill flow pattern map, showing the original map (continuous line, three regimes) and the revised one (dashed lines, four regimes)

calculating the parameters for a horizontal pipeline and then correct them for the actual angle of orientation.

The equations describing the original curves are,

$$L_1 = e^{-4.62-3.757\cdot\chi-0.481\cdot\chi^2-0.0207\cdot\chi^3} \tag{7.20a}$$

$$L_2 = e^{1.061-4.602\cdot\chi-1.609\chi^2-0.179\cdot\chi^3+0.635\times10^{-3}\cdot\chi^5} \tag{7.20b}$$

$$\chi = \ln C_l \tag{7.20c}$$

The revised lines are defined as,

$$L_1 = 316C_l^{0.302} \tag{7.21a}$$

$$L_2 = 9.252\times10^{-4}C_l^{-2.468} \tag{7.21b}$$

$$L_3 = 0.1C_l^{-1.4516} \tag{7.21c}$$

$$L_4 = 0.5C_l^{-6.738} \tag{7.21d}$$

**Table 7.3** Table showing the different conditions for the revised map

| Type of flow | Conditions |
|---|---|
| Segregated | $C_l < 0.01$ and $Fr_m < L_1$, or |
| | $C_l \geq 0.01$ and $Fr_m \leq L_2$ |
| Intermittent | $0.01 \leq C_l < 0.4$ and $L_3 < Fr_m \leq L_1$, or |
| | $C_l \geq 0.4$ and $L_3 \leq Fr_m \leq L_4$ |
| Distributed | $C_l < 0.4$ and $Fr_m > L_1$, or |
| | $C_l \geq 0.4$ and $Fr_m > L_4$ |
| Transition | $C_l \geq 0.01$ and $L_2 < Fr_m < L_3$, or |

**Table 7.4** Constants used to calculate the liquid hold-up in the Beggs and Brill calculation

| Type of flow | $A$ | $B$ | $C$ |
|---|---|---|---|
| Segregated | 0.980 | 0.4846 | –0.0868 |
| Intermittent | 0.845 | 0.5351 | –0.0173 |
| Distributed | 1.065 | 0.5824 | –0.0609 |

Based on this, one determines the pattern according to the following reasoning (Table 7.3).

### Liquid Hold-Up

After the patter has been determined, the next step consists in calculating the liquid hold-up; this step has been subdivided by Beggs and Brill in two parts: the first consists in calculating the valur for a horizontal pipe, $E_l(0)$; subsequently, this value is corrected for the actual angle of the pipeline, $E_l(\alpha)$. The horizontal hold-up should always be higher than then initial liquid hold-up. Thus, if $E_l(0) < C_l$, then the value will be set to be equal, $E_l(0) = C_l$. The value for the horizontal liquid hold-up can be written as a function depending on the type of flow encountered as,

$$H_l(0) = AC_l^B Fr^C \tag{7.22}$$

The second step in this procedure consists in correcting the liquid hold-up with a factor that considers the inclination of the pipeline (Table 7.4).

$$E_l(\alpha) = B(\alpha) \cdot E_l(0) \tag{7.23}$$

$$B(\alpha) = 1 + \beta\left[\sin(1.8\alpha) - \frac{1}{3}\sin^3(1.8\alpha)\right] \tag{7.24}$$

**Table 7.5** Different constants to correct the liquid hold-up as a function of the pipeline angle

| Type of Flow | D | E | F | G |
|---|---|---|---|---|
| Uphill flow | | | | |
| Segregated | 0.011 | –3.768 | 3.539 | –1.614 |
| Intermittent | 2.96 | 0.305 | –0.4473 | 0.0978 |
| Distributed | 1 | 0 | 0 | 0 |
| Downhill flow | | | | |
| All patterns | 4.7 | –0.3692 | 0.1244 | –0.5056 |

$$\beta = (1 - C_l) \ln D \cdot C_l^E N_{LV}^F Fr^G \tag{7.25}$$

$$N_{LV} = v_{SL} \left( \frac{\rho_l}{g\sigma} \right)^{\frac{1}{4}} \tag{7.26}$$

In the case of having a transition flow, the following approach should be followed, based on an interpolation of the values found for segregated and intermittent flows (Table 7.5),

$$E_l(0) = \mathcal{K} E_l(0)_{seg} + (1 - \mathcal{K})\, E_l(0)_{int} \tag{7.27}$$

$$\mathcal{K} = \frac{L_3 - Fr}{L_3 - L_2} \tag{7.28}$$

## Pressure Drop

Once the flow pattern has been determined and the liquid hold-up calculated and corrected, the last step consists in calculating the pressure drop. As mentioned in the beginning of the analysis, the pressure drop contains three components, namely: elevation, friction and acceleration. The friction element requires the calculation of a two-phase friction factor, based on the no-slip Reynolds number:

$$Re_{ns} = \frac{\rho_{ns} v_m D}{\mu_{ns}} \tag{7.29}$$

Beggs and Brill carried out a number of experiments and developed the following relationship between the two-phase friction factor ($f_{tp}$) and a no-slip friction factor ($f_{ns}$), resulting in the following equation,

$$\frac{f_{tp}}{f_{ns}} = e^s \tag{7.30}$$

The no-slip friction factor can be obtained from the Moody diagram. Beggs and Brill used a formula for smooth pipes, however, different and better formulae can be used to estimate this value. The exponential coefficient is a function of the liquid hold-up,

$$s(y) = \begin{cases} \ln(2.2y - 1.2), \quad \text{if} \quad 1.0 < y < 1.2 \\ \dfrac{\ln y}{-0.0523 + 3.182 \ln y - 0.8725 \ln^2 y + 0.01853 \ln^4 y}, \quad \text{if} \quad y \geq 1.2 \end{cases} \tag{7.31}$$

$$y = \frac{C_l}{H_l^2} \tag{7.32}$$

Finally one can calculated the pressure drop with Eq. 7.33,

$$-\frac{dp}{dz} = \frac{g\left[\rho_l H_l + \rho_g (1 - H_l)\right] \sin\alpha + f_{tp} \rho_{ns} \frac{v_m^2}{2} \frac{1}{D_{int}}}{1 - \frac{\left[\rho_l H_l + \rho_g (1 - H_l)\right] v_m v_{sg}}{p}} \tag{7.33}$$

### Conclusions over the Method

The previous procedure can be used in each pipe segment up and down, and add them to obtain the total head loss of the section under analysis. The original approach helps bridge previous work and provides a useful model to work from, but the specific coefficients are limited by the small size of the diameters and lengths used for the tests, as well as the use of water and air as test fluids, although tests have continued using other fluids.

One of the largest errors inherent in the method is the use of polynomial equations or logarithmic functions to fit the test data, whose errors are magnified when executing the design calculations. It must be recognized that Beggs and Brill and all preceding methods used liquids that give an essentially constant liquid head as it transits the line. In the case of natural gas condensates, whose volumes increase as the gas flow cools, the behavior of the phases must be incorporated into the calculation. For each section analyzed, the temperature must be known, estimating a temperature profile. With this average segment temperature and pressure known, a specific "flash" calculation must be performed to find the liquid head of the segment, which makes the calculation even more complex.

When major temperature changes occur in the line (e.g. underwater crossings simply resting on the bed) the liquid loading can change dramatically enough to produce different types of regimes across the line. Additionally, under changing pressures and temperatures, a normal oil will absorb or expel some gas. This situation will change its density and volume. There are many examples in the literature for performing relatively accurate "flash" calculations.

## 7.6 Comprehensive Review of the Technology

The above attempts to present a very tight summary of the engineering advances for the resolution of two-phase flows. It has been seen that the relative quantities of the two phases and their velocities are of fundamental importance. The basic calculation method depends on this data to largely define in which phases interact (Zahid et al. 2020; Usama 2009).

Although many researchers do not agree on how to discriminate the different flow regimes, the most logical thing is to define three main flow regimes: Distributed, Segregated and Intermittent, as indicated at the beginning. As a design factor $H_l$ is a very general term that tries—but not completely—to define the localized behavior within a segment and its flow regime. For example, if the continuous phase is greater than 80% and that phase is in a developed turbulent regime, it is very difficult to imagine anything other than the distributed regime as the predominant regime.

**Fluid Properties**: It is routine to calculate density, viscosity, and surface tension for a given phase, but energy balances require knowledge of compound properties. The possible thing to do is to calculate some average number that fits the chosen correlation.

**Pipe Size**: A predominant factor since those with large diameter have lower pressure losses and also the waves become larger since the geometry counts. However, it is possible to apply the data for small diameter lines to large diameter lines if the model is applied properly. Tilt: Upward tilted sections can exhibit all flow regimes. In descending sections, segregated flows normally predominate if the volume of the section is comparable to that of the preceding ascending section. Density will separate the phases and if this is true the pressure recovery can probably be reasonably represented by the static gas head involved.

**Pressure**: This effect is shown in both flow rate calculations and fluid properties calculations. However, scaling from low pressures to high pressures introduces a conceptual problem. In fact, the density of a gas changes more rapidly with pressure than that of a liquid. At 70 bars a typical natural gas is approximately 80 to 85% denser than at atmospheric pressure, but the density of the liquid changes very little and therefore the impact loads change a lot, among other issues.

**Temperatures**: Like pressure, it has an obvious effect. Temperature profile calculations are at best an approximation. Offshore, the lines are lined and covered with concrete, and the water currents and characteristics of the surrounding terrain vary. The best method to predict temperatures is by applying real data to similar systems (preferably under the same basic surrounding conditions).

**PVT behavior**: a prediction is necessary for the real system. In all research studies choosing table liquids eliminates this problem. It is of concern the change in liquid/gas ratios and fluid properties. The line segments must be selected so that the change in these values is not so abrupt as to compromise the calculations made on that section. The segments do not have to be the same length.

## 7.7 Approximate Optimization Methodology

A specific methodology has not been developed for the optimization of a pipeline with two-phase flow, however a series of guidelines will be given below to approach the topic. In general, these two-phase systems start from known volumes to be transported, that is, liquid flow and gas flow. To proceed to approximate an optimal solution, it is proposed:

1. Solve the pipeline as if all the fluid to be transported were only the liquid, and based on the liquid flow rate and its characteristics, calculate the optimal diameter as explained in the corresponding chapter.
2. Solve the duct as if all the fluid to be transported were only gas and based on the gas flow and its characteristics, calculate the optimal diameter as explained in the corresponding chapter.
3. Calculate the area corresponding to each optimal diameter, square these areas, add both areas (composition of areas) and extract the square root obtaining an equivalent area, and consequently an equivalent "theoretical optimal" diameter.
4. Solve the two-phase flow for the closest diameters (two larger and two smaller than the theoretical optimum) of the commercial series, varying the pressures and corresponding thicknesses, to achieve the established arrival pressure ($p_2$).
5. Carry out the economic calculation for each alternative analyzed—similar to what has already been explained for gas or liquid pipelines—and choose the installation with the lowest total cost.

## References


Aziz K, Govier GW. Pressure drop in wells producing oil and gas. J Canadian Petroleum Technol. 1972;11(03).

Babu C. Cfd analysis of multi-phase flow and its measurements. IOSR -JMCE. 2013;9(30–7):11. https://doi.org/10.9790/1684-0943037.

Barnea D. A unified model for predicting flow-pattern transitions for the whole range of pipe inclinations. Int J Multiph Flow. 1987;13(1):1–12.

Beggs DH, Brill JP. A study of two-phase flow in inclined pipes. J Petroleum Technol. 1973;25(05):607–17.

Cao Y, Xu Q, Yu H, Huang B, Liu T, Guo L. Experimental investigation on interface characteristics of gas-liquid two-phase flow in a kilometer-scale pipeline. Exper Thermal Fluid Sci. 2023;149:111017. ISSN 0894-1777. https://doi.org/10.1016/j.expthermflusci.2023.111017. https://www.sciencedirect.com/science/article/pii/S0894177723001735.

Cheng L, Xia G. Flow patterns and flow pattern maps for adiabatic and diabatic gas liquid two phase flow in microchannels: fundamentals, mechanisms and applications. Exper Thermal Fluid Sci. 110988.

Chen L, Tian YS, Karayiannis TG. The effect of tube diameter on vertical two-phase flow regimes in small tubes. Int J Heat Mass Transf. 2006;49(21):4220–30. ISSN 0017-9310. https://doi.org/10.1016/j.ijheatmasstransfer.2006.03.025. https://www.sciencedirect.com/science/article/pii/S0017931006002250.

Chisholm D. A theoretical basis for the lockhart-martinelli correlation for two-phase flow. Int J Heat Mass Transf. 1967;10(12):1767–78.

De Schepper SCK, Heynderickx GJ, Marin GB. Cfd modeling of all gas-liquid and vapor-liquid flow regimes predicted by the baker chart. Chem Engin J. 2008;138(1–3):349–57.

Duns H Jr, Ros NCJ. Vertical flow of gas and liquid mixtures in wells. In: World petroleum congress. WPC; 1963. p. WPC–10132.

Dziubinski M, Fidos H, Sosno M. The flow pattern map of a two-phase non-newtonian liquid-gas flow in the vertical pipe. Int J Multiph Flow. 2004;30(6):551–63.

Ganapathy H, Shooshtari A, Choo K, Dessiatoun S, Alshehhi MMMO, Ohadi M. Volume of fluid-based numerical modeling of condensation heat transfer and fluid flow characteristics in microchannels. Int J Heat Mass Transf. 2013;65:62–72.

Grassi B, Strazza D, Poesio P. Experimental validation of theoretical models in two-phase high-viscosity ratio liquid-liquid flows in horizontal and slightly inclined pipes. Int J Multiphase Flow. 2008;34(10):950–65. ISSN 0301-9322. https://doi.org/10.1016/j.ijmultiphaseflow.2008.03.006. https://www.sciencedirect.com/science/article/pii/S0301932208000608.

Hagedorn AR, Brown KE. Experimental study of pressure gradients occurring during continuous two-phase flow in small-diameter vertical conduits. J Petroleum Technol. 1965;17(04):475–84.

Hewitt GF. Churn and wispy annular flow regimes in vertical gas–liquid flows. Energy Fuels. 2012;26(7):4067–77. ISSN 0887-0624. https://doi.org/10.1021/ef3002422. https://doi.org/10.1021/ef3002422.

Kaya AS, Sarica C, Brill JP. Mechanistic modeling of two-phase flow in deviated wells. SPE Product Facilit. 2001;16(03):156–65.

Kim H-G, Kim S-M. Slug flow characteristics of air-liquid two-phase flow in horizontal pipes over a wide range of liquid viscosities. Int J Heat Mass Transf. 2023;208:124005. ISSN 0017-9310. https://doi.org/10.1016/j.ijheatmasstransfer.2023.124005. https://www.sciencedirect.com/science/article/pii/S0017931023001576.

Krumrick EA, López EJ, Camacho AG. Prediction of the flow regime in liquid-gas flows through straight pipes using computational fluid dynamics. Mecánica Comput. 2016;34(30):2101–14.

Ligus G, Zając D, Masiukiewicz M, Anweiler S. A new method of selecting the airlift pump optimum efficiency at low submergence ratios with the use of image analysis. Energies. 2019;12(4):735.

López J, Pineda H, Bello D, Ratkovich N. Study of liquid-gas two-phase flow in horizontal pipes using high speed filming and computational fluid dynamics. Exper Thermal Fluid Sci. 2016;76:126–34. ISSN 0894-1777. https://doi.org/10.1016/j.expthermflusci.2016.02.013. https://www.sciencedirect.com/science/article/pii/S0894177716300371.

Ma R, Ma T, Kang J, Yang K, Li L, Wang L. Prediction model of critical liquid-carrying gas velocity for high gas-to-liquid ratio gathering pipelines. J Pipeline Sci Engin. 2023;3(1):100093. ISSN 2667-1433. https://doi.org/10.1016/j.jpse.2022.100093. https://www.sciencedirect.com/science/article/pii/S2667143322000658.

Najmi K, McLaury BS, Shirazi SA, Cremaschi S. Experimental study of low concentration sand transport in wet gas flow regime in horizontal pipes. J Nat Gas Sci Engin. 2015;24:80–8.

Ozon PM, Ferschneider G, Chwetzoff A. A new multiphase flow model predicts pressure and temperature profiles in wells. In: SPE offshore Europe conference and exhibition. SPE; 1987. p. SPE–16535.

Petalas N, Aziz K. Development and testing of a new mechanistic model for multiphase flow in pipes. In: The 1996 ASME fluids engineering division summer meeting. Part 1(of 2), San Diego, CA, USA, 07/07-11/96. 1996. p. 153–9.

Petalas N, Aziz K. A mechanistic model for multiphase flow in pipes. J Can Pet Technol. 2000;39(6):04.

Poesio P, Strazza D, Sotgia G. Very-viscous-oil/water/air flow through horizontal pipes: pressure drop measurement and prediction. Chem Engin Sci. 2009;64(6):1136–42. ISSN 0009-2509. https://doi.org/10.1016/j.ces.2008.10.061. https://www.sciencedirect.com/science/article/pii/S0009250908005551.

Revellin R, Dupont V, Ursenbacher T, Thome JR, Zun I. Characterization of diabatic two-phase flows in microchannels: flow parameter results for r-134a in a 0.5 mm channel. Int J Multiphase Flow. 2006;32(7):755–74.

Schmid D, Verlaat B, Petagna P, Revellin R, Schiffmann J. Flow pattern observations and flow pattern map for adiabatic two-phase flow of carbon dioxide in vertical upward and downward direction. Exper Thermal Fluid Sci. 2022;131:110526. ISSN 0894-1777. https://doi.org/10.1016/j.expthermflusci.2021.110526. https://www.sciencedirect.com/science/article/pii/S0894177721001692.

Shi H, Holmes JA, Diaz LR, Durlofsky LJ, Aziz K. Drift-flux parameters for three-phase steady-state flow in wellbores. SPE J. 2005;10(02):130–7.

Sun L, Mishima K. An evaluation of prediction methods for saturated flow boiling heat transfer in mini-channels. Int J Heat Mass Transf. 2009;52(23–24):5323–9.

Taitel Y, Dukler AE. A model for predicting flow regime transitions in horizontal and near horizontal gas-liquid flow. AIChE J. 1976;22(1):47–55.

Triplett KA, Ghiaasiaan SM, Abdel-Khalik SI, Sadowski DL. Gas-liquid two-phase flow in microchannels part i: two-phase flow patterns. Int J Multiphase Flow. 1999a;25(3):377–94. ISSN 0301-9322. https://doi.org/10.1016/S0301-9322(98)00054-8. https://www.sciencedirect.com/science/article/pii/S0301932298000548.

Triplett KA, Ghiaasiaan SM, Abdel-Khalik SI, LeMouel A, McCord BN. Gas-liquid two-phase flow in microchannels: part ii: void fraction and pressure drop. Int J Multiphase Flow. 1999b;25(3):395–410. ISSN 0301-9322. https://doi.org/10.1016/S0301-9322(98)00055-X. https://www.sciencedirect.com/science/article/pii/S030193229800055X.

Usama K. Long liquid slugs in stratified gas/liquid flow in horizontal and slightly inclined pipes. PhD thesis. Netherlands: Delft University; 2009.

Zahid AA, Ur Rehman SR, Rushd S, Hasan A, Rahman MA. Experimental investigation of multiphase flow behavior in drilling annuli using high speed visualization technique. Front Energy. 2020;14:635–43.

Zhang J, Liu L, Liu S, Hanyang G. A method for predicting the breakup of gas column in a decaying swirling liquid field. Ann Nucl Energy. 2024;195: 110176.

# Chapter 8
# Transportation of Various Liquid or Gaseous and Solid Substances Simultaneously Through Pipelines

> If you really think that the environment is less important than the economy, try holding your breath while you count your money.
>
> *Guy McPherson*

## 8.1 Introduction

There are numerous applications where economics dictate the movement of more than one phase in a pipeline. The transportation of solids with liquids is used to move coal slurries, fracturing fluids, or others, and mixtures of air and coal are used in some large thermoelectric plants.

Heterogeneous flow is the flow of a fluid where the different phases of one or several different substances coexist within it. The phases can exist in solid, liquid and gaseous form. In general, the mixtures of greatest importance in engineering are solid-liquid, solid-gas. In the case of liquid-gas flows, they are called two-phase, and in some cases "three-phase" if they are two immiscible liquids and a gas (Jacobs 2003; Bratland 2010; Wang et al. 2023).

From a practical point of view, in solid-liquid and solid-gas mixtures what is important to transport is the mass of solid (Seitshiro et al. 2014; Reddy et al. 2023). Consequently, to do it like a pipe flow, the solid is fluidized into a liquid or a gas. In the case of two-phase flows, generally linked to primary production in oil fields, from a practical point of view, what is important to transport is both fluids, the liquid (e.g. crude oil) and the gas (e.g. natural gas). This chapter will be focused to heterogeneous solid-liquid and solid-gas flows (Fig. 8.1).



P. Druetta et al., *Transport by Pipelines and Its Optimization*, Petroleum Engineering,
https://doi.org/10.1007/978-3-031-66566-0_8

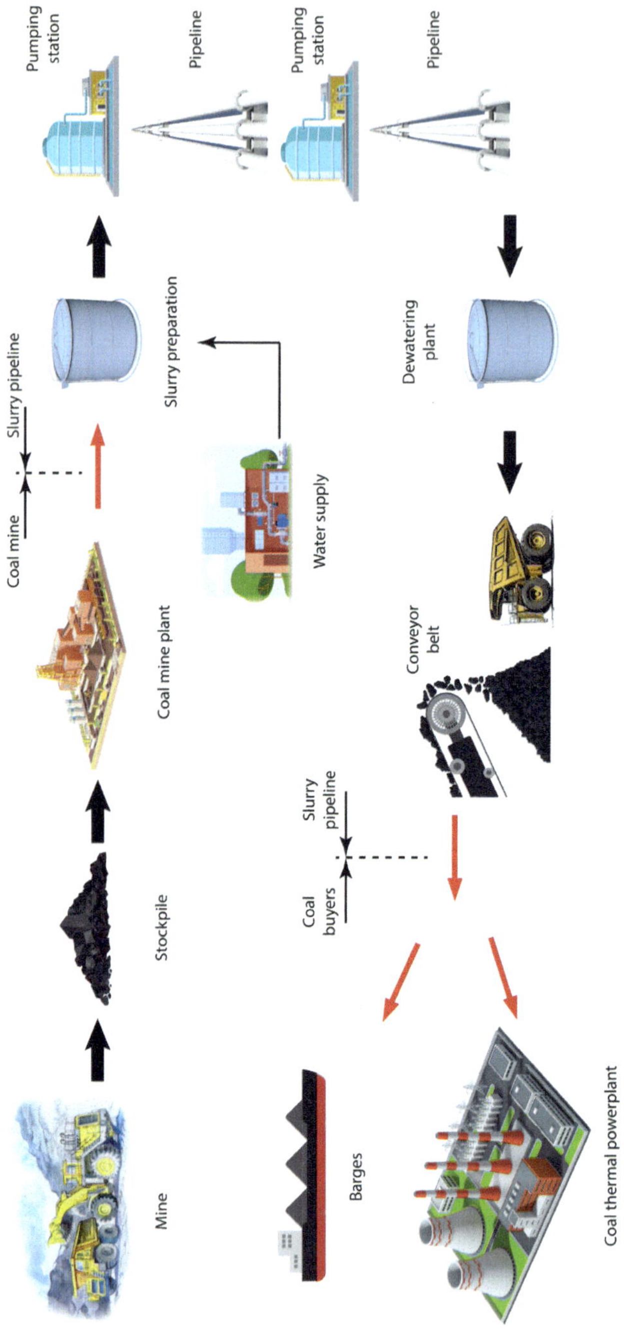


**Fig. 8.1** Schematic diagram with all the steps of a slurry pipeline, from the extraction of the ore to its final use or transportation

## 8.2 Solid-Liquid Heterogeneous Flow Transportation

For the solid to remain fluidized in the fluid, whether liquid or gas, a minimum transport speed is needed so that the solid does not settle along the way and clog the pipe, which defines the minimum transport speed (Fig. 8.2) (Hu et al. 2023).

One particle is relatively easy to entrain, several particles are more difficult and therefore it is important to know how many particles there are per mass of fluid, i.e., its concentration. On the other hand, so that the mixed flow does not get stuck in the pipe, the maximum amount of solid that can be introduced into the fluid vein must be known, which defines the concentration. The information necessary to start carrying out a proper design is: density of the solid being transported; density of the fluid (in this case, the cheapest fluid available and suitable for transport); the diameter of the particles, which can be changed by grinding, although there are practical limits to its modification).

The minimum transport speed and the maximum concentration of solid in fluid are experimental values or relationships specific to each fluid-solid mixture (Wang et al. 2023; Reddy et al. 2023). If the objective is to transport a solid inside a pipe, and therefore it is fluidized, it is advisable that its particles be small in size (i.e., powdery) and for this, it is previously passed through a grinding and screening process. The finer the solid, the easier it will be to transport and the less sedimentation and the consequent clogging of the pipe. The stream is passed through calibrated sieves, the grains that are rejected because they are larger than the size of the sieve are returned to the grinding process (Fig. 8.3).

A relative movement must be achieved between the fluid and the particle, the speed of the particle must be less than the speed of the flow for a pressure difference to

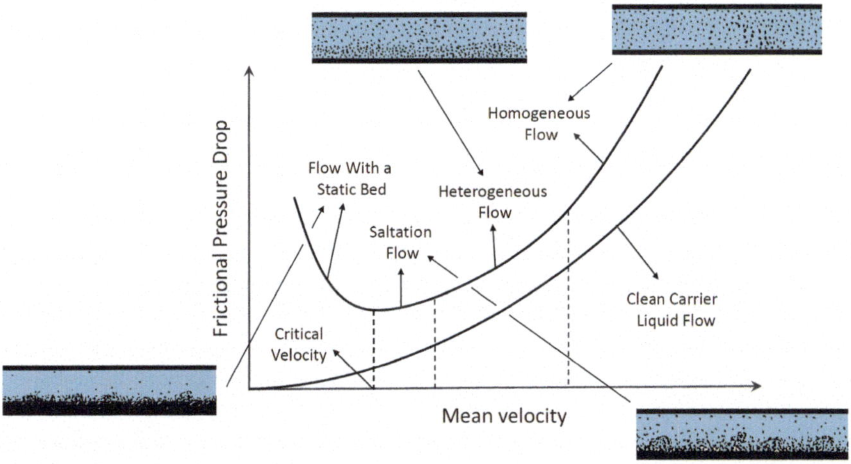


**Fig. 8.2** Plot showing the comparison between a pure fluid flow and a slurry flow, indicating the different flow patterns found

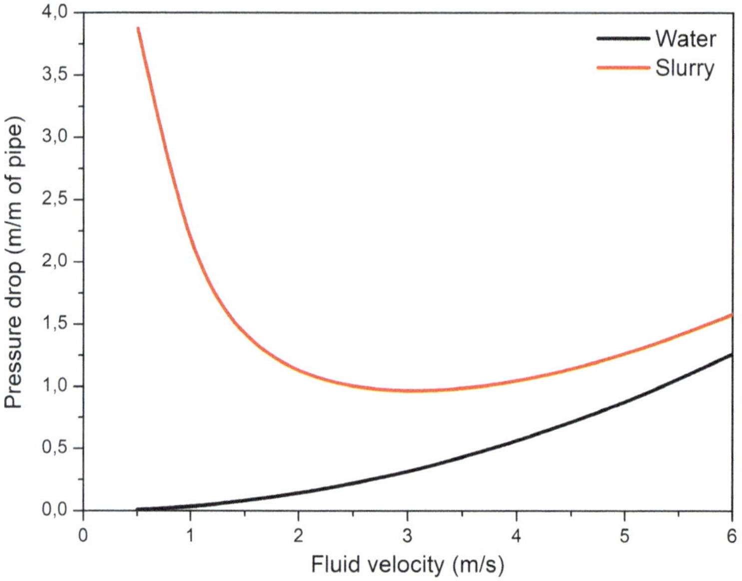


**Fig. 8.3** Plot showing the comparison between a pure fluid flow and a slurry flow, indicating the pressure drop as a function of the velocity

appear. If the powder is very fine, it can be dissolved in water (e.g., sodium chloride) and one transport a solution, that is, a homogeneous fluid; whereas if the grains are very large, dispersion is formed, which is not as stable and in a few days it decants (Jacobs 2003).

The most common carrier fluids used industrially are: fresh or salt water (it is important to consider that sometimes there are problems with salt water because it can contaminate the solid), or air.

In large diameter pipes, turbulent flows ($Re \approx 5000$) are achieved with low velocities, approximately less than 1 m/s, for water (Bratland 2010). It is necessary that there always be a difference in speed between the fluid and the particle, but sooner or later gravity acts and the particle descends to the lower intrados of the pipe. There it begins to be dragged and rotate, being worn or rounded by friction with the pipe to the extent that the pipe is harder than the solid, or by polishing and/or wearing away the inside of the pipe, if the particle is harder than the the pipe (Fig. 8.4).

As the particle rests on the lower intrados of the pipe, the fluid does not pass below and the current that passes above creates a pressure differential that causes the particle to rise and fall again. The minimum speed necessary for this to occur is the so-called *transport or hauling speed*. This speed is usually found initially by experimentation. The limiting or terminal velocity is the speed with which the particle settles and depends fundamentally on the density of the fluid and its viscosity. If there are many particles a type of internal dunes, i.e., with a *saw* shape, are produced. Since not all particles are the same, neither in shape nor size, and the limit velocity is difficult to

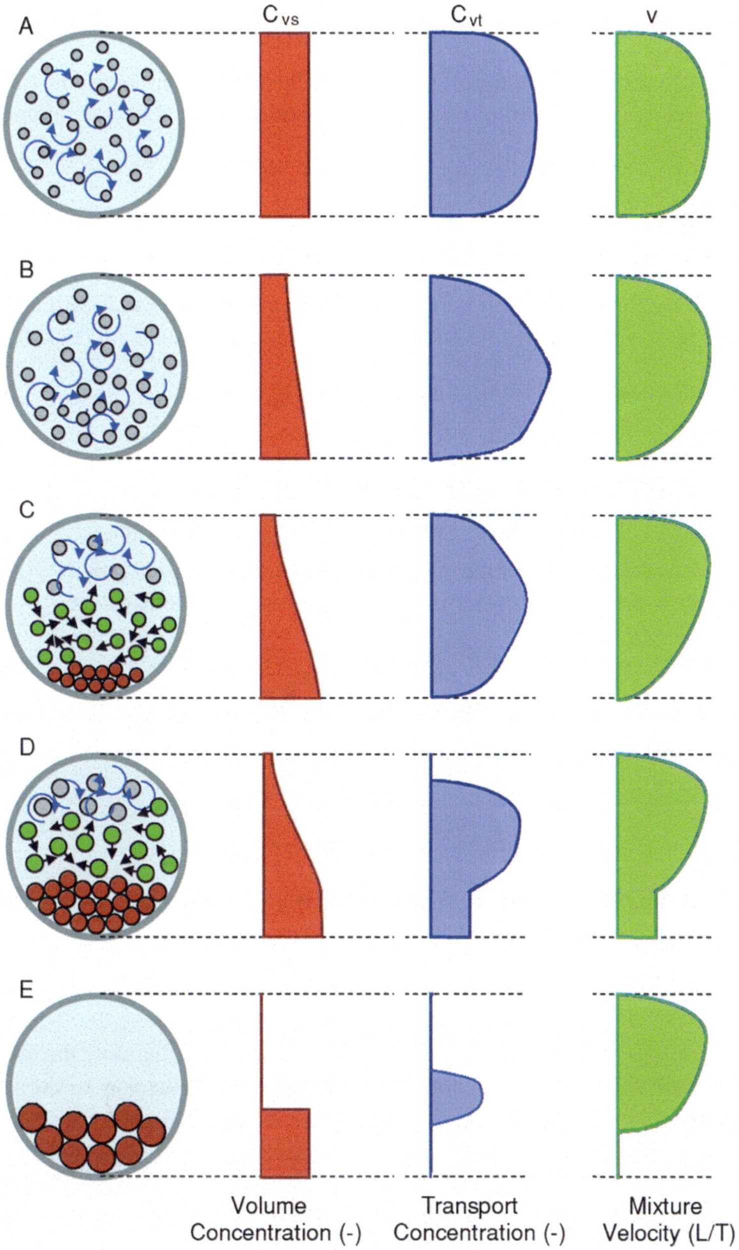


**Fig. 8.4** Scheme showing the different concentration profiles that can be found in pipes transporting solids as a function of the particles' sizes and concentration

estimate, except statistically, one must try to ensure that nothing settles to the bottom (Joshi et al. 2023; Jacobs 2003).

Since the objective is to transport the solid and the liquid is the carrier, its transport is useless and it is all "waste" of energy, transportation is not always economical.

## 8.3 Definitions

During this section several major concepts will be introduced, which will be useful during the presentation of the methodology and equations to solve this type of flow.

- $Q_s$ = solid volumetric flow rate [m$^3$/s];
- $Q_f$ = fluid volumetric flow rate [m$^3$/s];
- $Q$ = total flow rate [m$^3$/s];
- $\rho_s$ = solid density [kg/m$^3$];
- $\rho_f$ = fluid density [kg/m$^3$].

With these one can then define the mass flow rates, $\dot{m}_s = Q_s \cdot \rho_s$ and $\dot{m}_f = Q_f \cdot \rho_f$. The volumetric concentration of solid in the slurry is then calculated as,

$$c_{v,s} = \frac{\textbf{volume of solid}}{\text{volume of liquid} + \text{volume of solid}} = \frac{V_s}{V_f + V_f} \tag{8.1}$$

One can also define the volumetric concentration based on the transport properties,

$$c_{v,t} = \frac{Q_s}{Q_f + Q_s} \tag{8.2}$$

The relationship between these two can be determined by the slip velocity,

$$c_{v,t} = \left(1 - \frac{v_{sl}}{v_p}\right) \cdot c_{v,s} \tag{8.3}$$

From now on, it will be considered no slip in the calculations, therefore the transport and volumetric concentrations become the same value. Another way to express the concentration,

$$\epsilon = \frac{\text{mass of solid}}{\text{mass of fluid}} = \frac{\rho_s}{\rho_f} \cdot \frac{c}{1 - c} \tag{8.4}$$

### 8.3.1 Viscosity

The first study proposing the modification of the fluid viscosity as a function of the solid concentration was presented by Einstein (1906), which renders good results up to concentrations of 6%.

$$\mu_{red} \equiv \frac{\mu_s}{\mu_l} = 1 + 2.5c_v \tag{8.5}$$

In order to consider the influence in higher concentrations, Thomas (1965) in 1965 expanded Einstein's formula. A point relevant to this is the particle diameter: for small particles (less than 10 microns), the colloidal forces become important, rendering a non-Newtonian behavior of the fluid. For larger particles, the inertial effects due to rotations and collisions increase the energy (viscous) dissipation and therefore it renders higher relative viscosities. The expanded formula is then (Fig. 8.5),

$$\mu_{red} \equiv \frac{\mu_s}{\mu_l} = 1 + 2.5c_v + 10.05c_v^2 + 0.00273e^{16.6c_v} \tag{8.6}$$

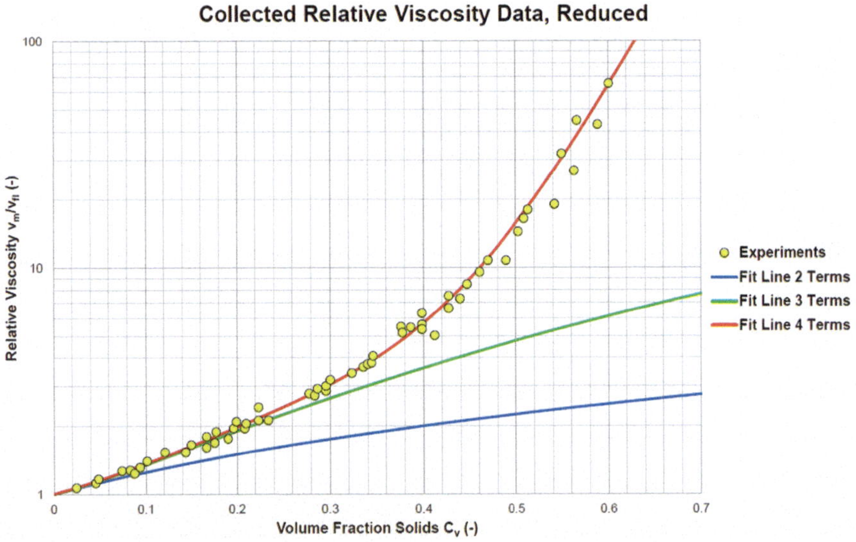


**Fig. 8.5** Relative viscosity of slurry samples compared to the Thomas viscosity equation 8.6. (Miedema 2013)

## 8.4 Sedimentation Velocity

The particles dragged by the carrier fluid eventually fall, but even against the lower part of the pipe a drag force will act on the particle (Zhiyao et al. 2008; Sarraf Shirazi and Frigaard 2021). A minimum speed is needed for it to settle. The minimum velocity for the transport is a function of the particle size, the densities of fluid/solid, and the particle shape. Performing a simple force balance over a particle, depicted in Fig. 8.6.

A solid particle that falls under the force of gravity reaches a maximum velocity of fall called limiting velocity or terminal velocity. With particles flowing in a pipeline, the drag force is based on the velocity difference between fluid and solid. There has to be a velocity difference for drag to appear, and it depends on the turbulence of the flow (Rajitha et al. 2006).

The rate of fall in the fluid according to the flow regime is based on the Reynolds number of the particle,

$$Re_p = \frac{v_t d_p \rho_f}{\mu_f} \tag{8.7}$$

and it can be divided in,

- Laminar regime: Stokes law;
- Intermediate or transition regime;
- Wake turbulent/Boundary layer laminar: Newton law;
- Wake and boundary layer turbulent: supercritical regime.

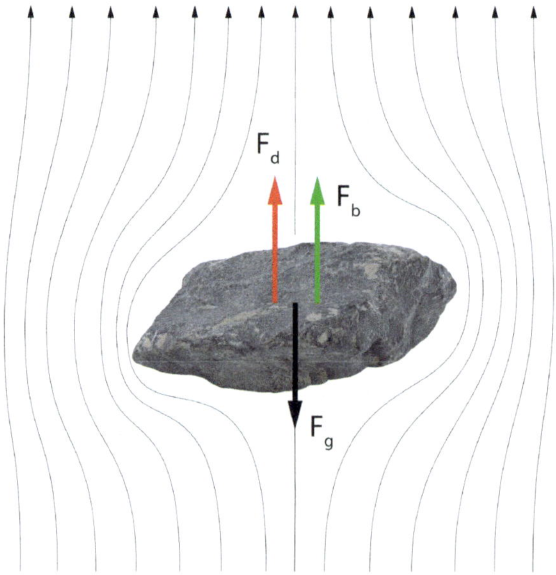


**Fig. 8.6** Force balance in a particle of a certain shape under the influence of the gravity field, showing as well the drag and buoyancy forces

The falling speed is limited to an experimental value, as the speed increases, the viscous resistance increases and the speed becomes constant. Performing the balance of forces, one can reach the terminal or sedimentation velocity for a single particle is,

$$v_t = \sqrt{\frac{4g\left(\rho_s - \rho_l\right) d\psi}{3\rho_l C_D}} \tag{8.8}$$

with $\psi$ being a factor considering the shape of the particles (equal to 1 for spherical particles) and the drag coefficient being a function of the Reynolds number (Figs. 8.7 and 8.8). For spheres,

$$C_D = \begin{cases} \frac{24}{Re_p}, & \text{if } Re_p \ll 1 \\ \frac{24}{Re_p} + \frac{3}{\sqrt{Re_p}} + 0.34, & \text{if } 1 < Re_p < 1000 \\ 0.445, & \text{if } 1000 < Re_p < 3 \times 10^5 \end{cases} \tag{8.9}$$

When considering non-spherical shapes, there are several methods reported in the literature, for instance, using dimensionless groups (e.g. Archimedes and Lyaschenko). The shape factor $\psi$ mentioned before, also known as sphericity, is equal to,

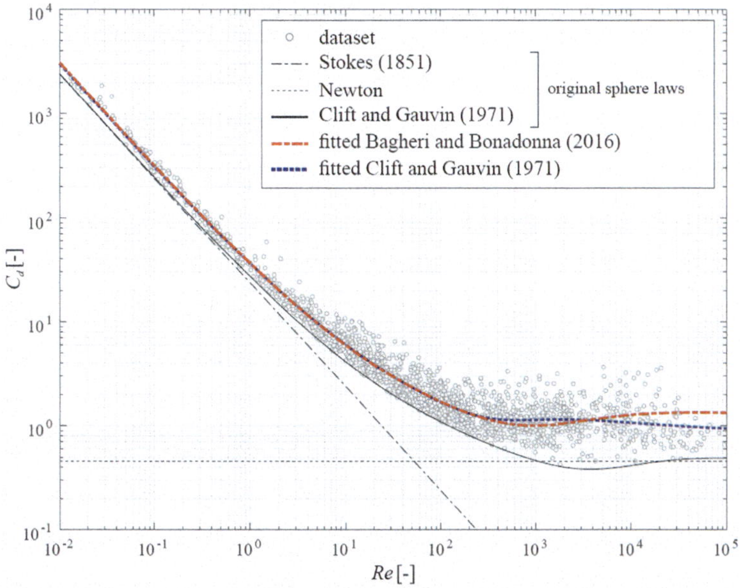


**Fig. 8.7** Drag coefficient as a function of the Reynolds number for spherical particles, showing the correspondence between experiments and Stokes and Newton's laws (Raffaele et al. 2020)

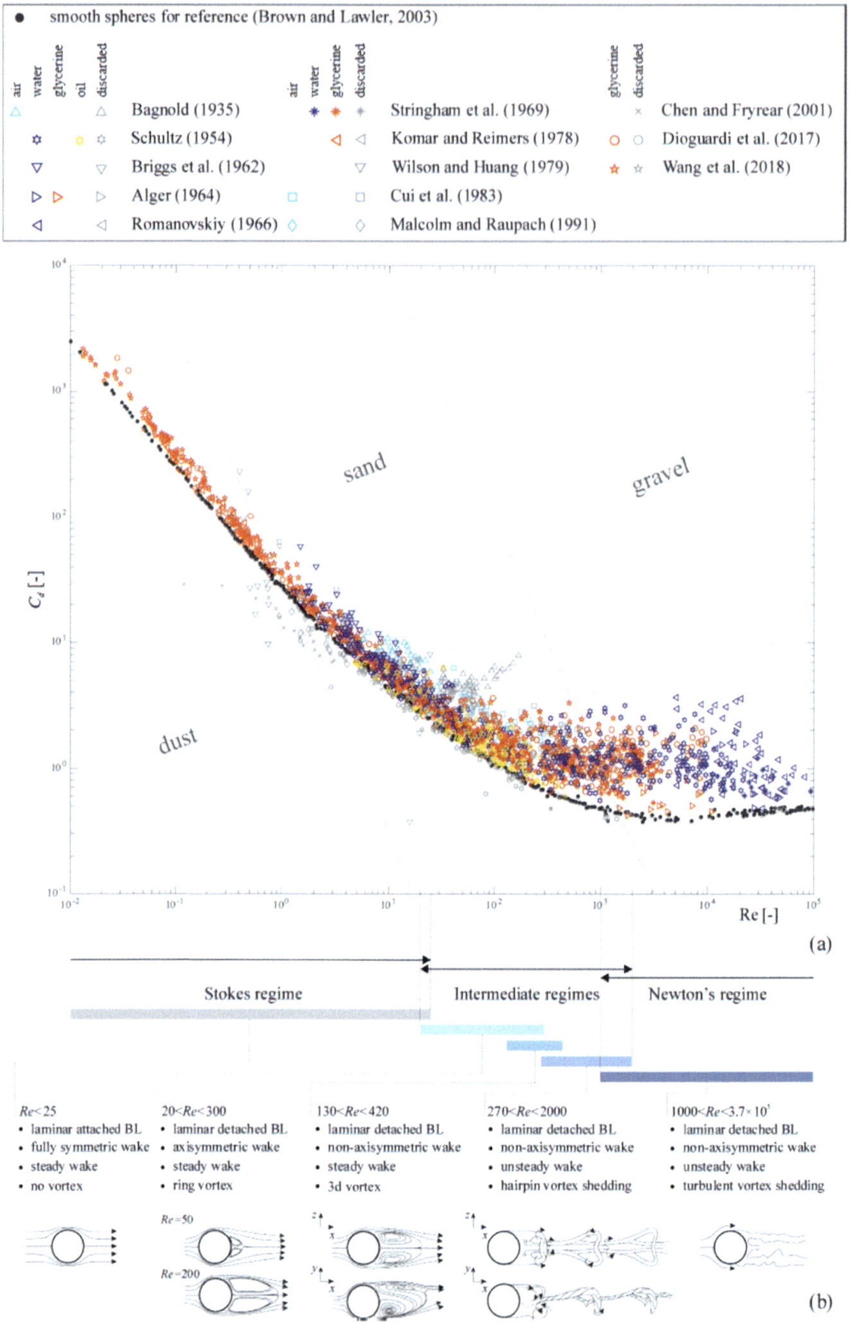


**Fig. 8.8** Drag coefficient as a function of the Reynolds number for natural particles and smooth spheres, depicting as well the different aerodynamic regimes and flow structure (Raffaele et al. 2020)

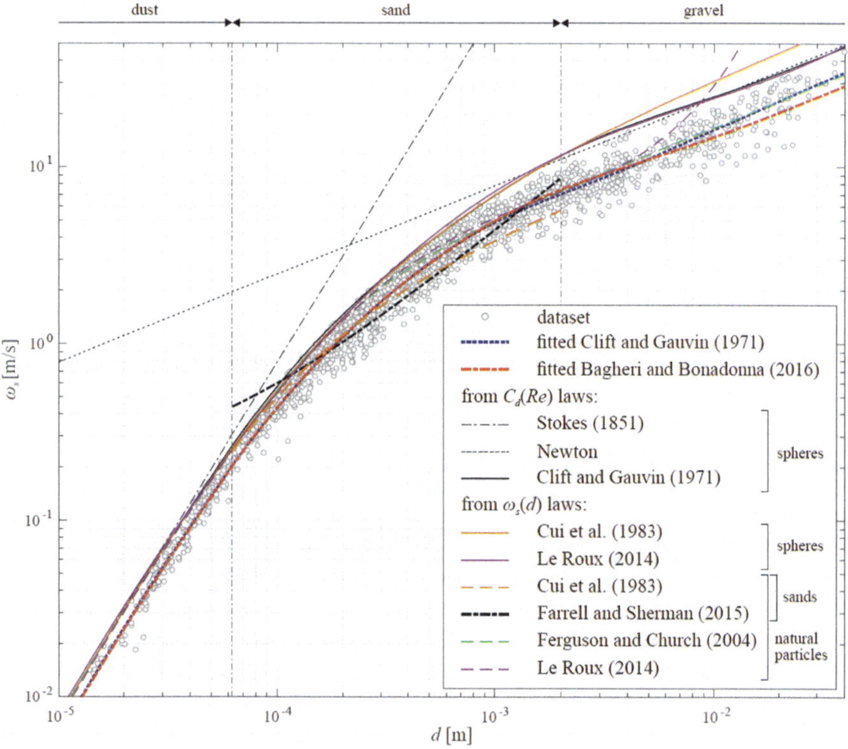


**Fig. 8.9** Comparison between mathematical fitting laws and experiments of the sedimentation velocity of quartz particles falling in air, for different shapes (Raffaele et al. 2020)

$$\psi = \frac{\pi d_{eq,p}^2}{SA_p} \tag{8.10}$$

This magnitude is then defined as the relationship of the surface area of a sphere, of the same volume as the particle, to the surface area of the real particle. Obviously, this value is equal to 1 for spheres and it decreases as the particles change their shape and become less spherical (Figs. 8.9 and 8.10).

Another technique is to use the concept of shape factors, which are simpler to measure than $\psi$. These involve measuring the longest (L), intermediate (I), and shortest (S) lengths of the particle. One can calculate the *flatness* $f = \frac{S}{I}$ and *elongation* $e = \frac{I}{L}$, as shown in Fig. 8.11.

Ganser (1993), using a similarity analysis, introduced a new formulation, considering that the drag coefficient for non-spherical particles can be calculated using the Reynolds number and two factors, dependent on the shape, called Stokes ($k_s$) and Newton ($k_N$) correction factors (Fig. 8.12),

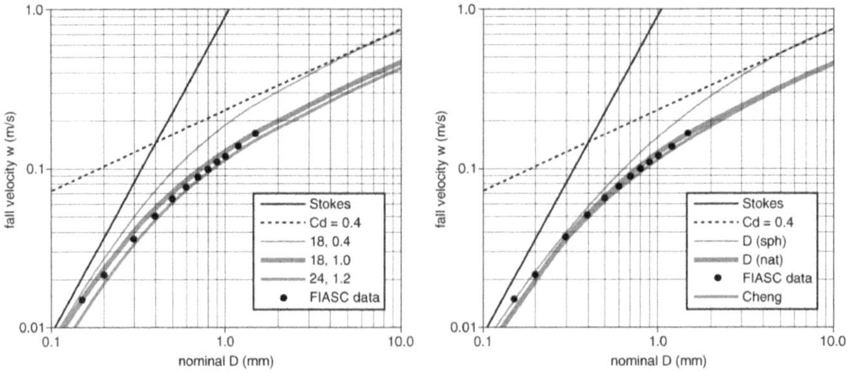


**Fig. 8.10** Predicted relation between sedimentation velocity and particle diameter for quartz grains, in water at 20 °C. The straight lines depict the expected asymptotic trends for smooth spheres, according to Stokes' law (Ferguson and Church 2004)

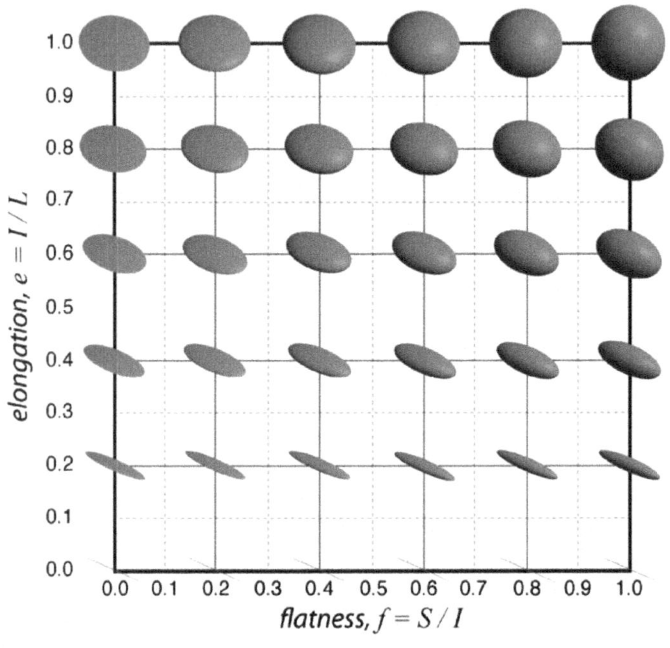


**Fig. 8.11** Effect of the particle shape, quantified by the *flatness* and *elongation* (Bagheri and Bonadonna 2016)

$$k_S \equiv \frac{C_D}{C_{D,s}} = \frac{C_D}{\frac{24}{Re_p}} \tag{8.11a}$$

$$k_N \equiv \frac{C_D}{C_{D,s}} = \frac{C_D}{0.445} \tag{8.11b}$$

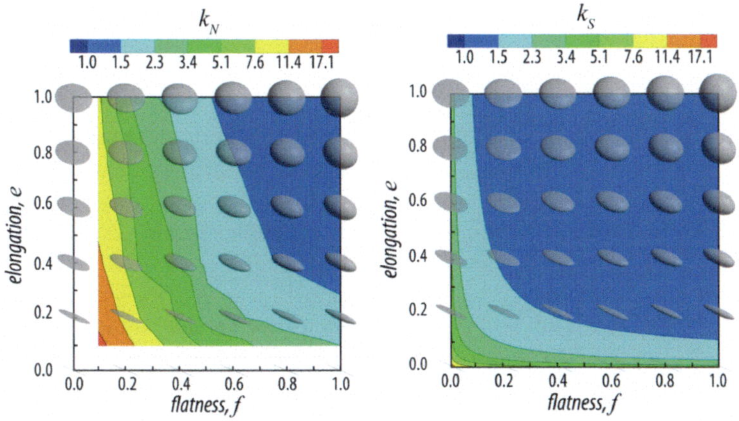


**Fig. 8.12** Impact of the flatness and elongation on the Stokes (left) and Newton (right) correction factors (Bagheri and Bonadonna 2016)

### 8.4.1 Hindered Settling

The previous equations are valid when calculating terminal velocities for individual particles. In a slurry mixture, the number of particles falling and settling is far larger than 1, and therefore the particles affect the settlement of other particles. This is, evidently, a function of the particles concentration (Fig. 8.13). A relationship to calculate the new terminal velocity is proposed by Richardson and Zaki (1954), valid for concentrations between 0 and 0.3. The equation is,

$$\frac{v_{t,h}}{v_t} = (1 - c)^{\beta} \tag{8.12}$$

The value of $\beta$ depends on the Reynolds number.

$$\beta = \begin{cases} 4.65, & \text{if } Re_p < 0.1 \\ 4.35Re^{-0.03}, & \text{if } 0.1 < Re_p < 1 \\ 4.45Re^{-0.1}, & \text{if } 1 < Re_p < 400 \\ 2.39, & \text{if } Re_p > 400 \end{cases} \tag{8.13}$$

## 8.5 Pressure Drop

An experimental relationship is sought in order to know the increase in pressure loss due to the presence of solid particles. The pressure loss of the heterogeneous flow in terms of variables can be expressed as $\Delta p = f\left(\Delta p_f; \rho_f; v_t; \rho_s; shape; v\right)$. Generally speaking, there are two possible approaches, which will be discussed in this

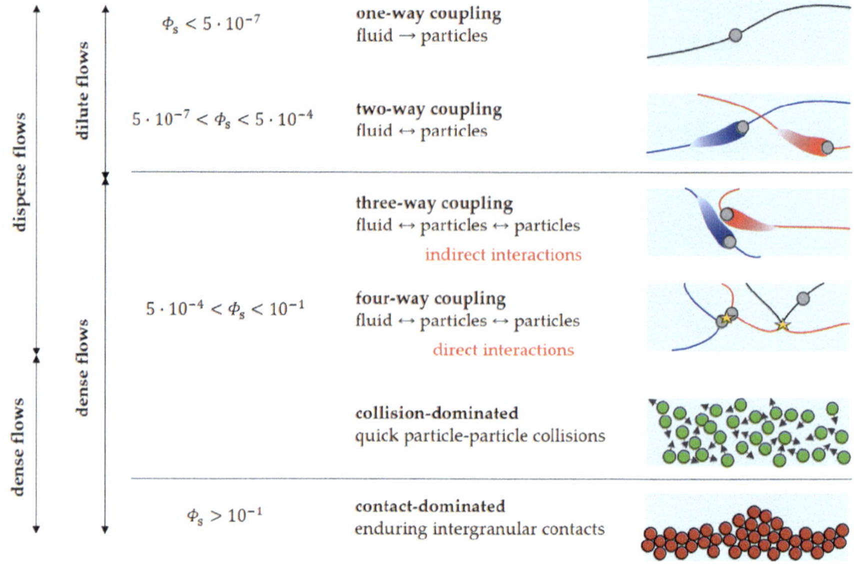


**Fig. 8.13** Classification of the different type of particle flows, according to the coupling regime (Messa et al. 2023)

Chapter. The first one involves the estimation of the pressure gradient as an addition of the fluid and solid components (Messa et al. 2021; Miedema 2013; Miedema and Ramsdell 2016; Jacobs 2003; Reddy et al. 2023).

$$\frac{\Delta p}{L} = \left(\frac{\Delta p}{L}\right)_f + \left(\frac{\Delta p}{L}\right)_s \tag{8.14}$$

$$\Delta p_f = f_f \rho_f \frac{v_f^2}{2} \frac{L}{D_{int}} = f_f \rho_f \frac{v_m^2 \left(1 - c^2\right)^2}{2} \frac{L}{D_{int}} = f_f \rho_f g \left(1 - c\right)^2 \frac{Fr^2}{2} L \tag{8.15}$$

The solid pressure loss on the other hand can be expressed similarly as,

$$\Delta p_s = c^2 K \rho_s \frac{v_m^2}{2} \frac{L}{D_{int}} = \frac{c K \rho_m g}{2} Fr^2 L, \quad Fr = \frac{v_m}{\sqrt{g D_{int}}} \tag{8.16}$$

Being $K$ an equivalent to the friction factor for solids, obtained experimentally as a function of the Froude number.

$$K = f\left(\rho_s; \rho_f; Fr; Fr_s\right), \quad Fr_s = \frac{v_t}{\sqrt{gd}} \tag{8.17}$$

The authors propose the following relationship for the constant $K$,

$$K = Y_1 R_{sd}^{Y_2} Fr^{Y_3} Fr_s^{Y_4} \tag{8.18}$$

rendering the total pressure drop as,

$$\begin{aligned}\frac{\Delta p}{L} = \left(\frac{\Delta p}{L}\right)_f + \left(\frac{\Delta p}{L}\right)_s &= f_f \rho_f g (1-c)^2 \frac{Fr^2}{2} + \frac{c Y_1 R_{sd}^{Y_2} Fr^{Y_3} Fr_s^{Y_4} \rho_m g}{2} Fr^2 \\ &= \left[ f_f \frac{\rho_f}{\rho_m} (1-c)^2 + cK \right] \rho_m \frac{v_m^2}{2} \frac{1}{D_{int}} \end{aligned} \tag{8.19}$$

The term between brackets can be considered as an equivalent friction factor, which is reduced to the Darcy factor when there is a pure fluid circulating. Minimizing the previous equation one could find an "optimum" Froude number for a minimum pressure drop, which would allow to estimate a diameter for the transport minimizing the pressure drop, not all the economical parameters.

$$\frac{\partial}{\partial Fr}\left(\frac{\Delta p}{L}\right) = 0 \tag{8.20}$$

The other approach, used by many authors (Miedema 2013; Miedema and Ramsdell 2016), consists in establishing a relationship between the pressure drop of a "clean" fluid ($i_f$) with the pressure drop of the slurry ($i_m$), as a function of the volumetric concentration. In the case of a pure fluid (i.e., no solids), it renders a quadratic curve obtained from the balance of energies. This approach can be used in both Newtonian (Fig. 8.14) and ineslatic non-Newtonian (Fig. 8.15) fluids.

### 8.5.1 *Durand*

Durand et al. (1953) and subsequently Durand (1953) carried out experiments in solids with an average diameter ranging from 0.18 mm to 22.5 mm in pipes with a diameter from DN 40 (NPS 1½) to DN 700 (NPS 28), and volumetric concentrations from 2% to 22%. They proposed an extra pressure loss due to solids calculated as,

$$\Phi = \frac{i_m - i_f}{i_f c_{v,t}} \tag{8.21}$$

$$\Phi = C\left(Fr^2 Fr_s^{-1}\right)^{-\frac{3}{2}} \tag{8.22}$$

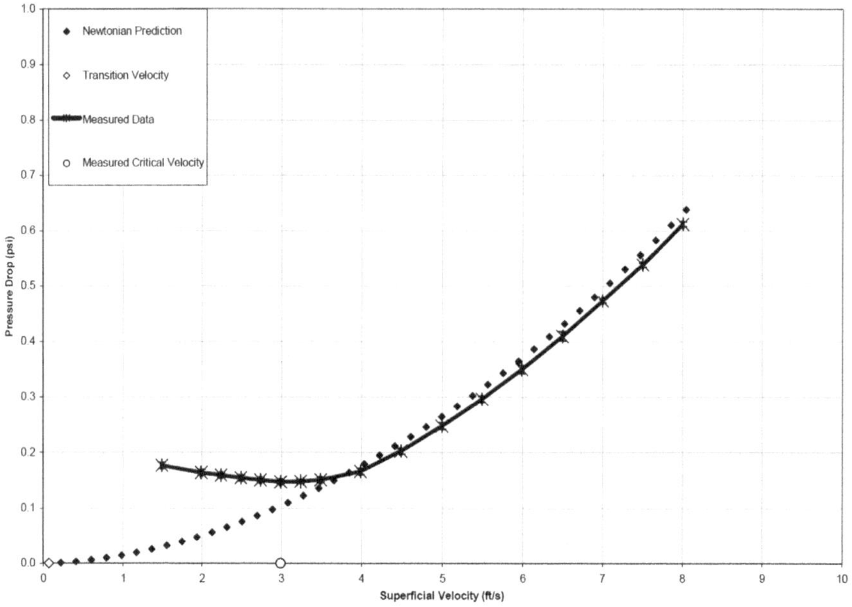


**Fig. 8.14** Pressure drop for a slurry mixture of 50-μm alumina in water, as a function of the fluid velocity (Poloski et al. 2009)

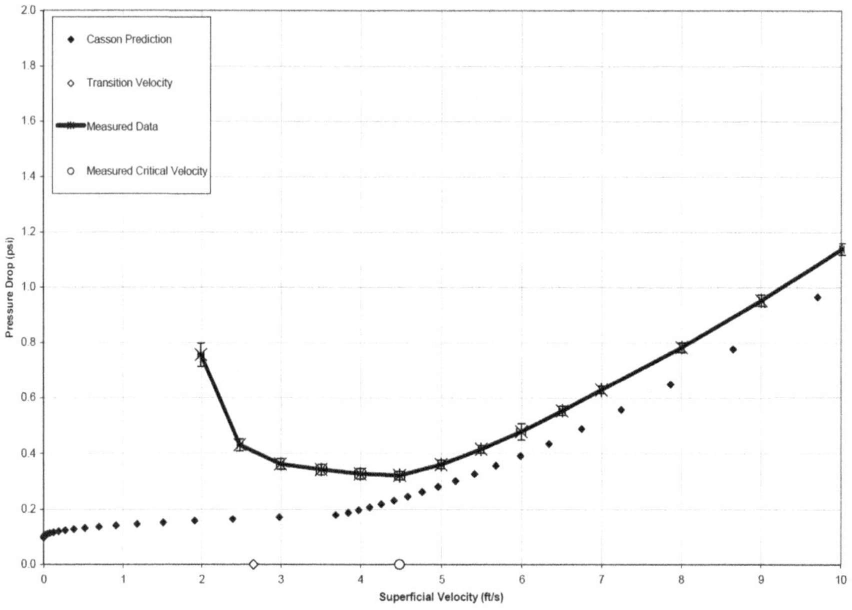


**Fig. 8.15** Pressure drop for a slurry mixture of 100-μm stainless steel in thin clay (modelled as a Casson non-Newtonian fluid), as a function of the fluid velocity (Poloski et al. 2009)

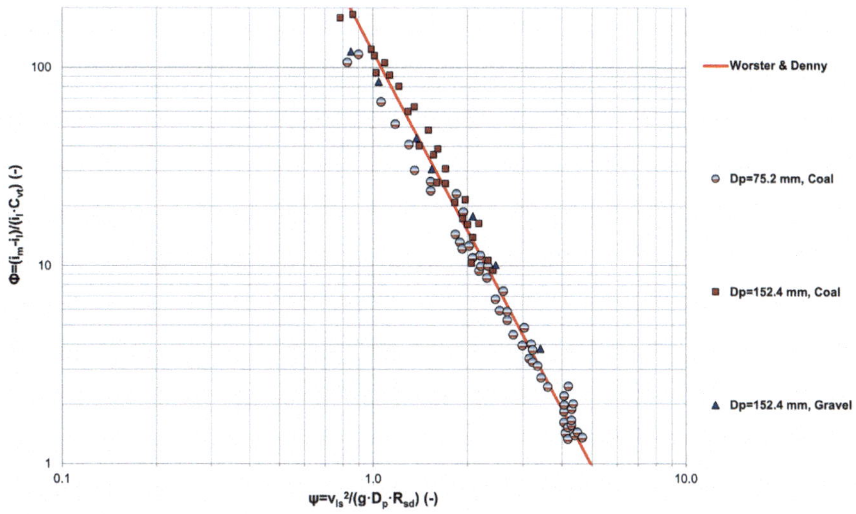


**Fig. 8.16** Relative pressure loss according to Worster & Denny, as a of the original Durand parameter $\psi$ (Miedema and Ramsdell 2016)

The constant of 176 was proposed originally in the paper of Durand and Condolios (1952). In order to consider the importance of the relative density between fluid and solid, Gibert (1960) presented a set of experiments with different densities of solids, and then he proposed an extension of the Froude number including the relative submerged density ($R_{sd}$). The liquid flow Froude number includes a new constant equal to 83.

$$Fr = \frac{v}{\sqrt{gD_{int}R_{sd}}} \tag{8.23}$$

Similar to Durand and Condolios, Worster and Denny (1955) presented the analysis of large particles in slurry flows, presenting a similar correlation to Durand's (Fig. 8.16).

$$\Phi = 120\left(\frac{v^2}{gD_{int}R_{sd}}\right)^{-\frac{3}{2}} \tag{8.24}$$

## 8.5.2 Babcock

Babcock (1971) adopted the approach to the previous authors and also of Charles (1970). He presented the simulations carried out in NPS 1 (DN 25) pipes, with coarse sand (20/30 mesh), and the proposed equation to consider the influence of solids is,

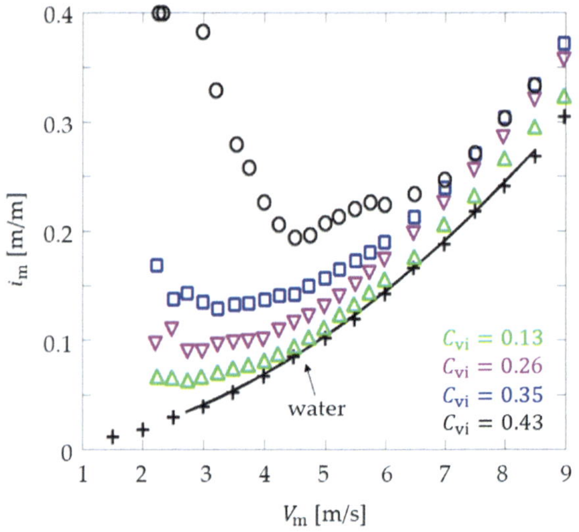


**Fig. 8.17** Pressure drop curves for flow of settling slurry at different volumetric concentrations (Messa et al. 2021)

$$\Phi = \frac{i_m - i_f}{i_f R_{sd} c_{v,t}} = 60.6\left(\frac{v^2}{gD_{int}}\right)^{-1} \quad (8.25)$$

Being $R_{sd}$ the relative submerged density, expressed as,

$$R_{sd} = \frac{\rho_s - \rho_l}{\rho_l} \quad (8.26)$$

According to some authors, the most important conclusion of this study is the verification of the influence of the volumetric concentration on the pressure drop (Fig. 8.17).

### 8.5.3 *Turian and Yuan*

The last study presented in this chapter was reported by Turian and Yuan (1977), who developed a a pressure drop model based on four different flow regimes for slurry pipelines, namely: stationary bed; saltation; heterogeneous; and homogeneous flows. The experiments consisted in 2848 data points to find correlations for each of these four regimes. These experiments were carried out in pipes ranging from NPS ½ (DN 15) to NPS 28 (DN 700), with relative submerged densities going from 0.16 to 10.3 particle sizes from 0.03 mm to 38 mm, and volumetric concentrations going from 0% (water) to 42%. The velocities tested ranged from 0 to 6.7 m/s. The excess pressure drop was proposed to be a function of four different parameters: the volumetric concentration, the friction factor, the drag coefficient and the Froude number.

**Table 8.1** Values of the different constants used for the different flow patterns in the Turian and Yuan method

| Type of flow | $\mathcal{K}$ | $\alpha$ | $\beta$ | $\gamma$ | $\delta$ |
|---|---|---|---|---|---|
| Fixed/sliding bed | 12.13 | 0.7389 | 0.7717 | –0.4054 | –1.096 |
| Saltation | 107.1 | 1.018 | 1.046 | –0.4213 | –1.3554 |
| Heterogeneous | 30.11 | 0.8687 | 1.2 | –0.1677 | –0.6938 |
| Homogeneous | 8.538 | 0.5028 | 1.428 | 0.1516 | –0.3531 |

$$\Delta p = f_s \rho_l \frac{v_l^2}{2} \frac{L}{D_{int}} \tag{8.27}$$

In this equation, the friction factor for the solid part is calculated as,

$$\frac{f_s}{4} = \frac{f_l}{4} + \mathcal{K} c_v^{\alpha} \left(\frac{f_l}{4}\right)^{\beta} \left(\frac{4}{3}\frac{R_{sd} g d_p}{v_t^2}\right)^{\gamma} \left(\frac{v_l^2}{R_{sd} g D_{int}}\right)^{\delta} = \mathcal{K} c_v^{\alpha} \left(\frac{f_l}{4}\right)^{\beta} C_{d,*}^{\gamma} Fr_*^{\delta} \tag{8.28}$$

The (*) indicates that it is a modified Froude or drag coefficient number. The four coefficients in the equations can be calculated according to the type of regime present (Table 8.1).

### *8.5.4 Delft Head Loss and Limit Deposit Velocity Framework*

Miedema in 2017 (Miedema 2013; Miedema and Ramsdell 2016) presented the a comprehensive study of all the previous slurry transport models and proposed a new framework, the DHLLDV, dividing the flow in five regimes, namely: fixed bed (FB); sliding bed (SB); heterogeneous (He); homogeneous (Ho); and sliding flow (SF) (Figs. 8.18 and 8.19).

## 8.6 Wear Verification for Steel Pipes

A vital part in the design of a slurry pipeline is also understand how the pipe material will wear as a function of time due to the interaction with the solid particles. This is defined as the cumulative loss (usually in mm/year) of material in the inner surface of the pipe (Gupta et al. 1995). One of the relationships was presented by Gupta et al. (1995) based on the concentration in weight,

$$E_r = 0.223 v^{2.145} D_s^{0.344} c_w^{0.556} \tag{8.29}$$

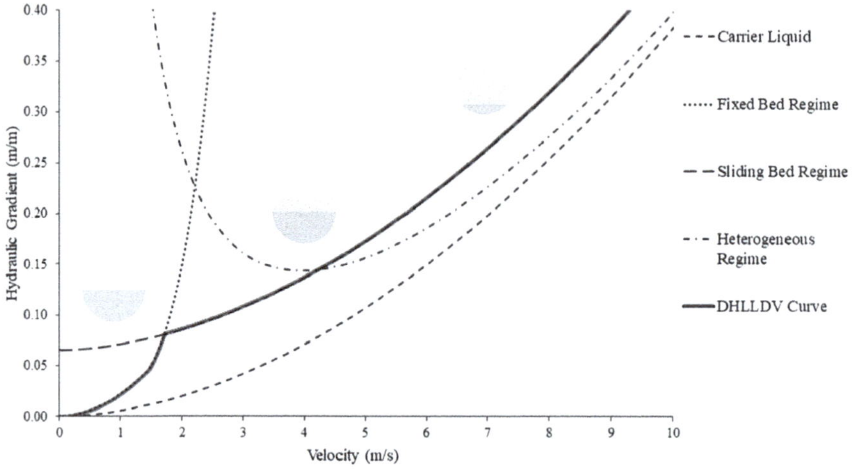


**Fig. 8.18** Hydraulic loss as a function of the fluid velocity in the pipeline for different fluid conditions and/or regimes, showing the path of the DHLLDV method (Ting et al. 2019)

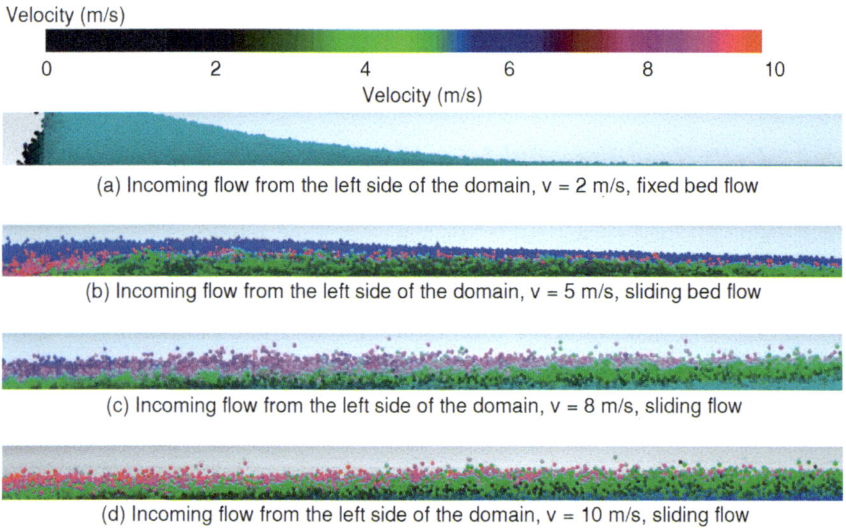


**Fig. 8.19** CFD simulation of the different flow regimes based on the DHLLDV method (Ting et al. 2019)

The exponents are valid for mild steel, commonly used for pipelines. If one considers brass:

$$E_r = 0.178 v^{2.4882} D_s^{0.291} c_w^{0.516} \tag{8.30}$$

$$\rho_m = \frac{1}{\frac{c_w}{\rho_s} + \frac{(1-c_W)}{\rho_l}} \tag{8.31}$$

The velocity is expressed in m/s and the particle diameter in mm. This equation assumes spherical particles; in the case of non-spherical or rounded particles, the diameter of the equivalent sphere must be calculated. For example, for an oblong-shaped particle 10 mm long and 3 mm in diameter with an approximate volume of $2.25 \times 10^{-8}$ m$^3$ gives an equivalent spherical diameter of 3.5 mm. It is noteworthy to mention that the velocity in these equations is approximately squared. Therefore, it is important to carry out a proper design to minimize the velocities, without promoting deposition, minimizing both the pressure drop and the wear (van den Boogaard 2020; Zhang et al. 2023; Sadeghi et al. 2023; Joshi et al. 2023).

## 8.7 Optimization of a Slurry Pipeline for the Transport Solid-Liquid

### *8.7.1 Definition of the Problem*

When the study of heterogeneous solid-fluid flow is focused, the input data of the problem are:

- Diameter of the solid particle (m);
- Density of the solid/fluid material (kg/m$^3$);
- Flow rate of the solid material to be transported (m$^3$/s);
- Geometry of the pipeline, this is, length (L) and height difference (H) (m).

With these data, the sedimentation velocity $v_t$ is determined, and then with experimental relationships, the following values are determined,

- Maximum volumetric concentration;
- Minimum transport speed to avoid sedimentation;
- Pipe diameter;
- Discharge pressure of the pump stations;
- Number of pumping stations.

However, these values are generally not optimal values, as discussed in previous chapter, which is why one do not obtain minimum cost transportation. In this section, a system of general equations is proposed to obtain the optimal values of $c$, $D_{int}$, $p_1$, and $\chi$, in a manner analogous to that developed for liquid and gaseous flows in the previous chapters.

### 8.7.2 Cost Functions

Analogously to what was developed for the flow of liquids and gases through pipes, one can define a number of costs representing the pipeline, energy, and pumping stations. The cost $c_4$ is not considered assuming that the fluid, whether air or water, does not have a significant cost. Regarding the solid, which can have a significant cost, it is recovered by circulating more fluid. Otherwise, if these assumptions are not met, the cost of the material should be added to the total (van den Boogaard 2020; Sadeghi et al. 2023; Reddy et al. 2023).

The Lagrange method can then be applied according to the Total Cost and Condition equations, and depending on the variables: diameter ($D_{int}$), maximum pressure ($p_1$), solid concentration ($c$), and number of plants ($\chi$). The four Lagrange equations plus the condition equation form the system of 5 equations with 5 unknowns, so that the system is perfectly defined and has a unique solution. Solving in the first instance the partial derivatives of the cost with respect to the unknowns, one obtains,

$$c_T = c_1 + c_2 + c_3(+c_4) \tag{8.32}$$

$$\frac{\partial c_T(p_1, \chi, D, c)}{\partial D} + \lambda \frac{\partial \phi(p_1, \chi, D, c)}{\partial D} = 0 \tag{8.33a}$$

$$\frac{\partial c_T(p_1, \chi, D, c)}{\partial p_1} + \lambda \frac{\partial \phi(p_1, \chi, D, c)}{\partial p_1} = 0 \tag{8.33b}$$

$$\frac{\partial c_T(p_1, \chi, D, c)}{\partial c} + \lambda \frac{\partial \phi(p_1, \chi, D, c)}{\partial c} = 0 \tag{8.33c}$$

$$\frac{\partial c_T(p_1, \chi, D, c)}{\partial \chi} + \lambda \frac{\partial \phi(p_1, \chi, D, c)}{\partial \chi} = 0 \tag{8.33d}$$

The condition equation can be obtained from the methods previously described, for instance using the Turian and Yuan model,

$$\frac{f_s}{4} = \mathcal{K} c_v^{\alpha} \left(\frac{f_l}{4}\right)^{\beta} C_{d,*}^{\gamma} Fr_*^{\delta} \tag{8.34}$$

$$\Delta p = 4\mathcal{K} c_v^{\alpha} \left(\frac{f_l}{4}\right)^{\beta} C_{d,*}^{\gamma} Fr_*^{\delta} \rho_l \frac{v_l^2}{2} \frac{L}{D_{int}} \tag{8.35}$$

Finally, the calculation proposed sequence is,

1. The diameter $D_{int}$ is estimated;
2. Calculate the Lagrange variable ($\lambda$);
3. Calculate the transport concentration ($c$);

4. Calculate the number of pumping stations ($\chi$);
5. Calculate the discharge pressure ($p_1$);
6. Verify the diameter assumed. If the verification fails, assume a different diameter and restart the calculation.
7. Once the optimum values have been found, round up to the commercial values and find the new concentration used in the transport.

## References


Babcock HA. Heterogeneous flow of heterogeneous solids. In: Advances in solid liquid flow in pipes and its applications. 1971. p. 125–48.

Bagheri G, Bonadonna C. On the drag of freely falling non-spherical particles. Powder Technol. 2016;301:526–44. ISSN 0032-5910. https://doi.org/10.1016/j.powtec.2016.06.015. https://www.sciencedirect.com/science/article/pii/S0032591016303539.

Bratland O. Pipe flow 2: multi-phase flow assurance. 2010.

Durand RJ. Basic relationships of the transportation of solids in pipes-experimental research. In: Proceedings of the IAHR 5th congress, minneapolis. 1953. p. 89–103.

Durand R, Condolios E, Gibert M. Etude expérimentale du refoulement des matériaux en conduites, en particulier des produits de dragage et des schlamms. Journées de l'hydraulique. 1953;2(1):27–55.

Einstein A. A new determination of molecular dimensions. Annln Phys. 1906;19:289–306.

Ferguson RI, Church M. A simple universal equation for grain settling velocity. J Sediment Res. 2004;74(6):933–7.

Ganser GH. A rational approach to drag prediction of spherical and nonspherical particles. Powder Technol. 1993;77(2):143–52.

Gupta R, Singh SN, Sehadri V. Prediction of uneven wear in a slurry pipeline on the basis of measurements in a pot tester. Wear. 1995;184(2):169–78. ISSN 0043-1648. https://doi.org/10.1016/0043-1648(94)06566-7. https://www.sciencedirect.com/science/article/pii/0043164894065667.

Hu H, Song K, Xu Y, Geng X, Wang W, Li Y, Zhang W, An C. Study on a numerical model of hydrate bed critical velocity in solid–liquid two-phase flow pipelines. Energy Fuels. 2023;37(7):4960–72. ISSN 0887-0624. https://doi.org/10.1021/acs.energyfuels.2c04105.

Jacobs BEA. Design of slurry transport systems. Taylor & Francis. 2003. ISBN 9780203215913. https://books.google.nl/books?id=YoLZLJAf40cC.

Joshi T, Parkash O, Krishan G. Slurry flow characteristics through a horizontal pipeline at different prandtl number. Powder Technol. 2023;413:118008. ISSN 0032-5910. https://doi.org/10.1016/j.powtec.2022.118008. https://www.sciencedirect.com/science/article/pii/S0032591022008890.

Messa GV, Yang Q, Adedeji OE, Chára Z, Duarte CAR, Matoušek V, Rasteiro MG, Sanders RS, Silva RC, de Souza FJ. Computational fluid dynamics modelling of liquid-solid slurry flows in pipelines: State-of-the-art and future perspectives. Processes. 2021;9(9):1566.

Messa GV, Yang Q, Rasteiro MG, Faia P, Matoušek V, Silva RC, Garcia F. Computational fluid dynamic modelling of fully-suspended slurry flows in horizontal pipes with different solids concentrations. KONA Powder Particle J. 2023;40:219–35.

Miedema SA. An overview of theories describing head losses in slurry transport: a tribute to some of the early researchers. In: International conference on offshore mechanics and arctic engineering, vol. 55362. American Society of Mechanical Engineers; 2013. p. V04AT04A038.

Miedema S, Ramsdell RC. Slurry transport: fundamentals, a historical overview and the delft head loss and limit deposit velocity framework. 2016.

Poloski AP, Adkins HE, Abrefah J, Casella AM, Hohimer RE, Nigl F, Minette MJ, Toth JJ, Tingey JM, Yokuda ST. Deposition velocities of newtonian and non-newtonian slurries in pipelines. Technical report, Pacific Northwest National Lab.(PNNL), Richland, WA (United States). 2009.

Raffaele L, Bruno L, Sherman DJ. Statistical characterization of sedimentation velocity of natural particles. Aeolian Res. 2020;44:100593. ISSN 1875-9637. https://doi.org/10.1016/j.aeolia.2020.100593. https://www.sciencedirect.com/science/article/pii/S1875963719302010.

Rajitha P, Chhabra RP, Sabiri NE, Comiti J. Drag on non-spherical particles in power law non-newtonian media. Int J Mineral Process. 2006;78(2):110–21. ISSN 0301-7516. https://doi.org/10.1016/j.minpro.2005.09.003. https://www.sciencedirect.com/science/article/pii/S0301751605001122.

Reddy NVK, Pothal JK, Barik R, Senapati PK. Pipeline slurry transportation system: an overview. J Pipeline Syst Engin Pract. 2023;14(3):03123001. https://doi.org/10.1061/JPSEA2.PSENG-1391. https://ascelibrary.org/doi/abs/10.1061/JPSEA2.PSENG-1391

Richardson JF, Zaki WN. The sedimentation of a suspension of uniform spheres under conditions of viscous flow. Chem Eng Sci. 1954;3(2):65–73.

Sadeghi M, Sontti SG, Zheng E, Zhang X. Computational fluid dynamics (cfd) simulation of three-phase non-newtonian slurry flows in industrial horizontal pipelines. Chem Engin Sci. 2023;270:118513. ISSN 0009-2509. https://doi.org/10.1016/j.ces.2023.118513. https://www.sciencedirect.com/science/article/pii/S0009250923000696.

Seitshiro IT, Katende J, Sato H. Slurry pipeline design of multi-sized solids: application of innovated models. Int J Soc Mater Engin Resourc. 2014;20(2):201–6.

Shirazi AS, Frigaard I. Slurrynet: predicting critical velocities and frictional pressure drops in oilfield suspension flows. Energies. 2021;14(5):1263.

Thomas DG. Transport characteristics of suspension: Viii. a note on the viscosity of newtonian suspensions of uniform spherical particles. J Colloid Sci. 1965;20(3):267–77.

Ting X, Xinzhuo Z, Miedema SA, Xiuhan C. Study of the characteristics of the flow regimes and dynamics of coarse particles in pipeline transportation. Powder Technol. 2019;347:148–58. ISSN 0032-5910. https://doi.org/10.1016/j.powtec.2019.02.031. https://www.sciencedirect.com/science/article/pii/S0032591019301408.

Turian RM, Yuan T-F. Flow of slurries in pipelines. AIChE J. 1977;23(3):232–43.

van den Boogaard T. Centrifugal dredge pumps: a study on particle trajectories near impeller blades. 2020.

Wang L, Cheng L, Yin S, Yan Z, Zhang X. Multiphase slurry flow regimes and its pipeline transportation of underground backfill in metal mine: mini review. Construct Build Mater. 2023;402:133014. ISSN 0950-0618. https://doi.org/10.1016/j.conbuildmat.2023.133014. https://www.sciencedirect.com/science/article/pii/S0950061823027319.

Worster RC, Denny DF. Hydraulic transport of solid material in pipes. Proc Inst Mech Engin. 1955;169(1):563–86.

Zhang R, Xu K, Liu Y, Liu H. A general numerical method for solid particle erosion in gas-liquid two-phase flow pipelines. Ocean Engin. 2023;267:113305. ISSN 0029-8018. https://doi.org/10.1016/j.oceaneng.2022.113305. https://www.sciencedirect.com/science/article/pii/S0029801822025884.

Zhiyao S, Tingting W, Fumin X, Ruijie L. A simple formula for predicting settling velocity of sediment particles. Water Sci Engin. 2008;1(1):37–43. ISSN 1674-2370. https://doi.org/10.1016/S1674-2370(15)30017-X. https://www.sciencedirect.com/science/article/pii/S167423701530017X.

# Chapter 9
# Transient in Pipelines

> A common mistake that people make when trying to design something completely foolproof is to underestimate the ingenuity of complete fools.
>
> *Douglas Adams*

## 9.1 Introduction

The analysis so far in this book has assumed that pipelines were operating in steady-state conditions, i.e., no significant variations in time of the operational parameters such as pressures or flow rates. However, it is well-known from practice that pipelines operate under conditions or regimes that are far away from steady-state conditions. The analysis of the transient flow in pipelines is one of the most important parts of its design, but for different reasons in liquid and gas lines since the compressibility of fluids affects the flow parameters and also the mechanical design. This chapter is subdivided, presenting firstly the analysis of incompressible fluids under transient conditions, stressing the critical points to consider; finally, the second part includes a mathematical derivation of a model to calculate pressure, temperature and flow rate variations in compressible pipelines (Ellul et al. 2004; Mohitpour et al. 2007).

This analysis is an important part of their design, but for different reasons if they are liquids or gases. The main reason for this is how fluid compressibility affects flow parameters and mechanical design. In nearly incompressible fluids, transient flows are generated by the closing of valves, creating a sudden increase in pressure. In gas pipelines the causes and effects are completely different. In general, over-pressures do not appear in gas pipelines, but due to this behavior, the amount of gas mass inside the pipe will be a function of time (Larock et al. 1999).



P. Druetta et al., *Transport by Pipelines and Its Optimization*, Petroleum Engineering,
https://doi.org/10.1007/978-3-031-66566-0_9

The gas pipeline design process depends on many factors, some more obvious than others. The final result of the process will aim to determine the diameter, the MAPO, and the number of compressor stations and their location. In general, gas transmission systems have been designed with steady-state simulations and under stable demand scenarios. Transient analyzes are recommended when anticipated during system operation: large load factors (Qmax/Qmin); significant variations in mass flow; and frequent entry/exit of compressors to the system.

- In nearly incompressible fluids ($\nabla \cdot \mathbf{u} \approx 0$), transient flows are generated by the closure of line valves, creating a pressure surge which can provoke the blow-out of the pipeline (aka *water hammer*). In these cases it is quite important to achieve **very low** velocities of closure in valves.
- For gas pipelines the causes and effects are completely different (since $\nabla \cdot \mathbf{u} \neq 0$). Usually overpressures are not found in gas lines. However, due to this behavior, the amount of mass inside the gas pipeline will be a function of time, acting as some kind of huge vessel. During this chapter this system will be studied.

The classification can also be done based on the type of pipes/pipelines in which the transient takes place, namely: in closed conduits; in open channels; and, a combined free-surface and pressurized transient flow. Another possible classification is to divide the systems in *distributed* and *lumped* ones. In the latter it is assumed that changes in the operating conditions are transmitted instantaneously through the system, they depend on time, but not on the position itself; for instance, the oscillations in a surge tank after a change in an adjacent turbomachinery. In the former, one considers the fluid compressible, and the transient is expressed as the transmission of waves in the fluid; an example of this is the change in the conditions of a gas pipeline with variations in the consumption. In this case the operating parameters depend both on time and position.

A question one may ask is what the nature of these transients is? These changes, either planned or accidental, provoke the onset of waves travelling through the fluid (and pipe), altering both the velocity and pressure. Some of these causes are listed below.

1. Opening and/or closing of lines valves;
2. Chattering of relief valves;
3. Failure of hydraulic dams;
4. Pumps (reciprocating/centrifugal) operation in any pumping facility;
5. Start-up of hydraulic turbines or change in their power demand;
6. Vibrations that may take place in, for instance, impellers or blades;
7. Formation of waves on a reservoir;

## 9.2 Transport of Incompressible Fluids

The objective thus far in this book has been to introduce and explain a way to carry out the design of pipelines facilities by combining engineering with economical tools. A series of formulations have been discussed and derived allowing to find, based on a certain input data, an optimum diameter for a project. Nevertheless, this approach has been based on a static analysis of the pipeline, i.e., no changes in time. It is evident that normal operation of any pipeline comes with a certain number of operations involving time dependant functionalities (McNeeley 2021).

As mentioned before, there are two possible types of transients: quasi-steady flow, in which inertial and elastic effects are considered negligible; and real transients, in which the previously mentioned effects cannot be neglected. This last case comprises what is known in industry as *water hammer* condition (Bergant et al. 2006). There is no golden rule to determine the borderline between one and the other, and each case should be carefully analyzed. In case of doubts, one should always use the real transient analysis, since it uses the complete equations.

### *9.2.1 Water Hammer*

Water hammer is a well-known phenomenon in pipeline engineering taking place when the change in the velocities take place so fast that the fluid and pipe's elastic properties cannot be neglected in the study, for instance, the sudden closure of a valve. In order to present this analysis, the classical example of the pipeline connected to a reservoir is used, as shown in Figs. 9.1 and 9.2. The system is working under steady state conditions until the valve at the end of the line is operated, creating a transient flow. Considering that the valve is completely closed almost instantaneously, the following events will take place (Bergant et al. 2006, 2008; Kaneko et al. 2014).

1. The velocity of the fluid upstream the valve becomes zero; however, not all the fluid in the line reduces its momentum to zero at the same time. Thus, the fluid still moving increases the pressure of the vicinity of the valve. Due to the elastic behavior of pipe and fluid, the pipe deforms increasing it diameter, and the fluid increases its density.
2. This creates a pressure wave moving backwards with a speed **a** which depends on the pipe and fluid properties. The pipe deforms and the fluid comes at a stop;
3. After a time $L/a$, all the pipe has been deformed and expanded and the fluid came to a full stop in the pipeline with an increase pressure.
4. In the vicinity of the reservoir, an unbalance of pressure is taking place: the pipeline contains fluid at a higher pressure than the reservoir, creating a backflow of the fluid in the line back to the reservoir and the elastic pipeline deforming back to the original conditions. This takes place during a period of time also $L/a$ until the front reaches the valve.

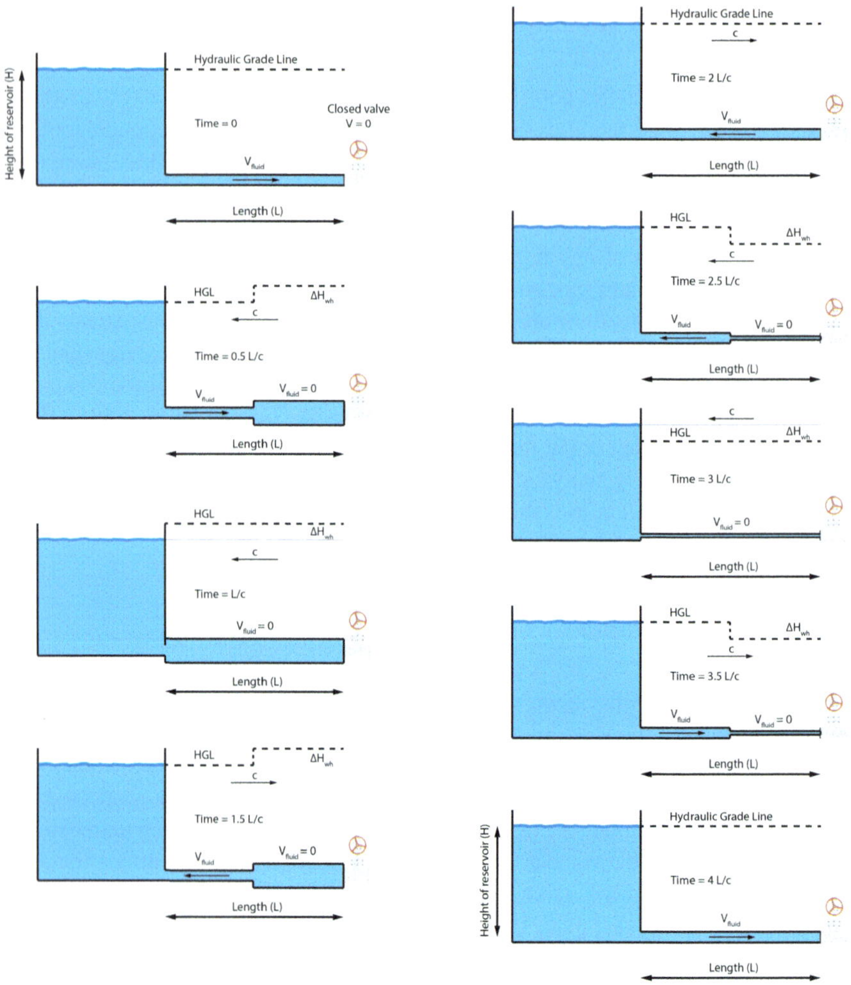


**Fig. 9.1** Scheme showing the different stages of a water hammer process caused by the closure of a valve, assuming an ideal flow

5. After $2L/a$, the pipe is in original state but the fluid is moving backwards. This creates a lower pressure region near the valve, shrinking the pipe while the fluid comes again at a stop.
6. The wave travels back to the reservoir and at a time $3L/a$ all the pipe is shrunk, with the fluid at zero velocity inside the pipe. This originates one more time an unbalance between pipe and reservoir, this time allowing the fluid to enter back in the pipeline, expanding the pipe back to its original dimensions with the fluid moving one more time towards the valve.

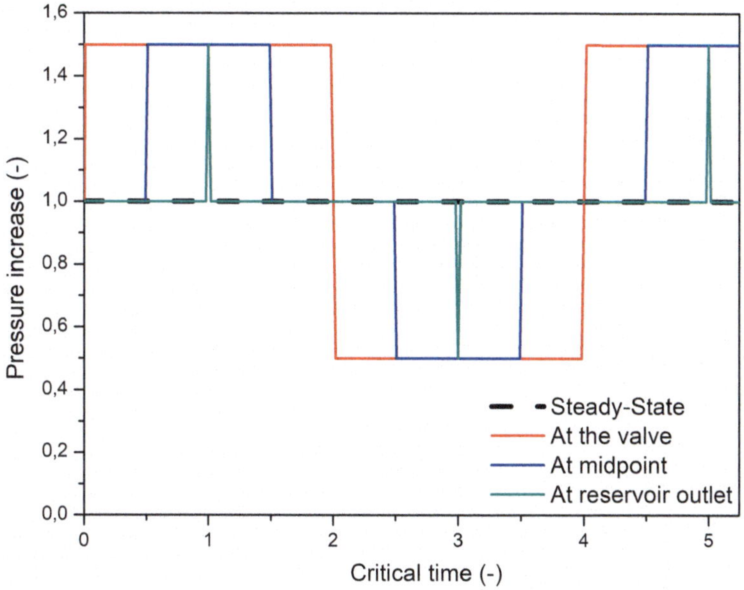


**Fig. 9.2** Plot showing the duration and intensity of the peaks during a water hammer phenomenon described in Fig. 9.1

7. The cycle continues until $4L/a$ in which one goes back to the original conditions at the beginning of this cycle.

### *9.2.2 Joukowski Equation*

The phenomenon of water hammer, its causes and how it is developed in a system with elastic (i.e. deformable) materials, has been introduced and explained qualitatively in the previous section. The next step consists in developing a mathematical formulation to explain and predict what happens in these cases. The phenomenon of transients in pipelines has been widely studied during the last 100 years, with the pioneering work of Allievi in 1925 as a starting point in this matter (Allievi 1925).

The study of this starts with the differential analysis of the momentum equation in a deformable pipeline, in which a mechanical wave (of unknown origin) is travelling at a certain speed, as shown in the Fig. 9.3.

The Reynolds Transport Theorem, also known as the Leibniz-Reynolds Transport Theorem, or simply RTT, is a generalization of the Leibniz integral rule that is commonly used in Fluid Mechanics to describe the conservation laws acting on a fluid. It can be simply written as,

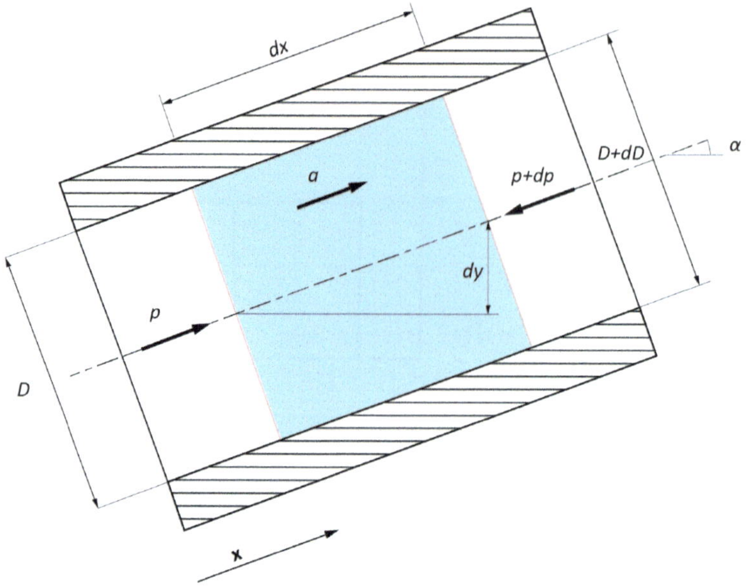


**Fig. 9.3** RTT analysis of a control volume inside a pipeline investigating the effect of the velocity change on the pressure

*what was already there + what enters – what goes out = what is there now*

$$\frac{d}{dt}\left(B_{syst}\right) = \int_{CV} \frac{\partial}{\partial t}\left(\beta\rho\right) dV + \int_{CS} \beta\rho\left(\mathbf{v}\cdot\mathbf{n}\right) dA \tag{9.1}$$

$$\beta = \frac{d}{dm}\left(B_{syst}\right) \tag{9.2}$$

For a balance of forces, it is necessary to replace the dummy variable for a linear momentum.

$$B_{syst} = m\mathbf{v} \Rightarrow \beta = \mathbf{v} \tag{9.3}$$

$$\frac{d}{dt}\left(m\mathbf{v}\right)_{syst} = \sum \mathbf{F} = \frac{d}{dt}\iiint_{CV} \mathbf{v}\rho dV + \iint_{CS} \mathbf{v}\rho\left(\mathbf{v}\cdot\mathbf{n}\right) dA \tag{9.4}$$

$$pA - \left(p + dp\right)\left(A + dA\right) = \left(V + a\right) A\rho_f dV \tag{9.5}$$

The left-hand side term can be expanded and rewritten, considering that the variations in the area can be neglected in comparison with the variations of the pressure ($dA \ll dp$),

$$-Adp = -A\rho_f g dH = (V + a)\, A\rho_f dV \tag{9.6}$$

$$dH = -\frac{dV}{g}(V + a) = -\frac{dVa}{g}\left(1 + \frac{V}{a}\right) \tag{9.7}$$

The velocity of propagation of mechanical waves in the system ($a$) is usually several orders of magnitude higher than the fluid speed ($V$), hence, the equation can be then written as what is known as the *Joukowski equation*,

$$dH = -\frac{dVa}{g} \tag{9.8}$$

The Joukowski equation allows to calculate a variation in the pressure due to a variation in the fluid's velocity. Any decrease in the velocity, such as the closure of a valve, will provoke an increase in the pressure, as described previously. Nonetheless, there is a critical aspect of this formula, which is not yet known, the wave speed $a$, or the velocity of propagation of mechanical waves in the system.

### 9.2.3 Thin-Walled Pipes and the Speed of Sound

In order to reach an expression to calculate the velocity of propagation, a mass balance analysis will be introduced, similar to the momentum used in the previous section. Thus, using a similar control volume to that of Fig. 9.3,

$$B_{syst} = m \Rightarrow \beta = 1 \tag{9.9}$$

$$\left(\frac{dm}{dt}\right)_{syst} = 0 = \frac{d}{dt}\iiint_{CV} \rho dV + \iint_{CS} \rho\,(\mathbf{v}\cdot\mathbf{n})\,dA \tag{9.10}$$

$$0 = \frac{d}{dt}\left(\rho_f \forall\right) - \rho_f AV + \left(\rho_f + d\rho_f\right)(V + dV)(A + dA) \tag{9.11}$$

$$\frac{dm}{dt} = -\left(\rho_f VdA + \rho_f dVA + d\rho_f VA\right) \approx -\rho_f AdV \tag{9.12}$$

$$dm \approx -\rho_f AdV\frac{dL}{a} \tag{9.13}$$

The elastic properties of both fluid and solid have been mentioned during this chapter to play a major role in true transients (Garg and Kumar 2020; Kandil et al. 2020). Therefore, the next step consists in establishing a formulation to understand how these materials change under the influence of a changing pressure. For almost incompressible fluid, the bulk modulus of elasticity defines the relationship between the pressure and the change in density,

$$K = -\frac{dp}{\frac{d\forall}{\forall}} \tag{9.14}$$

With respect to the change or deformation of an elastic pipe, the analysis more complex, since it involved the deformation of an isotropic, elastic material in two dimensions under the influence of an internal pressure, with a certain number of physical constraints.

$$\sigma_L = \frac{\epsilon_L + \mu_p \epsilon_C}{1 - \mu^2} E \tag{9.15a}$$

$$\sigma_C = \frac{\epsilon_C + \mu_p \epsilon_L}{1 - \mu^2} E \tag{9.15b}$$

Where $L$ stands for the longitudinal axis of the pipeline and $C$ represents the circumferential (or girth) axis. The coefficient $\mu_p$ is the Poisson's coefficient. The pipe will change its diameter under the influence of an internal pressure, changing therefore its volume,

$$d\forall_C = \frac{\pi}{4}(D + dD)^2 L - \frac{\pi}{4} D^2 L = \frac{\pi}{2} D dD L \tag{9.16}$$

$$d\epsilon_C = \frac{dD}{D} \tag{9.17}$$

$$d\forall_C = \frac{\pi}{2} D^2 d\epsilon_C L \tag{9.18}$$

Similarly, the change of volume due to the longitudinal expansion is,

$$d\forall_L = \frac{\pi D^2}{4} dL = \frac{\pi D^2}{4} d\epsilon_L L \tag{9.19}$$

Combining both equations, one can obtain the total volume change as,

$$d\forall = \frac{\pi D^2}{4} L \left(d\epsilon_L + 2 d\epsilon_C\right) \tag{9.20}$$

With the study of the deformations complete, now these must be related to a specific stress condition, such as the change in the internal pressure. The circumferential stress for thin walled vessels is well-known from the literature and the simplification of Lamé's equations as,

$$d\sigma_C = \frac{dpD}{2t} \tag{9.21}$$

with $t$ begin the pipeline wall thickness. Relating this formula to Eq. 9.15b,

$$d\sigma_C = \frac{dpD}{2t} = \frac{d\epsilon_C + \mu_p d\epsilon_L}{1-\mu^2} E \tag{9.22}$$

While the relationship between stresses and deformations in the circumferential axis is well determined, this is not the case for the longitudinal axis, which depends on the existing restraints on the pipe.

- Pipeline anchored at one end;
- Pipeline fully restrained from axial displacements;
- Pipeline with expansion joints.

The derivation of the speed of sound will be applied to the second case as example. The longitudinal deformation is then zero and the total volume change is,

$$d\forall = \frac{\pi D^2}{2} L d\epsilon_C \tag{9.23}$$

$$\sigma_L = \frac{\mu_p \epsilon_C}{1-\mu^2} E = \mu_p \sigma_C \tag{9.24}$$

$$\frac{dpD}{2t} = \frac{\epsilon_C}{1-\mu^2} E \tag{9.25}$$

$$d\forall = \frac{\pi D^2}{2} L \frac{dpD}{2tE} \left(1-\mu^2\right) \tag{9.26}$$

$$dm = \rho_f d\forall + \forall d\rho_f = \rho_f A dV \frac{L}{a} \rightarrow d\forall + \forall \frac{d\rho_f}{\rho_f} = A dV \frac{L}{a} \tag{9.27}$$

$$d\forall + \forall \frac{dp}{K} = A dV \frac{L}{a} \tag{9.28}$$

$$\frac{dVa}{a^2} = \frac{\rho_f g \Delta H}{K} + 2 \frac{\rho_f g \Delta H D}{2tE} \left(1-\mu^2\right) \tag{9.29}$$

$$\frac{1}{a^2} = \frac{\rho_f}{K} + \frac{\rho_f D}{tE}\left(1 - \mu^2\right) \tag{9.30}$$

$$a^2 = \frac{\frac{K}{\rho_f}}{1 + \frac{KD}{tE}\left(1 - \mu^2\right)} \tag{9.31}$$

This equation considers the influence of all the elastic elements in the speed of sound within the pipeline. For instance, a pipe of infinite rigidity would render the classical speed of sound in liquids, $\sqrt{\frac{K}{\rho_f}}$. Furthermore, one can observe an extra factor, which depends exclusively on the constraints in the system, as listed before. These extra factors are,

- Pipeline anchored at one end = $1 - \frac{mu_p}{2}$ or $\frac{5}{4} - \mu_p$
- Pipeline fully restrained from axial displacements = $1 - \mu_p^2$
- Pipeline with expansion joints = 1

## 9.2.4 Thick-Walled Pipes and Other Conduits

The vast majority of the pipelines installed in the World can be modelled as thin-walled pipes, in which the relationship $D/t > 40$. However, there are a reduced number of cases where the thin-walled theory cannot be used. In this case, the complete Lamé's equations of elasticity must be solved in order to obtain the complete three dimensional state of stresses and deformations in the elastic pipe material (see Chapter 2). As in the previous case, the restraint conditions affect the value of the speed of sound, obtaining also in this case a series of pipe restraining factors,

- Pipeline anchored at one end = $\frac{2t}{D}\left(1 + \mu_p\right) + \frac{D}{D+t}\left(1 - \frac{\mu_p}{2}\right)$ or $\frac{2t}{D(1+\mu_p)+\frac{D}{D+t}}\left(\frac{5}{4} - \mu_p\right)$
- Pipeline fully restrained from axial displacements = $\frac{2t}{D}\left(1 + \mu_p\right) + \frac{D}{D+t}\left(1 - \mu_p^2\right)$
- Pipeline with expansion joints = $\frac{2t}{D(1+\mu_p)+\frac{D}{D+t}}$
- Circular tunnel (wall thickness $\rightarrow \infty$) = $\frac{2t}{D}\left(1 + \mu_p\right)$

## 9.2.5 Entrapment of Compressible Fluids in the Flow

The previous sections aimed at studying the influence on the speed of sound of the materials used in the construction of the lines. However, a question remains: what is the effect, sometimes undesired, of a compressible fluid, in the form of bubbles for instance, in the speed of propagation of mechanical waves (Zhang et al. 2018). It was seen that any addition of "elasticity" in the system will reduce the speed of propagation. In the case of bubbles, its effect is major: even a small percentage of trapped bubbles will decrease substantially the speed of sound. The first step to calculate this consists in finding an equivalent bulk modulus circulating in the pipeline,

$$K_{mix} = \frac{K_{if}}{1 + \upsilon_{f,cf}\left(\frac{K_{if}}{K_{cf}} - 1\right)} \tag{9.32}$$

where $K_{ic}$ and $K_{cf}$ are the bulk modulus of the incompressible and compressible fluids, respectively and $\upsilon_{f,cf}$ is the volume concentration of the compressible fluid. The density of the mixture can be also obtained using the volume fraction, as seen in previous chapters (Yan et al. 2018). Thus, the final formula for the speed of sound is,

$$a = \frac{\sqrt{\frac{K_{if}}{\rho_{mix}}}}{1 + \frac{K_{icD}}{Et} C_{pipe} + \upsilon_{f,cf}\frac{K_{if}}{K_{cf}}} \tag{9.33}$$

It is important to note that the bulk modulus of the compressible fluid is strongly dependant on the thermodynamic evolution of the flow: for an isothermal process, $K_{cf} = p$; for an isentropic process, $K_{cf} = kp$; and for a polytropic process, $K_{cf} = np$.

### 9.2.6 *Differential Equations of Transient Flow of Liquids*

Previous sections in this chapter were aimed at introducing the concept of transient flows and describing mathematically a number of concepts necessary to study these phenomena in pipelines. The next step consists on making a differential analysis of a control volume of the pipeline and establishing relationships for the conservation of mass and momentum, which take place within the line. This analysis renders the following system of partial differential equations for an inclined pipe with friction,

$$\frac{Dp}{Dt} + \rho_f a^2 \frac{\partial V}{\partial x} = 0 \tag{9.34a}$$

$$\frac{DV}{Dt} + \frac{1}{\rho_f}\frac{\partial p}{\partial x} + g\frac{dy}{dx} + \frac{f}{2D}V\,|V| = 0 \tag{9.34b}$$

where $\frac{D\ldots}{Dt}$ is the material derivative, comprising partial derivatives of the magnitude with respect to time and space. Expanding these terms,

$$\frac{\partial p}{\partial t} + V\frac{\partial p}{\partial x} + \rho_f a^2 \frac{\partial V}{\partial x} = 0 \tag{9.35a}$$

$$\frac{\partial V}{\partial t} + V\frac{\partial V}{\partial x} + \frac{1}{\rho_f}\frac{\partial p}{\partial x} + g\frac{dy}{dx} + \frac{f}{2D}V\,|V| = 0 \tag{9.35b}$$

Considering an horizontal, frictionless flow and discarding the non-linear convective terms of the material derivative, it renders,

$$\frac{\partial p}{\partial t} + \rho_f a^2 \frac{\partial V}{\partial x} = 0 \tag{9.36a}$$

$$\frac{\partial V}{\partial t} + \frac{1}{\rho_f}\frac{\partial p}{\partial x} = 0 \tag{9.36b}$$

One can approximate the non-linear convective term derivative as $V\frac{\partial V}{\partial x} \approx \frac{V}{a}\frac{\partial V}{\partial t}$; as long as the speed of sound is several order of magnitude higher than the circulation velocity, these terms can be neglected without introducing significant errors in the calculation. Replacing the pressure for its geometric expression, $p = \rho_f g H$, taking the second cross derivatives and applying the Schwarz's theorem of symmetry of second derivatives, one obtains,

$$\frac{\partial^2 H}{\partial t^2} + \frac{a^2}{g}\frac{\partial^2 V}{\partial x \partial t} = 0 \tag{9.37a}$$

$$\frac{\partial^2 V}{\partial t \partial x} + g\frac{\partial^2 H}{\partial x^2} = 0 \tag{9.37b}$$

Combining both equations,

$$\frac{\partial^2 H}{\partial t^2} - a^2\frac{\partial^2 H}{\partial x^2} = 0 \tag{9.38}$$

Rendering the well-known, hyperbolic in nature, wave partial differential equation.

### 9.2.7 Numerical Solution—Method of Characteristics

The partial differential equations described before do not have, as it happens usually in engineering, an analytical solution; thus, an appropriate numerical scheme should be chosen in order to find the approximate solution. In the case of the wave equation, a good starting technique is the Method of Characteristics (MoC) (Khalighi et al. 2017; Li et al. 2023). In this chapter only the simplified equations will be presented; the analysis for the complete system of equations can be found in the literature. Thus, starting with the two simplified differential equations,

$$\frac{\partial p}{\partial t} + V\frac{\partial p}{\partial x} + \rho_f a^2 \frac{\partial V}{\partial x} = 0 \tag{9.39a}$$

$$\frac{\partial V}{\partial t} + V\frac{\partial V}{\partial x} + \frac{1}{\rho_f}\frac{\partial p}{\partial x} + g\frac{dy}{dx} + \frac{f}{2D}V\left|V\right| = 0 \tag{9.39b}$$

The idea in the MoC is to replace a pair of PDE's by an equivalent set of ODE's. Then, it is assumed that both equations may be written as a linear combination of themselves. Using $\lambda$ as a constant linear factor, the *Lagrange multiplier*,

$$\frac{\partial p}{\partial t} + V\frac{\partial p}{\partial x} + \rho_f a^2 \frac{\partial V}{\partial x} + \lambda\left(\frac{\partial V}{\partial t} + V\frac{\partial V}{\partial x} + \frac{1}{\rho_f}\frac{\partial p}{\partial x} + g\frac{dy}{dx} + \frac{f}{2D}V\left|V\right|\right) = 0 \tag{9.40}$$

$$\left(\lambda\frac{\partial V}{\partial t} + a^2\frac{\partial V}{\partial x}\right) + \left(\frac{1}{\rho_f}\frac{\partial p}{\partial t} + \frac{\lambda}{\rho_f}\frac{\partial p}{\partial x}\right) + \lambda g\frac{dy}{dx} + \lambda\frac{f}{2D}V\left|V\right| = 0 \tag{9.41}$$

In order to replace this by an ordinary differential equation, the following conditions must be met,

$$\lambda\frac{dx}{dt} = a^2 \tag{9.42a}$$

$$\frac{\lambda}{\rho_f} = \frac{1}{\rho_f}\frac{dx}{dt} \tag{9.42b}$$

The solution to this system is then $\lambda = \pm a$. Then, one can rewrite this as a system of ODE's,

$$\frac{dV}{dt} + \frac{1}{\rho_f a}\frac{dp}{dt} + g\frac{dy}{dx} + \frac{f}{2D}V\left|V\right| = 0 \tag{9.43a}$$

$$\frac{dV}{dt} - \frac{1}{\rho_f a}\frac{dp}{dt} + g\frac{dy}{dx} + \frac{f}{2D}V\left|V\right| = 0 \tag{9.43b}$$

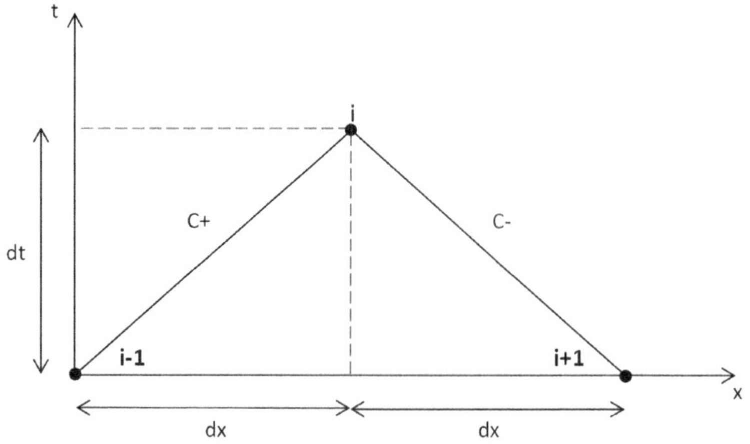


**Fig. 9.4** Diagram showing the graphical representation of the MoC indicating the positive and negative characteristic lines and their evolution in time through the nodes of the system (Kandil et al. 2020)

Considering an horizontal line and replacing the pressure for its equivalent in piezometric height,

$$\frac{dV}{dt} + \frac{g}{a}\frac{dH}{dt} + \frac{f}{2D}V\,|V| = 0, \quad \text{valid for} \quad \frac{dx}{dt} = +a \tag{9.44a}$$

$$\frac{dV}{dt} - \frac{g}{a}\frac{dH}{dt} + \frac{f}{2D}V\,|V| = 0, \quad \text{valid for} \quad \frac{dx}{dt} = -a \tag{9.44b}$$

Integrating the equations on the right, it yields a series a straight lines with the same slope but different integration constants (Fig. 9.4),

$$t = \pm\frac{x}{a} + C^{I} \tag{9.45}$$

To compute the values of $H$ and $V$ at the different points of the pipeline as a function of time, the initial conditions must be known along the x-axis and boundary conditions for all time at the pipe ends. The process is depicted in Fig. 9.5.

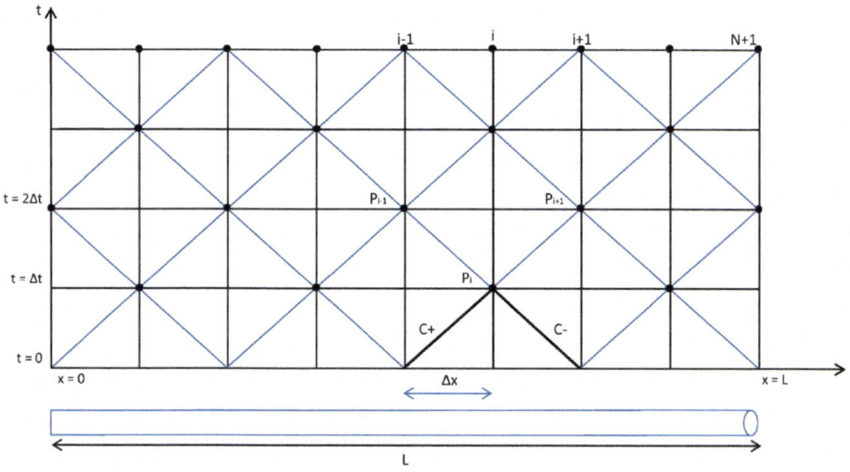


**Fig. 9.5** General scheme of the MoC for several time-steps in a discretized domain (Kandil et al. 2020)

And the finite difference approximation for the conservation equation is,

$$\frac{V_p - V_{l,p}}{\Delta t} + \frac{g}{a}\frac{H_p - H_{l,p}}{\Delta t} + \frac{f}{2D}V_{l,p}\left|V_{l,p}\right| = 0, \quad \text{valid for} \quad \frac{\Delta x}{\Delta t} = +a \tag{9.46a}$$

$$\frac{V_p - V_{p,r}}{\Delta t} - \frac{g}{a}\frac{H_p - H_{r,p}}{\Delta t} + \frac{f}{2D}V_{r,p}\left|V_{r,p}\right| = 0 \quad , \quad \text{valid for} \quad \frac{\Delta x}{\Delta t} = -a \tag{9.46b}$$

## Boundary and Initial Conditions

The initial conditions are the steady-state solutions of the Bernoulli equation within the pipeline, with a constant flow rate and inlet pressure (Wan et al. 2018). Generally speaking, there are three different boundary conditions used in water hammer, representing the operational behavior at the inlet/outlet of the pipeline (Figs. 9.6 and 9.7). These are,

- Upstream reservoir $\rightarrow H_{p,1}(t) = H_{res}$;
- Downstream valve condition $\rightarrow \begin{cases} V_{p,N}(t) = V_0\left(1 - \frac{t}{T_{clo}}\right) & \forall t < T_{clo} \\ V_{p,N}(t) = 0 \quad \forall t \leq T_{clo} \end{cases}$ ;
- Upstream centrifugal pump $\rightarrow H_{p,1} = AV_{p,1}^2 + BV_{p,1} + C$.

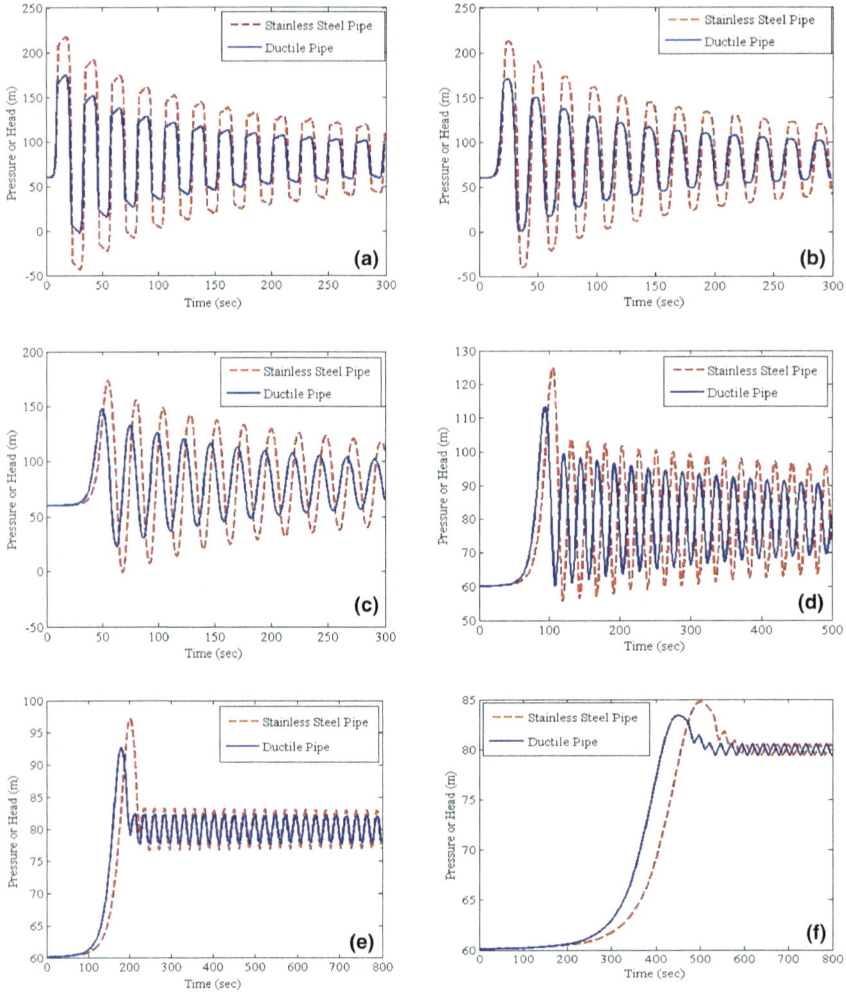


**Fig. 9.6** Transient pressure fluctuations according to a linear valve closure scheme, considering the following times: **a** t = 2L/c; **b** t = 4L/c; **c** t = 10L/c; **d** t = 20L/c; **e** t = 40L/c; and **f** t = 100L/c (Yuce and Omer 2019)

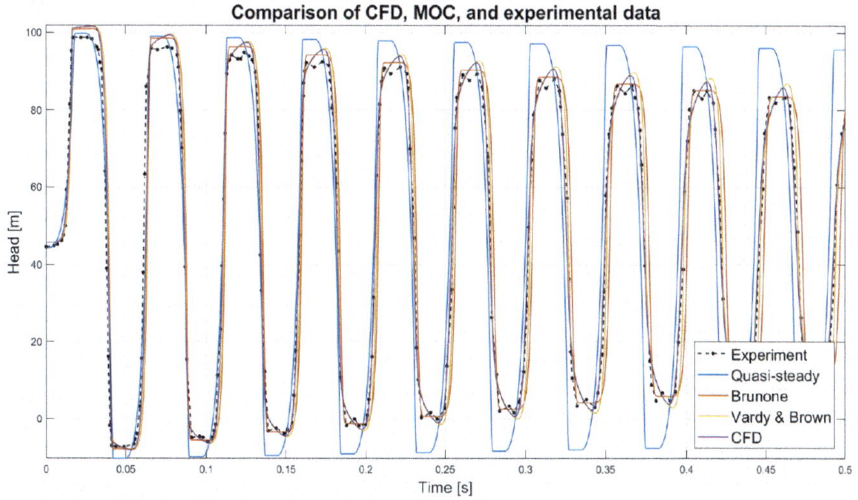


**Fig. 9.7** Diagram showing the pressure as a function of time for a valve closure, comparing different simulation techniques (Saidani et al. 2022)

## Pseudocode for the MoC

The method described in this chapter can be easily implemented in any programming tool in order to simulate and optimize a system which undergoes a transient period with incompressible fluids (Carlsson 2016).

---

**Pseudocode 9.1:** Water hammer

---

Set $\Delta t$, $L$, $N_{ele}$, $p_1^n$, $D$, $t$, $\varepsilon$, $max_{time}$ and $T_{clo}$;
Set $K$, $\rho_f$, $E$, $t$, $\mu_p$;
Set Boundary Conditions;
$A = \frac{\pi D_{int}^2}{4}$;
Speed of sound $a$ based on restraining conditions;
Friction factor $f = f\left(Re, \frac{\epsilon}{D}\right)$;
Define initial parameters $H(n)^0$ and $V(n)^0$ with steady state formulation;
$t = \Delta t$;
**while** $t \leq max_{time}$ **do**
    **for** i=2:1:N-1 **do**
        Set equation valid for $+a$;
        Set equation valid for $-a$;
        Solve the two equations to find $H(i)^t$ and $V(i)^t$;
    **end for**
    Using equation valid for $+a$ calculate unknown parameter in (N,t);
    Using equation valid for $-a$ calculate unknown parameter in (1,t);
    $t = t + \Delta t$;
**end while**

---

## 9.3 Transport of Compressible Fluids

The most important factor to consider before proceeding with the design of a gas pipeline is the flow requirement (i.e., demand), and the process begins precisely with the supply and demand forecast. To do this, it is necessary to survey both the potential consumers along the pipeline route and their projection over time, as well as the eventual producers who can deliver volumes of gas for transportation (Menon 2005; Mohitpour et al. 2007).

Unlike the transport of liquids, for the transport of gas in general there are no storage tanks, and the tank is somehow the pipeline itself. All variable consumption must be absorbed with the storage in the gas pipeline, up to the limits of the minimum and maximum pressures. However, peak control plants called *peak-shaving* have been developed, these are large cryogenic LNG storage, or on a smaller scale, propane+air injection plants from LPG cylinders or spheres; and also underground storage (caverns or deposits), where pressurized natural gas is injected in summer and extracted in winter (Clegg and Mancarella 2015; van der Hoeven 2007).

The design process of a (gas) pipeline (or a pipeline network—Fig. 9.8) depends on many factors, some of them more evident than the others. The final outcome of this process will be aimed at determining the pipe size, the amount of compression stations, their physical locations, their power (and how they will be powered) and the operating pressures, which may be sometimes limited by the country's legislation (Liu et al. 2019). Thus, in order to determine all these factors one should pay special attention to,

- the total length: this is usually considered from the supply points (e.g. gas fields or distribution centers) to the final users;
- the geographical location (see Fig. 9.9): the gas fields may be located in such a way that the pipeline must cross bodies of water to reach the final users;
- the location of the consumption: in some cases one could have a concentrated point of consumption (e.g. Argentina), whilst in others (e.g. the Netherlands) the distribution of gas in the country can be more balanced;
- the prices of the different commodities (e.g. steel, gas, oil, electricity);
- the availability of energy, water, etc. Building a gas pipeline near the Randstad area in the Netherlands may be a bit different than doing it in the Amazonian jungle or in the Atacama desert.

However, it also may depend on some other factors, less evident than the previous ones,

- interest rates;
- gas composition;
- local/regional legislation;
- protected areas (e.g. National Parks, Natura 2000);
- weather conditions;
- mating period of endangered species;

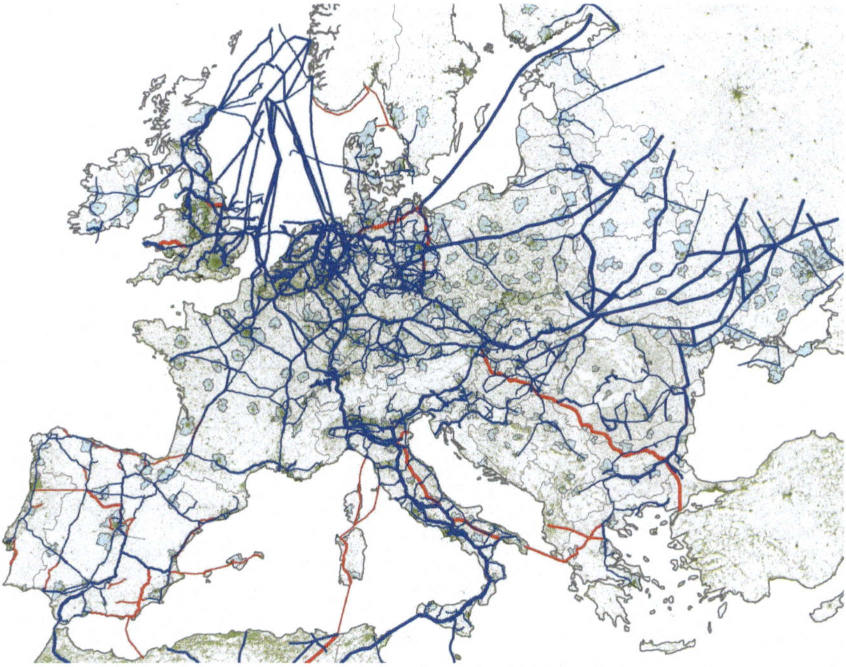

**Fig. 9.8** European Gas Network, showing the main gas pipelines in the continent, including the Mediterranean Sea crossings (Zurich 2024)

- political interests.

Still, the most important parameter has to be determined: the **flowrate**, i.e. how much gas companies want to transport. In gas pipelines that value might be hard to adopt. What flow rate should be chosen? Obviously, the amount of gas consumed changes with the seasons, especially for residential usage since the industrial/energy should remain approximately constant, and also every week, every day, and every hour (Fig. 9.10). Depending on time, these consumption-time curves can be obtained.

The question of what flow rate one should use for the design is one of the most difficult to answer. As engineers, one can choose not to perform a transient analysis and then perform for a steady state design. However, this may introduce certain risks, namely:

- considering the maximum: then the pipe will be over-designed, the project will not be optimum since a large diameter will be used (and therefore, thicknesses), for a condition that occurs less than a month per year.
- deciding to go for some average flow rate, then during winter there may be a problem of low pressure at the entrance of the PCV's (pressure control valves) in the cities or power plants, unless you count with a peak-shaving plant.

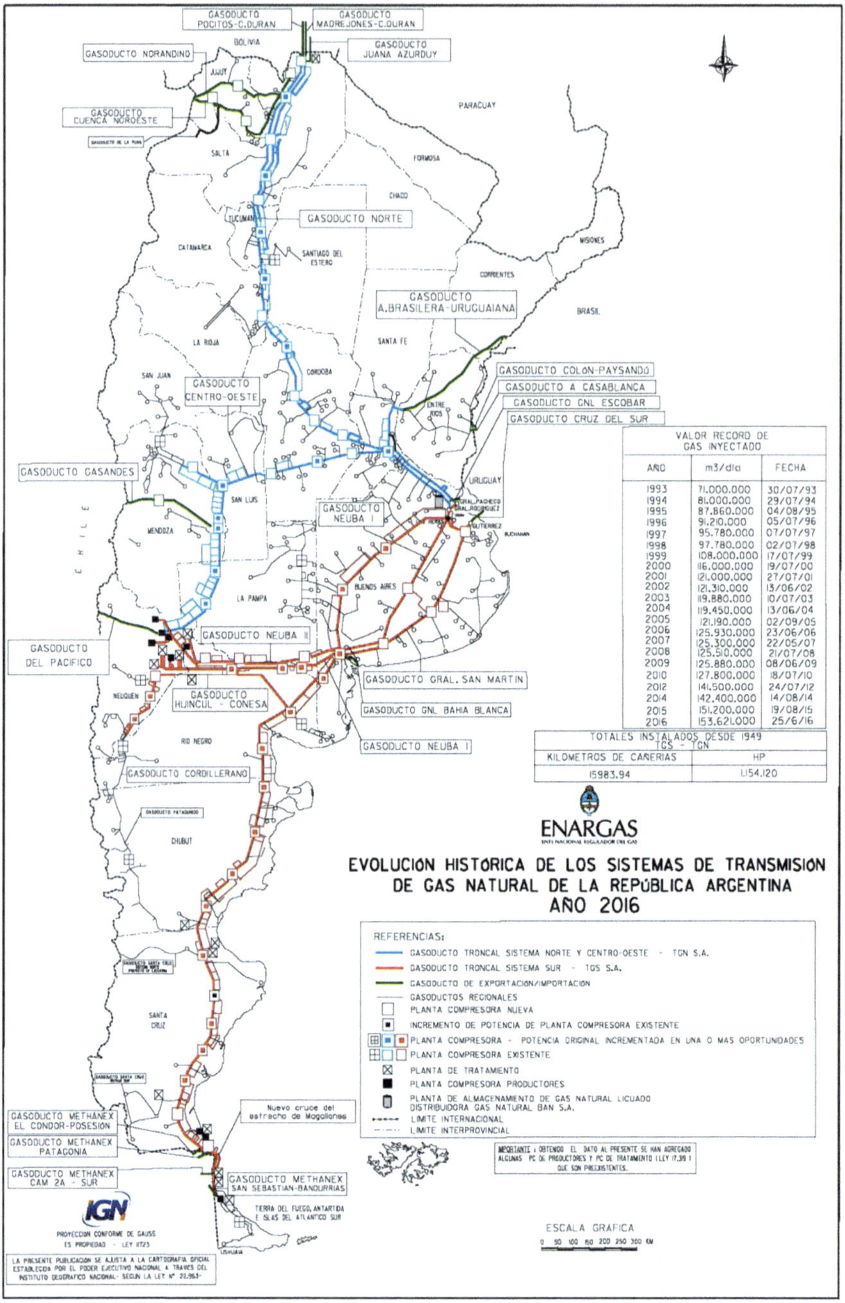


**Fig. 9.9** Argentinian Gas Network, showing the main gas pipelines coming from the north and south of the country to the capital, Buenos Aires (ENARGAS 2017)

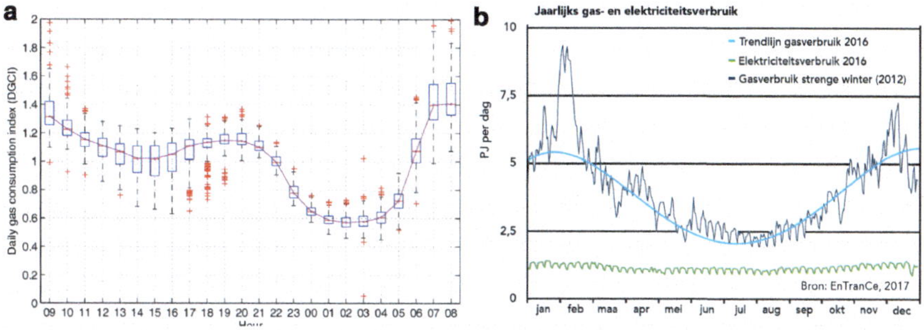


**Fig. 9.10** Plots showing the variability of the gas consumption: **a** daily gas consumption index as a function of the time of the day (Potocnik and Govekar 2010); and **b** daily energy demand of natural gas in the Netherlands, indicating also a peak due to one of the coldest winters registered in February. In green the electricity and in blue the natural gas demands with its trend line also depicted (Mertens 2022)

All these variations, caused by the constant change in the demand, take place due to the compressibility of the gas. The density (i.e., the total mass) inside the pipeline changes both with space and time. Thus, the pipe acts as some kind of large, large vessel. The capacity of the pipe to respond to the variations in the demand based on the volume of gas stored inside is known as **linepack**.

## *9.3.1 Linepack*

The most important factor to consider before proceeding with the design of a gas pipeline is the flow requirement (demand) and the process begins precisely with the supply and demand forecast. Steady-state simulations are limited to know a realistic response of the gas pipeline system to flow fluctuations and to appreciate the differences between both modeling it is necessary to relate the pressure drops to the flow rates and the linepack, i.e., the effect of storage of pressurized gas in the gas pipeline (Fig. 9.11). The linepack has a damping effect on the transient flow (Clegg and Mancarella 2015; Krishnaswami et al. 2004). Using exclusively the analysis of steady-state flows can lead to very pronounced erroneous conclusions if demand is very cyclical.

The pipeline, in addition to its essential function of transporting natural gas, is normally used to meet consumption peaks and be used as buffer or storage tank to cope up with this variations. The volume of gas contained in the line is a function of pressure and the gas physical properties, so in order to develop an expression that allows calculating the total volume contained, the pressure conditions must be known. Thus, it is possible to know the gas content of the line in the "compacted" and "non-compacted" conditions.

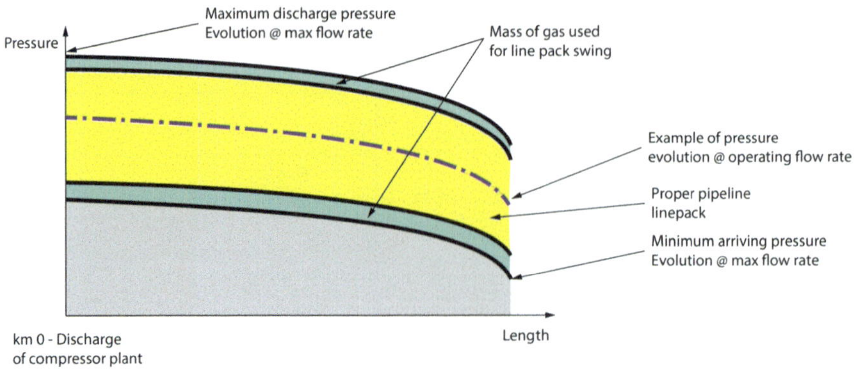


**Fig. 9.11** Scheme showing the different pressures that can be found in a compressible flow pipeline, indicating as well the linepack area

$$V = \frac{2}{3}\frac{A}{p_0}\left[\left(p_1^3 - p_2^3\right)\frac{L}{p_1^2 - p_2^2}\right] \tag{9.47}$$

with $A$ as the internal area of the pipeline and the average pressure being,

$$p_m = \frac{2}{3}\frac{p_1^3 - p_2^3}{p_1^2 - p_2^2} \tag{9.48}$$

This renders,

$$V = \frac{AL}{p_0}p_m = \mathcal{K}\frac{D_i^2 L}{p_0}\left(p_1 + p_2 - \frac{p_1 p_2}{p_1 + p_2}\right) \tag{9.49}$$

The linepack is then defined as the difference of the volume of gas stored in the *compacted* conditions minus the volume of gas at the pipeline's minimum working conditions, *non-compacted* (Fig. 9.12). Thus,

$$S = \frac{\pi}{4p_0}\left(p_1^* - p_1 + p_2^* - p_2 - \frac{p_1^* p_2^*}{p_1^* + p_2^*} + \frac{p_1 p_2}{p_1 + p_2}\right) D_i^2 L \quad [Sm^3] \tag{9.50}$$

where,

- $p_1$ = Maximum allowable operating pressure (MAPO);
- $p_1^*$ = Minimum admissible supply pressure;
- $p_2$ = Final pressure of the section (compacted) calculated with nominal flow;
- $p_2^*$ = Initial pressure of the section (non-compacted) calculated with the nominal flow;
- $p_0$ = Reference pressure;

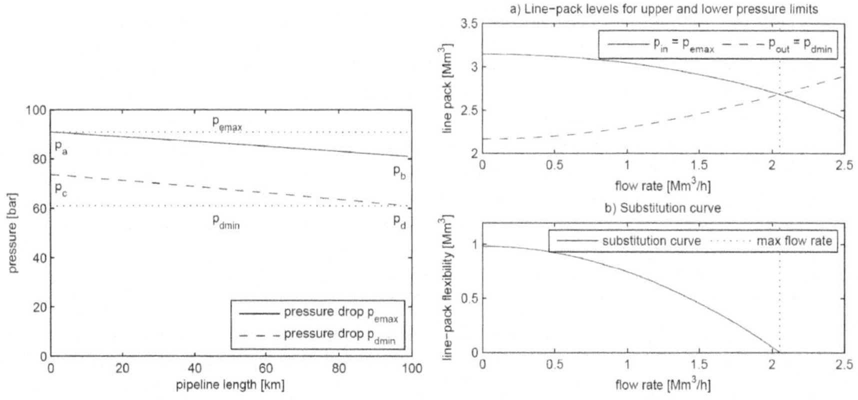


(a) (b)

**Fig. 9.12** Graphs showing the line-pack flexibility, with the area between the maximum and minimum pressures representing the potential storage in the line (Keyaerts et al. 2011)

- $L$ = Pipeline length;
- $D_i$ = Internal diameter.

## 9.3.2 *Differential Equations of Transient Flow of Gases*

As it was done before, one need to make some assumptions before starting the analysis to obtain a set of Partial Differential Equations to solve the transient, non-isothermal flow of compressible fluids in pipelines. Firstly, an analysis of all the differential, conservation equations that are applied in the system is introduced. Then, some simplifications will be made before discussing solution strategies (Gugat et al. 2023; Yin et al. 2023). As starting assumption, it is considered that the flow inside the pipeline can be well represented using a 1D system. As with liquids, the Reynolds Transport Theorem can be applied to derive all the conservation equations, adding in this case the energy, also known as the First Law of Thermodynamics.

- Mass conservation ($B_{syst} = m \Rightarrow \beta = 1$);
- Momentum ($B_{syst} = m\mathbf{v} \Rightarrow \beta = \mathbf{v}$);
- Energy ($B_{syst} = \mathbf{E} \Rightarrow \beta = e$).

**Conservation of Mass**

In order to study this, a control volume in the pipe is considered as depicted in Fig. 9.13.

$$\rho Au - \left[\rho Au + \frac{\partial}{\partial x}(\rho Au)\,dx\right] = \frac{\partial}{\partial t}(\rho Adx) \tag{9.51}$$

$$\frac{\partial \rho}{\partial t} + \frac{\partial}{\partial x}(\rho u) = 0 \tag{9.52}$$

### Conservation of Momentum

The momentum analyzes the balance of forces in the control volume shown in Fig. 9.14. Since one considered the friction (due to the viscosity), a stress/force in the pipe walls appears.

$$pA - \left[pA + \frac{\partial}{\partial x}(pA)\,dx\right] - \tau_w \pi Ddx - \rho gAdx\left(\frac{dy}{dx}\right) = \rho Adx\left(u\frac{\partial u}{\partial x} + \frac{\partial u}{\partial t}\right) \tag{9.53}$$

The wall shear stress can be expressed as a function of the velocity and the friction factor,

$$\tau_w = \frac{f\rho u \mid u \mid}{8} \quad \text{and} \quad f = fun\left(Re, \frac{\epsilon}{D}\right) \tag{9.54}$$

$$\rho\frac{\partial u}{\partial t} + \rho u\frac{\partial u}{\partial x} + \frac{\partial p}{\partial x} = -\frac{f\rho u \mid u \mid}{8A}\pi D - \rho g \sin\alpha \tag{9.55}$$

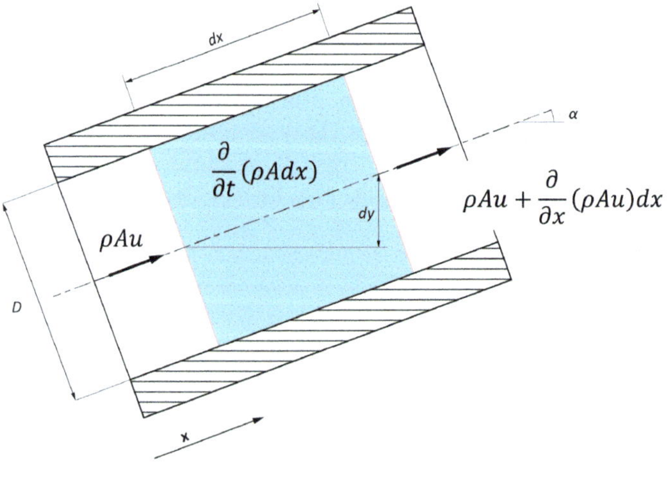


**Fig. 9.13** RTT analysis of the conservation of mass

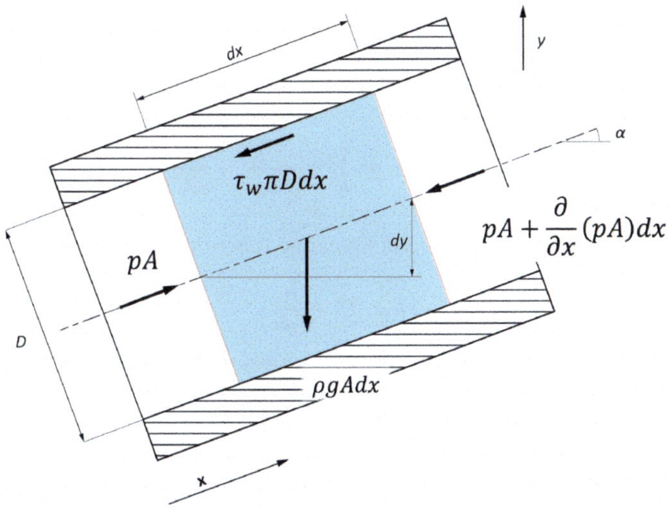


**Fig. 9.14** RTT analysis of the conservation of momentum

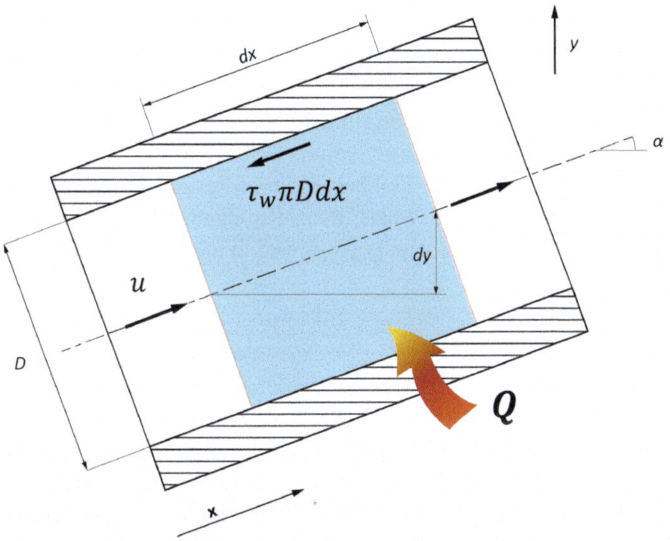


**Fig. 9.15** RTT analysis of the conservation of energy

## Conservation of Energy

The last part is to study also how the energy is transferred/transformed in the system, and how this affect the thermodynamics parameters (Christie 2007). Using the same control volume (Fig. 9.15).

$$Q = q\rho A dx = \frac{\partial}{\partial t}\left[\rho A dx\left(U + \frac{u^2}{2} + gy\right)\right] + \frac{\partial}{\partial x}\left[\rho u A\left(U + \frac{u^2}{2} + gy + \frac{p}{\rho}\right)\right] dx \quad (9.56)$$

$$q = \frac{D}{Dt}\left(U + \frac{u^2}{2} + gy\right) + \frac{u}{\rho}\frac{\partial p}{\partial x} + \frac{p}{\rho A}\frac{\partial}{\partial x}(uA) \quad (9.57)$$

Introducing now the enthalpy, a thermodynamic function which is equal to,

$$h = U + \frac{p}{\rho} \quad (9.58)$$

$$dh = c_p dT + \left[1 + \frac{T}{\rho}\left(\frac{\partial \rho}{\partial T}\right)_p\right]\frac{dp}{\rho} \quad (9.59)$$

$$q\rho + \frac{f\rho u^2 \mid u \mid}{8A}\pi D = \rho\left(\frac{\partial h}{\partial t} + u\frac{\partial h}{\partial x}\right) - \frac{\partial p}{\partial t} - u\frac{\partial p}{\partial x} \quad (9.60)$$

**Equation of State**

Finally, a thermodynamic equation describing a system given a set of specific physical conditions (e.g., temperature, pressure, volume, enthalpy, or other state properties) is introduced, considering also the real behaviour of the gas. Many formulations are used in engineering (e.g., van de Waals, Peng-Robinson, Redlich-Kwong), but the compressibility factor version is usually employed in the oil industry to quantify this deviation from the ideal gas.

$$\frac{p}{\rho} = ZRT \quad (9.61)$$

**Speed of Sound**

In order to obtain the main system of equations, another concept is introduced such as in the study of water hammer, which is the speed of propagation of a mechanical wave in an elastic medium (Smith and Clancy 2010). This velocity, also known as the speed of sound ($c$), depends on many factors, mainly the state of matter of the medium (e.g., solid, water or gas) and in the case of compressible gases, it depends majorly on the temperature. The analysis begins by considering the entropy as a function of the pressure and temperature and taking its differential considering an isentropic evolution (Fig. 9.16),

$$s = f(p, t) \quad \rightarrow \quad ds = 0 = \frac{\partial s}{\partial p}_T dp + \frac{\partial s}{\partial T}_p dT \quad (9.62)$$

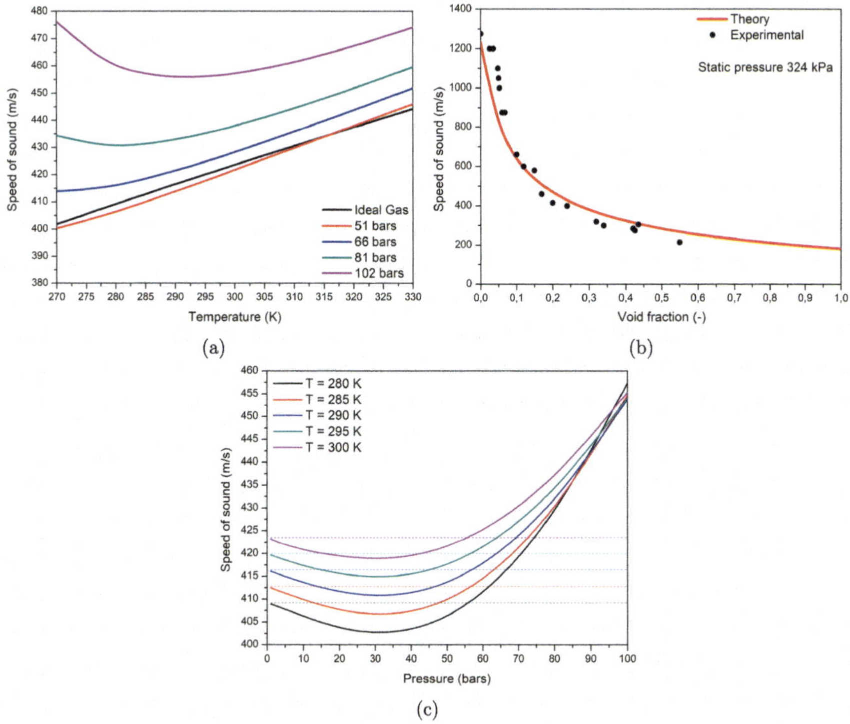


**Fig. 9.16** Plots showing the behavior of the speed of sound in real gases: **a** as a function of the temperature for different pressures; **b** comparison between experiments and theoretical results to work as validation of the model; and **c** as a function of the pressure for different temperatures

Based on this equation, one can obtain the speed of sound for a real gas,

$$c = \sqrt{\frac{ZRT}{\left\{1 - \frac{p}{Z}\frac{\partial Z}{\partial p} - \frac{p}{\rho c_p T}\left[1 + \frac{T}{Z}\frac{\partial Z}{\partial T}\right]^2\right\}}} \tag{9.63}$$

## 9.3.3 *Final System of Equations*

The system of equations resulting of applying the conservation laws, the equation of state and the speed of sound theory presented in this chapter can be written as a function of the pressure, temperature and velocity (Chapman et al. 2005).

$$\frac{\partial p}{\partial t} + v\frac{\partial p}{\partial x} + \rho c^2 \frac{\partial v}{\partial x} = \frac{c^2}{c_p T}\left(1 + \frac{T}{Z}\frac{\partial Z}{\partial T}\right)\frac{q\rho A + \frac{f\rho v^2 |v|}{8}\pi D}{A} \tag{9.64}$$

$$\frac{\partial v}{\partial t} + v\frac{\partial v}{\partial x} + \frac{1}{\rho}\frac{\partial p}{\partial x} = -g\sin\alpha - \frac{1}{\rho A}\frac{f\rho u \mid u \mid}{8}\pi D \tag{9.65}$$

$$\frac{\partial T}{\partial t} + v\frac{\partial T}{\partial x} + \frac{c^2}{c_p}\left[1 + \frac{T}{Z}\frac{\partial Z}{\partial T}\right]\frac{\partial v}{\partial x} = \frac{c^2}{c_p p}\left[1 - \frac{p}{Z}\frac{\partial Z}{\partial p}\right]\frac{q\rho A + \frac{f\rho v^2|v|}{8}\pi D}{A} \tag{9.66}$$

The final system of equations can be slightly modified in order to take into account the operational parameters usually measured. Pressure and temperature can be easily measured and are also important for the line operation. Nonetheless, the velocity presents no interest to the system, since it is the flow rate the one considered at the injection and discharge of the line. Therefore, the equations are written as a function of the pressure, temperature and mass flow rate ($\dot{m} = \rho v A$). Introducing this into the system of PDE's, one can obtain the final set of equations to be solved.

$$\frac{\partial p}{\partial t} + \frac{\phi_m ZRT}{pA}\left[1 - \frac{c^2}{ZRT}\left(1 - \frac{p}{Z}\frac{\partial Z}{\partial p}\right)\right]\frac{\partial p}{\partial x} + \phi_m c^2\left[\frac{1}{\phi_m}\frac{\partial \phi_m}{\partial x} + \frac{1}{T}\left(1 + \frac{T}{Z}\frac{\partial Z}{\partial T}\right)\frac{\partial T}{\partial x}\right]$$
$$= \frac{c^2}{c_p T}\left(1 + \frac{T}{Z}\frac{\partial Z}{\partial T}\right)\left(q\rho + \frac{\phi_m ZRT}{pA^2}\frac{f\rho u \mid u \mid}{8}\pi D\right) \tag{9.67}$$

$$\frac{ZRT}{p^2 A}\left[p\left(\frac{\partial \phi_m}{\partial t} + \frac{\phi_m ZRT}{pA}\frac{\partial \phi_m}{\partial x}\right) - \phi_m\left(1 - \frac{p}{Z}\frac{\partial Z}{\partial p}\right)\left(\frac{\partial p}{\partial t} + \frac{\phi_m ZRT}{pA}\frac{\partial p}{\partial x}\right) + \ldots\right.$$
$$\left.\frac{p\phi_m}{T}\left(1 + \frac{T}{Z}\frac{\partial Z}{\partial T}\right)\left(\frac{\partial T}{\partial t} + \frac{\phi_m ZRT}{pA}\frac{\partial T}{\partial x}\right)\right] + \frac{ZRT}{p}\frac{\partial p}{\partial x} =$$
$$-g\sin\alpha - \frac{ZRT}{pA}\frac{f\rho u \mid u \mid}{8}D \tag{9.68}$$

$$\frac{\partial T}{\partial t} + \frac{c^2}{c_p}\frac{ZRT}{p^2 A}\left(1 + \frac{T}{Z}\frac{\partial Z}{\partial T}\right)\left[p\frac{\partial \phi_m}{\partial x} - \phi_m\left(1 - \frac{p}{Z}\frac{\partial Z}{\partial p}\right)\frac{\partial p}{\partial x}\right] + \ldots$$
$$\frac{\phi_m ZRT}{pA}\left[1 + \frac{c^2}{c_p T}\left(1 + \frac{T}{Z}\frac{\partial Z}{\partial T}\right)^2\right]\frac{\partial T}{\partial x} =$$
$$\frac{c^2}{c_p T}\left(1 - \frac{p}{Z}\frac{\partial Z}{\partial p}\right)\left(q\rho + \frac{\phi_m ZRT}{pA^2}\frac{f\rho u \mid u \mid}{8}\pi D\right) \tag{9.69}$$

**Example: Simplified Model**

The equations in this chapter provided a complete tool to model the real, non-isothermal, transient flow of compressible fluids by pipelines. This rendered a set of highly non-linear partial differential equations. One can start with a simplified model of the line by considering the following assumptions:

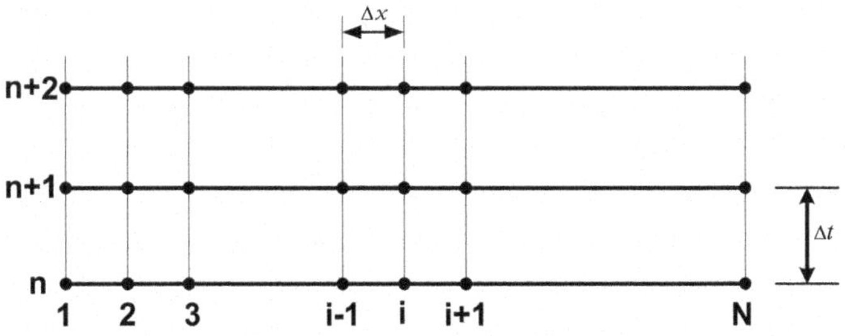


**Fig. 9.17** Discretized pipeline domain to solve the non-linear system of equations

- Ideal gas ($Z = 1$);
- No heat transfer;
- Horizontal pipeline;
- Isothermal system;
- The change in the pipe diameter due to thermal or mechanical reasons is neglected.

These simplifications render the following system of non-linear equations,

$$\frac{\partial p}{\partial t} + \frac{\phi_m RT}{pA}(1-k)\frac{\partial p}{\partial x} + c^2\frac{\partial \phi_m}{\partial x} = \frac{kR}{c_p}\frac{\phi_m}{A^2}\frac{fu^2}{8}\pi D \tag{9.70}$$

$$\frac{RT}{p^2A}\left[p\left(\frac{\partial \phi_m}{\partial t} + \frac{\phi_m RT}{pA}\frac{\partial \phi_m}{\partial x}\right) - \phi_m\left(\frac{\partial p}{\partial t} + \frac{\phi_m RT}{pA}\frac{\partial p}{\partial x}\right)\right] + \frac{RT}{p}\frac{\partial p}{\partial x} = -\frac{1}{A}\frac{fu^2}{8}D \tag{9.71}$$

$$\frac{c^2}{c_p}\frac{RT}{p^2A}\left(p\frac{\partial \phi_m}{\partial x} - \phi_m\frac{\partial p}{\partial x}\right) = \frac{c^2}{c_pT}\frac{\phi_m ZRT}{pA^2}\frac{f\rho u \mid u \mid}{8}\pi D \tag{9.72}$$

It is evident that it is impossible to find an analytical solution, so a numerical approach is needed. In order to do so, the continuum or PDE's are replaced by a discretized domain with a finite number of points (see Appendix A), which evolve in time (Fig. 9.17).

The values of the variables (and their derivatives) are therefore replaced by for approximations, obtained by the Taylor expansion of the derivative terms. Using the pressure as example,

$$p \approx p_i^n \quad \text{or} \quad p \approx \frac{p_i^n + p_{i+1}^n}{2} \tag{9.73}$$

$$\frac{\partial p}{\partial x} \approx \frac{p_{i+1}^n - p_i^n}{\Delta x} \tag{9.74}$$

$$\frac{\partial p}{\partial t} \approx \frac{p_i^{n+1} - p_i^n}{\Delta t} \quad \text{or} \quad \frac{p_i^{n+1} + p_{i+1}^{n+1} - p_i^n - p_{i+1}^n}{2\Delta t} \tag{9.75}$$

### 9.3.4 Numerical Solution—Newton's Method for Non-Linear Systems

The system presented above is composed by a set of non-linear PDE's. The definition a non-linear function is when it does not satisfy the superposition principle, i.e., cannot be written as a linear combination of separate functions of each variable.

$$f_1(x_1, x_2, \cdots, x_n) \neq f(x_1) + f(x_2) + \cdots + f(x_n) \tag{9.76}$$

Standard linear techniques (e.g. Gauss, Jacobi, SOR) do not work with this system of equations, thus a different approach is used. Considering the following system as example,

$$\begin{aligned} f_1(x_1, x_2, \cdots, x_n) &= 0 \\ f_2(x_1, x_2, \cdots, x_n) &= 0 \\ f_3(x_1, x_2, \cdots, x_n) &= 0 \\ &\vdots \\ f_n(x_1, x_2, \cdots, x_n) &= 0 \end{aligned} \tag{9.77}$$

Or, in its vectorial form,

$$\mathbf{F}(\vec{\mathbf{x}}) = \vec{0}, \quad \mathbf{F}(\vec{\mathbf{x}}) : \mathbb{R}^n \to \mathbb{R}^n \tag{9.78}$$

Most of solvers for non-linear systems are based on iterative techniques, starting from an initial guess for the vector $\vec{\mathbf{x}}_0$. So, from some initial vector $\vec{\mathbf{x}}_0$, one can calculate the value in $\vec{\mathbf{x}}_1$ (see Appendix A for the Big-$\mathcal{O}$ notation).

$$\mathbf{F}(\vec{\mathbf{x}}_1) = \mathbf{F}(\vec{\mathbf{x}}_0) + \mathbf{J}(\vec{\mathbf{x}}_0)(\vec{\mathbf{x}}_1 - \vec{\mathbf{x}}_0) + \mathcal{O}\left(\|\Delta x\|^2\right) \tag{9.79}$$

$$\mathbf{F}(\vec{\mathbf{x}}_1) \approx \mathbf{F}(\vec{\mathbf{x}}_0) + \mathbf{J}(\vec{\mathbf{x}}_0)(\vec{\mathbf{x}}_1 - \vec{\mathbf{x}}_0) \tag{9.80}$$

Since the solution to the system is the desired value, the assumption is that that $\vec{\mathbf{x}}_1$ approaches the solution more than the initial guess, thus,

$$\mathbf{0} \approx \mathbf{F}\left(\vec{\mathbf{x}}_0\right) + \mathbf{J}\left(\vec{\mathbf{x}}_0\right)\left(\vec{\mathbf{x}}_1 - \vec{\mathbf{x}}_0\right) \tag{9.81}$$

Then, the **only** unknown in this **now linear** system is the variation between the two approximations $\Delta\vec{\mathbf{x}}$, which is calculated as,

$$\left(\vec{\mathbf{x}}_1 - \vec{\mathbf{x}}_0\right) \approx -\left[\mathbf{J}\left(\vec{\mathbf{x}}_0\right)\right]^{-1} \mathbf{F}\left(\vec{\mathbf{x}}_0\right) \tag{9.82}$$

In this case, the Jacobian is required to not lose its rank (i.e., to be non-singular). The advantages of the Newton's method is that it is easy to apply, and besides it converges quadratically. On the other hand, the main, major disadvantage is the *Jacobian* and its inversion, which has to be calculated for every iteration and, in case of time-dependent problems, repeat this every time-step. This can be computationally demanding and also cause problems in case of ill-conditioned Jacobians. A minor problem is the election of the initial guess. Poor choosing may cause the system to oscillate or fail to converge. Therefore, this technique is sometimes replaced by what is called *Quasi-Newton* methods (e.g. Broyden, Broyden–Fletcher–Goldfarb–Shanno or BFGS, SR1, DFP). These are defined as any numerical procedure, when solving a non-linear or optimization system requiring the calculation of the Jacobian (and/or its inverse), that replaces the latter for an approximation matrix. The main advantage of the QN techniques is that they usually do not need to calculate the inverse of the Jacobian in every step, being less susceptible to ill-conditioned matrices. On the other hand, QN methods converge at a rate slower than Newton, which means they take more steps to converge.

### 9.3.5 *Pseudocode for Newton's Method*

The method described in this chapter can be easily implemented in Matlab in order to simulate and optimize a system which undergoes a strong winter.

**Pseudocode 9.2:** Transient gas pipeline

Set $\Delta t$, $L$, $\phi_{m,N}(t)$, $p_1^n$, $Dia$, $R$, $min_{error}$, $max_{time}$ and $T$;
$\rho_0 = \frac{101325Pa}{RT}$;
$A = \frac{\pi D^2}{4}$;
$u = \frac{\phi_{m,N}}{\rho A}$;
Friction factor $f = f\left(Re, \frac{\epsilon}{D}\right)$;
Define initial parameters $p(n)^0$ with steady state formulation;
$t = \Delta t$;
**while** $t \leq max_{time}$ **do**
  **while** $error \geq min_{error}$ **do**
    Calculate functions of the Jacobian $J(i, j)$;
    Assembly Jacobian;
    Find the inverse $\left[\mathbf{J}\left(\vec{\mathbf{x}}_0\right)\right]^{-1}$;
    Calculate $\Delta\vec{\mathbf{x}}$;
    Calculate $error$ $\|\Delta\vec{\mathbf{x}}\|$;
  **end while**
  plot($p$,$x$) and plot($\phi_m$,$x$);
  $t = t + \Delta t$;
**end while**

# References


Allievi L. Theory of water-hammer, vol. 1. Garroni: Typography R; 1925.

Bergant A, Simpson AR, Tijsseling AS. Water hammer with column separation: a historical review. J Fluids Struct. 2006;22(2):135–71.

Bergant A, Tijsseling AS, Vítkovský JP, Covas DIC, Simpson AR, Lambert MF. Parameters affecting water-hammer wave attenuation, shape and timing-part 2: case studies. J Hydraul Res. 2008;46(3):382–91.

Carlsson J. Water hammer phenomenon analysis using the method of characteristics and direct measurements using a "stripped" electromagnetic flow meter. Master's thesis, KTH, Physics. 2016.

Chapman KS, Krishniswami P, Wallentine V, Abbaspour M, Ranganathan R, Addanki R, Sengupta J, Chen L. Virtual pipeline system testbed to optimize the us natural gas transmission pipeline system. Technical report, Kansas State University, Manhattan, KS (United States). 2005.

Christie B. Analysis of the frictional heating effect in pipelines. In: PSIG annual meeting. 2007. p. PSIG–0702. PSIG.

Clegg S, Mancarella P. Integrated electrical and gas network flexibility assessment in low-carbon multi-energy systems. IEEE Trans Sustain Energy. 2015;7(2):718–31.

Ellul IR, Saether G, Shippen ME. The modeling of multiphase systems under steady-state and transient conditions-a tutorial. In: PSIG annual meeting. 2004. p. PSIG–0403. PSIG.

ENARGAS. Gasoductos de argentina. 2017. https://cyt-ar.com.ar/cyt-ar/index.php/Gasoductos_de_Argentina.

ETH Zurich. Preventing natural gas crises, 2024. https://ethz.ch/en/news-and-events/eth-news/news/2014/03/preventing-natural-gas-crises.html.

Garg RK, Kumar A. Experimental and numerical investigations of water hammer analysis in pipeline with two different materials and their combined configuration. Int J Pressure Vessels Piping. 2020;188:104219. ISSN 0308-0161. https://doi.org/10.1016/j.ijpvp.2020.104219. https://www.sciencedirect.com/science/article/pii/S0308016120301964.

Gugat M, Krug R, Martin A. Transient gas pipeline flow: analytical examples, numerical simulation and a comparison to the quasi-static approach. Optimizat Engin. 2023;24(1):425–46. ISSN 1573-2924. https://doi.org/10.1007/s11081-021-09690-4

Kandil M, Kamal AM, El-Sayed TA. Effect of pipematerials on water hammer. Int J Pressure Vessels Piping. 2020;179:103996. ISSN 0308-0161. https://doi.org/10.1016/j.ijpvp.2019.103996. https://www.sciencedirect.com/science/article/pii/S0308016119303850.

Kaneko S, Nakamura T, Inada F, Kato M, Ishihara K, Nishihara T, Langthjem MA, editors. Chapter 5 - vibration induced by pressure waves in piping. In: Flow-induced vibrations (second edition). Oxford: Academic Press; 2014. p. 197–275. ISBN 978-0-08-098347-9. https://doi.org/10.1016/B978-0-08-098347-9.00005-9. https://www.sciencedirect.com/science/article/pii/B9780080983479000059.

Keyaerts N, Hallack M, Glachant JM, D'haeseleer W. Gas market distorting effects of imbalanced gas balancing rules: inefficient regulation of pipeline flexibility. Energy Policy. 2011;39(2):865–876.

Khalighi F, Ahmadi A, Keramat A. Water hammer simulation by explicit central finite difference methods in staggered grids. J Comput Appl Res Mech Engin (JCARME). 2017;6(2):69–77. ISSN 2228-7922. https://doi.org/10.22061/jcarme.2017.598. https://jcarme.sru.ac.ir/article_598.html.

Krishnaswami P, Chapman KS, Abbaspour M. Compressor station optimization for linepack maintenance. In PSIG annual meeting. 2004. p. PSIG–0410. PSIG.

Larock BE, Jeppson RW, Watters GZ. Hydraulics of pipeline systems. CRC Press; 1999. ISBN 9781420050318. https://books.google.nl/books?id=RAMX5xuXSrUC.

Li C, He J, Jia W, Yang F, Ban J, Qiu B. Modeling transient cavitating flow in large drop crude oil pipelines. J Petroleum Sci Engin. 2023;220:111241. ISSN 0920-4105. https://doi.org/10.1016/j.petrol.2022.111241. https://www.sciencedirect.com/science/article/pii/S0920410522010932.

Liu E, Kuang J, Peng S, Liu Y. Transient operation optimization technology of gas transmission pipeline: a case study of west-east gas transmission pipeline. IEEE Access. 2019;7:112131–41. https://doi.org/10.1109/ACCESS.2019.2934315.

McNeeley S. Pipeline hydraulics system: a beginner's guide to hydraulic systems: pipeline hydraulics solution manual. Independently Published; 2021. ISBN 9798542442372. https://books.google.nl/books?id=bwmOzgEACAAJ.

Menon ES. Gas pipeline hydraulics. Taylor & Francis, 2005. ISBN 9780849327858. https://books.google.nl/books?id=uFpHngEACAAJ.

Mertens S. Design of wind and solar energy supply, to match energy demand. Cleaner Engin Technol. 2022;6: 100402.

Mohitpour M, Golshan H, Murray MA. Pipeline design and construction: a practical approach. ASME Press: Pipelines and Pressure Vessels; 2007. https://books.google.nl/books?id=_d1SAAAAMAAJ.

Potocnik P, Govekar E. Practical results of forecasting for the natural gas market. In: Potocnik P, editor. Natural gas, chapter 16. IntechOpen, Rijeka; 2010. https://doi.org/10.5772/9853

Saidani A, Fourar A, Massouh F. Influence of temperature on transient flow with cavitation in copper pipe-rig. Model Earth Syst Environ. 2022;1–11.

Smith JP, Clancy J. Understanding aga reprot no. 10, speed of sound in natural gas and other related hydrocarbons gases. Technical report, Report. 2010.

van der Hoeven T. Gas transport pipe. In: PSIG annual meeting. 2007. p. PSIG–0710. PSIG.

Wan W, Zhang B, Chen X. Investigation on water hammer control of centrifugal pumps in water supply pipeline systems. Energies. 2018;12(1):108.

Yan T, Qu J, Sun X, Chen Y, Pan Y. Propagation velocity and time laws of backpressure wave in the wellbore during managed pressure drilling. Nat Gas Indust B. 2018;5(3):253–60. ISSN 2352-8540. https://doi.org/10.1016/j.ngib.2017.11.009. https://www.sciencedirect.com/science/article/pii/S2352854018300391.

Yin X, Wen K, Huang W, Luo Y, Ding Y, Gong J, Gao J, Hong B. A high-accuracy online transient simulation framework of natural gas pipeline network by integrating physics-based and data-driven methods. Appl Energy. 2023;333:120615. ISSN 0306-2619. https://doi.org/10.1016/j.apenergy.2022.120615. https://www.sciencedirect.com/science/article/pii/S0306261922018724.

Yuce MI, Omer AF. Hydraulic transients in pipelines due to various valve closure schemes. SN Appl Sci. 2019;1:1–11.

Zhang B, Wan W, Shi M. Experimental and numerical simulation of water hammer in gravitational pipe flow with continuous air entrainment. Water. 2018;10(7):928.

# Appendix A
# Discretization

In this Appendix (Thom Badings and Vakis 2022), a brief introduction is given to some discretization methods. The first point consists in underlying why discretization is relevant for many engineering problems. Then, the concept of the *mesh* is introduced, and a discussion of how it is applied to discretize different types of problems is presented, including different formulations of the *finite difference method.*

## A.1 Why Discretization?

As known by many scientists and professionals, many problems in engineering cannot be solved analytically. Especially problems involving Navier-Stokes equations, diffusion and waves are difficult to solve in multiple dimensions or with coupled physics as these are typically result in a complex system of non-linear PDE's. Analytical solutions are often only obtainable if one reduces the problem using restrictive assumptions, thereby ofoten rendering the problem devoid of practical interest.

When the interest is in the behavior of a physical system, which dependent on space and time and need to solve it numerically, one must employ discretization. This is in essence the transformation of a continuous set of equations describing this system into discrete elements, which can be evaluated using a numerical scheme. By calculating these elements in time and space one can then evaluate system behavior, whether that is steady state or transient.

Various discretization methods exist. This appendix will be limited to the discussion of the discretization of a partial differential equation using the finite difference method. The origin of this method can be found in the early works of the mathematicians Euler and Kutta, and is similar to the Runge-Kutta family of numerical methods employed in solving ODE's. For PDE's, the finite difference methods consist in approximating the differential operator with an equation using differential


P. Druetta et al., *Transport by Pipelines and Its Optimization*, Petroleum Engineering, https://doi.org/10.1007/978-3-031-66566-0

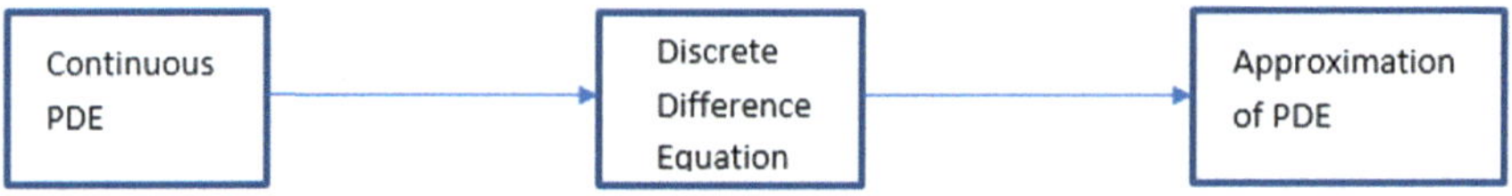


**Fig. A.1** General workflow of discretization methods

quotients to replace the derivatives in the equation. The difference between the differential operator and differential quotient is a measure for the error of the numerical solution compared to the analytical solution; this is often known as the *truncation error*. In essence, Figure Fig. A.1 describes the workflow of discretizing a PDE using the finite difference method.

**Example**

Suppose one wishes to approximate $u$ at a point $x$. Assuming $u(x)$ is a smooth differentiable function, the derivative of $u$ around $x$ for step size $h$ can be assessed as follows:

$$\frac{\partial u}{\partial x} \approx \lim_{h \to 0} \frac{u(x+h) - u(x)}{h}. \tag{A.1}$$

As one would expect, as $h$ becomes smaller the approximation on the right hand side of the equation becomes more accurate. Equation A.1 illustrates the basic concept behind the finite difference schemes, which shall be further explained later.

## A.2 Generating the Mesh

Discretizing the domain means that the domain will end up with a series of points in space and time which may be (or not, depending on the physics being modelled) evenly spaced. Supposing one is interested in the temperature distribution over time in a rod along its length $L$, which shall be called the $x$-direction, then it renders,

$$0 \leq x \leq L \tag{A.2}$$

$$x_i = (i-1)\Delta x \qquad \text{for } i = 1, 2, \cdots, N, \tag{A.3}$$

where $n$ represents the amount of grid points. Based on $L$ and $N$, $\Delta x$ can be then determined:

$$\Delta x = \frac{L}{N-1}. \tag{A.4}$$

Similar equations hold for discretizing time:

$$0 \leq t \leq t_{\text{final}} \tag{A.5}$$

$$t_k = (k-1)\Delta t \qquad \text{for } k = 1, 2, \cdots, K, \tag{A.6}$$

where $K$ denotes the number of time steps, and $\Delta t$ refers to the size of the time step given by

$$\Delta t = \frac{t_{\text{final}}}{K-1}. \tag{A.7}$$

The above can be visualized in a grid known as a mesh, as shown in Fig. A.2. In this figure, the horizontal axis represents the $x$-direction of the rod (along its length $L$), and the vertical axis represents steps through time. For each time step, all grid points $i$ must be calculated. In other words, at $t = 0$, also known as the *initial conditions*, the solution is completely known and given as information and for the next time step, all the values of the variable under study are calculated. The row $k$ is assessed using row $k-1$. As calculation of a grid point often depends on its neighbors in the $x$-direction in the previous time step, one needs to specify the values at the boundaries of the mesh. These are called the *boundary conditions* (indicated by the red rectangles), and need to be specified at both $x = 0$ and $x = L$, for each time $t > 0$.

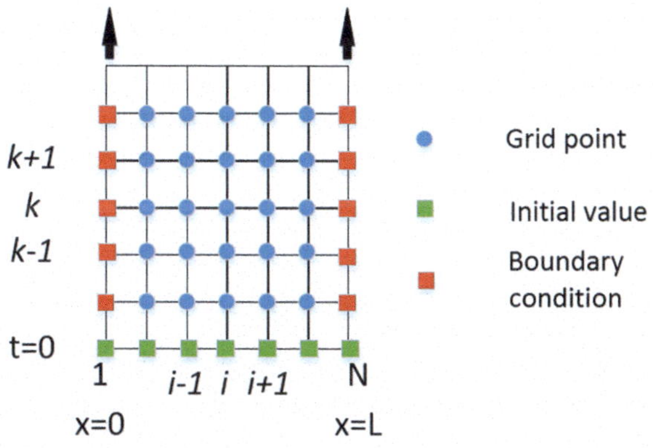


**Fig. A.2** Mesh of the heat equation example

## A.3 First Order Forward Difference

In this section the first order forward difference scheme will be covered, which is derived from a Taylor series expansion $u(x)$ around a point $x_i$:

$$\left.\frac{\partial u}{\partial x}\right|_{x_i} = \frac{u_{i+1} - u_i}{\Delta x} - \frac{\Delta x}{2}\left.\frac{\partial^2 u}{\partial x^2}\right|_{x_i} - \frac{\Delta x^2}{3!}\left.\frac{\partial^3 u}{\partial x^3}\right|_{x_i} + \mathcal{O}(\Delta x^3). \tag{A.8}$$

Replacing the higher order terms using the mean value theorem,

$$\left.\frac{\partial u}{\partial x}\right|_{x_i} \approx \frac{u_{i+1} - u_i}{\Delta x} + \frac{\Delta x^2}{2}\left.\frac{\partial^2 u}{\partial x^2}\right|_{\epsilon}, \tag{A.9}$$

where the terms to the right of the difference in $u$ over $\Delta x$ in Eq. A.9 represent the *truncation error*, which results from truncating (in other words, cutting off or neglecting) the higher order terms in Eq. A.8. In general,the value of $\epsilon$ is not known and also the second order term of $u$ with respect to $x$ cannot be computed, as the function $u(x, t)$ is not known. To deal with the truncation error, the term $\mathcal{O}(\Delta x^2)$ is often used. The $\mathcal{O}$ represents the dependence of the truncation error on the mesh, which we determine prior to simulation:

$$\frac{\Delta x^2}{2}\left.\frac{\partial^2 u}{\partial x^2}\right|_{\epsilon} = \mathcal{O}(\Delta x^2). \tag{A.10}$$

Rewriting Eq. A.9 then yields

**Definition: First order backward difference equation**

$$\left.\frac{\partial u}{\partial x}\right|_{x_i} = \frac{u_{i+1} - u_i}{\Delta x} + \mathcal{O}(\Delta x). \tag{A.11}$$

Note that $\Delta x^2$ was replaced with $\Delta x$ as the right hand side of Equation A.10 is not meant as a strict equality; it shows the dependence of the order of magnitude of the error, and how is approaches zero as function of the meshing. Equation A.11 is called the forward difference equation because it utilizes the nodes $i$ and $i + 1$.

## A.4 First Order Backward Difference

Replacing the use of $\Delta x$ in Eq. A.8 with $-\Delta x$, the backward difference equation is obtained:

$$\left.\frac{\partial u}{\partial x}\right|_{x_i} \approx \frac{u_i - u_i - 1}{\Delta x} + \frac{\Delta x}{2}\left.\frac{\partial^2 u}{\partial x^2}\right|_{x_i} - \frac{\Delta x^2}{3!}\left.\frac{\partial^3 u}{\partial x^3}\right|_{x_i} + \cdots . \tag{A.12}$$

$$u_{i-1} = u_i - \Delta x\left.\frac{\partial u}{\partial x}\right|_{x_i} + \frac{\Delta x^2}{2}\left.\frac{\partial^2 u}{\partial x^2}\right|_{x_i} - \frac{(\Delta x)^3}{3!}\left.\frac{\partial^3 u}{\partial x^3}\right|_{x_i} + \cdots . \tag{A.13}$$

Again, truncating the higher order terms and using $\mathcal{O}$ notation,

**Definition: First order backward difference equation**

$$\left.\frac{\partial u}{\partial x}\right|_{x_i} = \frac{u_i - u_i - 1}{\Delta x} + \mathcal{O}(\Delta x), \tag{A.14}$$

This is called the backward difference, as it depends on evaluating nodes $i$ and $i - 1$. The truncation error for both the forward and backward difference equation is the same.

## A.5 First Order Central Difference

The central difference equation is actually a combination of both the forward and backward difference equations.The main idea is to reduce the truncation error; in the central difference equation, this error approaches zero much faster as $\Delta x$ is decreased. First, recall the Taylor series expansions for $u_{i+1}$ and $u_{i-1}$:

$$u_{i+1} = u_i + \Delta x\left.\frac{\partial u}{\partial x}\right|_{x_i} + \frac{\Delta x^2}{2}\left.\frac{\partial^2 u}{\partial x^2}\right|_{x_i} + \frac{(\Delta x)^3}{3!}\left.\frac{\partial^3 u}{\partial x^3}\right|_{x_i} + \cdots , \tag{A.15}$$

$$u_{i-1} = u_i - \Delta x\left.\frac{\partial u}{\partial x}\right|_{x_i} + \frac{\Delta x^2}{2}\left.\frac{\partial^2 u}{\partial x^2}\right|_{x_i} - \frac{(\Delta x)^3}{3!}\left.\frac{\partial^3 u}{\partial x^3}\right|_{x_i} + \cdots . \tag{A.16}$$

Subtracting Eq. A.15 from A.16 yields

$$u_{i+1} - u_{i-1} = 2\Delta x\left.\frac{\partial u}{\partial x}\right|_{x_i} + \frac{2(\Delta x)^3}{3!}\left.\frac{\partial^3 u}{\partial x^3}\right|_{x_i} + \cdots . \tag{A.17}$$

Solving for $\left.\frac{\partial u}{\partial x}\right|_{x_i}$,

$$\left.\frac{\partial u}{\partial x}\right|_{x_i} = \frac{u_{i+1} - u_{i-1}}{2\Delta x} - \frac{(\Delta x)^3}{3!}\left.\frac{\partial^3 u}{\partial x^3}\right|_{x_i} + \cdots, \tag{A.18}$$

which, with the $\mathcal{O}$ notation, leads to the central difference equation:

**Definition: First order central difference equation**

$$\left.\frac{\partial u}{\partial x}\right|_{x_i} = \frac{u_{i+1} - u_{i-1}}{2\Delta x} + \mathcal{O}(\Delta x^2). \tag{A.19}$$

The truncation error will approach zero much faster than for the forward and back equations, Eqs. A.11 and A.14, respectively. However, there is a disadvantage to the first order central difference equation, as it does not contain the grid point $i$. Depending on the problem, this might cause complications, which are discussed in many texts of numerical analysis.

## A.6 Second Order Central Difference

Similar to the first order central difference, the second order central difference equation involves Eqs. A.15 and A.16 with the appropriate Taylor expansion. Contrary to the first order difference equation, the goal is not finding an equation to replace the *second order* derivative of the function. To this end, using Taylor expansions up to fourth order,

$$u_{i+1} = u_i + \Delta x\left.\frac{\partial u}{\partial x}\right|_{x_i} + \frac{\Delta x^2}{2}\left.\frac{\partial^2 u}{\partial x^2}\right|_{x_i} + \frac{(\Delta x)^3}{3!}\left.\frac{\partial^3 u}{\partial x^3}\right|_{x_i} + \frac{\Delta x^4}{4!}\left.\frac{\partial^4 u}{\partial x^4}\right|_{x_i} + \cdots, \tag{A.20}$$

$$u_{i-1} = u_i - \Delta x\left.\frac{\partial u}{\partial x}\right|_{x_i} + \frac{\Delta x^2}{2}\left.\frac{\partial^2 u}{\partial x^2}\right|_{x_i} - \frac{\Delta x^3}{3!}\left.\frac{\partial^3 u}{\partial x^3}\right|_{x_i} + \frac{\Delta x^4}{4!}\left.\frac{\partial^4 u}{\partial x^4}\right|_{x_i} + \cdots. \tag{A.21}$$

Instead of subtracting the two Taylor series from each other, these are added together, resulting in the following expression:

$$u_{i+1} + u_{i-1} = 2u_i + \Delta x^2\left.\frac{\partial^2 u}{\partial x^2}\right|_{x_i} + \frac{\Delta x^4}{4!}\left.\frac{\partial^4 u}{\partial x^4}\right|_{x_i} + \cdots. \tag{A.22}$$

Rearranging it yields,

$$\left.\frac{\partial^2 u}{\partial x^2}\right|_{x_i} = \frac{u_{i+1} - 2u_i + u_{i-1}}{\Delta x^2} - \frac{\Delta x^2}{4!}\left.\frac{\partial^4 u}{\partial x^4}\right|_{x_i} + \cdots . \tag{A.23}$$

After rearranging, the first error term in Eq. A.23 contains $\Delta x^2$. Hence, the conclusion is that the second order central difference equation has error $\mathcal{O}(\Delta x^2)$,

**Definition: Second order central difference equation**

$$\left.\frac{\partial^2 u}{\partial x^2}\right|_{x_i} = \frac{u_{i+1} - 2u_i + u_{i-1}}{\Delta x^2} + \mathcal{O}(\Delta x^2). \tag{A.24}$$

## Reference

Badings PDTKT, de Bruijn C, Vakis A. Dynamics of engineering systems: course reader. Rijksuniversiteit Groningen; 2022.

**SPRINGER NATURE**

# GPSR Compliance

*The European Union's (EU) General Product Safety Regulation (GPSR) is a set of rules that requires consumer products to be safe and our obligations to ensure this.*

*If you have any concerns about our products, you can contact us on ProductSafety@springernature.com*

In case Publisher is established outside the EU, the EU authorized representative is:

Springer Nature Customer Service Center GmbH
Europaplatz 3
69115 Heidelberg, Germany

The manufacturer's authorised representative in the EU is Springer Nature Customer Service Centre GmbH, Europaplatz 3, 69115 Heidelberg, Germany. If you have any concerns regarding our products, please contact ProductSafety@springernature.com

Printed and bound by CPI Group (UK) Ltd, Croydon, CR0 4YY
01/10/2026
02225965-0010